# SOURCES, EFFECTS AND RISKS OF IONIZING RADIATION

United Nations Scientific Committee on the Effects of Atomic Radiation

UNSCEAR 2024
Report to the General Assembly, with Scientific Annexes

VOLUME I
Scientific Annex A



UNITED NATIONS
New York, 2025

NOTE

The report of the Committee without its annexes appears as *Official Records of the General Assembly*, Seventy-ninth Session, Supplement No. 46 (A/79/46).

The designations employed and the presentation of material in this publication do not imply the expression of any opinion whatsoever on the part of the Secretariat of the United Nations concerning the legal status of any country, territory, city or area, or of its authorities, or concerning the delimitation of its frontiers or boundaries.

The country names used in this document are, in most cases, those that were in use at the time the data were collected or the text prepared. In other cases, however, the names have been updated, where this was possible and appropriate, to reflect political changes.

UNITED NATIONS PUBLICATION
Sales No. E.25.IX.1
ISBN: 978-92-1-003330-5
e-ISBN: 978-92-1-106873-3



This publication has not been formally edited.

Information on uniform resource locators and links to Internet sites contained in the present publication are provided for the convenience of the reader and are correct at the time of issue. The United Nations takes no responsibility for the continued accuracy of that information or for the content of any external website.

# CONTENTS

# Report of the United Nations Scientific Committee on the Effects of Atomic Radiation

## Contents

# Chapter I

## Introduction

1. The mandate of the United Nations Scientific Committee on the Effects of Atomic Radiation (UNSCEAR) was first established in General Assembly resolution 913 (X), adopted in 1955. In Assembly resolution 3154 A–C (XXVIII), adopted in 1973, the Committee was further requested to consider the risks of ionizing radiation from all sources.[1] In pursuit of its mandate, the Committee thoroughly reviews and evaluates global and regional exposures to radiation. The Committee also evaluates evidence of radiation-induced health effects in exposed groups and advances in the understanding of the biological mechanisms by which radiation-induced effects on human health or on non-human biota can occur. Those evaluations provide the scientific basis used, inter alia, by the relevant agencies of the United Nations system in formulating international standards for the protection of the public, workers and patients;[2] those standards, in turn, are linked to important legal and regulatory instruments.

2. Exposure to ionizing radiation arises from naturally occurring sources (such as radiation from outer space and radon gas emanating from rocks on the Earth) and from sources with an artificial origin (such as medical diagnostic and therapeutic procedures; radioactive material resulting from nuclear weapons testing; electricity generation, including by means of coal, natural gas, oil, nuclear power, and geothermal and other energy sources; unplanned events; and workplaces where there may be increased exposure to artificial or naturally occurring sources of radiation).

---

[1] The United Nations Scientific Committee on the Effects of Atomic Radiation (UNSCEAR) was established by the General Assembly at its tenth session, in 1955. The terms of reference of the Committee are set out in resolution 913 (X). The Committee was originally composed of the following Member States: Argentina, Australia, Belgium, Brazil, Canada, Czechoslovakia (later succeeded by Slovakia), Egypt, France, India, Japan, Mexico, Sweden, Union of Soviet Socialist Republics (later succeeded by the Russian Federation), United Kingdom of Great Britain and Northern Ireland and United States of America. The membership of the Scientific Committee was subsequently enlarged by the Assembly in its resolution 3154 C (XXVIII) to include the Federal Republic of Germany (later succeeded by Germany), Indonesia, Peru, Poland and the Sudan. By its resolution 41/62 B, the Assembly increased the membership of the Committee to 21 members and invited China to become a member. In its resolution 66/70, the Assembly further enlarged the membership of the Committee to 27 and invited Belarus, Finland, Pakistan, the Republic of Korea, Spain and Ukraine to become members. In its resolution 76/75, the Assembly further enlarged the membership of the Committee to 31 and invited Algeria, Iran (Islamic Republic of), Norway and the United Arab Emirates to become members.

[2] For instance, relevant international safety standards that take into account the findings of the Scientific Committee include: (a) the international Fundamental Safety Principles, which are jointly sponsored by the European Atomic Energy Community, the Food and Agriculture Organization of the United Nations (FAO), the International Atomic Energy Agency (IAEA), the International Labour Organization (ILO), the International Maritime Organization, the Nuclear Energy Agency of the Organisation for Economic Co-operation and Development (OECD/NEA), the Pan American Health Organization (PAHO), the United Nations Environment Programme (UNEP) and the World Health Organization (WHO); and (b) the *Radiation Protection and Safety of Radiation Sources: International Basic Safety Standards – General Safety Requirements Part 3*, which are co-sponsored by the European Commission, FAO, IAEA, ILO, OECD/NEA, PAHO, UNEP and WHO. Both international standards were established under the aegis of IAEA.

# Chapter II

## Deliberations of the United Nations Scientific Committee on the Effects of Atomic Radiation at its seventy-first session

3. The Scientific Committee held its seventy-first session in Vienna from 20 to 24 May 2024. The following were elected to serve as officers of the Committee for its seventy-first and seventy-second sessions: Sarah Baatout (Belgium) as Chair; Anssi Auvinen (Finland), Reiko Kanda (Japan) and Aayda Al Shehhi (United Arab Emirates) as Vice-Chairs; and Carol Robinson (Norway) as Rapporteur.

4. The Scientific Committee took note of General Assembly resolution 78/71 on the effects of atomic radiation. The Committee heard statements from five member States, Argentina, Belgium, Iran (Islamic Republic of), the Russian Federation and the United Arab Emirates, and from the following observers: the International Atomic Energy Agency (IAEA), the Scientific Advisory Group for the Treaty on the Prohibition of Nuclear Weapons and the World Health Organization. Assistant Secretary-General of the United Nations and Deputy Executive Director of the United Nations Environment Programme (UNEP), Elizabeth Mrema, also addressed the Committee. The recorded statements are available from the Committee's secretariat.

5. The Scientific Committee also noted and discussed other issues addressed in the resolution. Those discussions are summarized in chapter II, section E (Administrative issues), of the present report.

### A. Completed evaluations

6. The Scientific Committee discussed in detail the evaluation of second primary cancer after radiotherapy and the evaluation of public exposure to ionizing radiation, adopted the scientific reports based on the findings of those evaluations (see chapter III) and requested that the scientific annexes be published in the usual manner, subject to the agreed modifications.

#### 1. Second primary cancer after radiotherapy

7. At its sixty-sixth session, the Scientific Committee endorsed the plan to initiate an evaluation of second primary cancer after radiotherapy. The aim of the evaluation was to raise awareness in the scientific and medical communities and in national authorities with respect to the fact that cancer treatment by radiation, while contributing to the effective treatment of increasing numbers of patients, can involve off-target exposures which may, in some patients, result in second primary cancer several years later. Quantification of the risk of second primary cancer after radiotherapy and evaluations of the factors affecting that risk require data (e.g. on dose distributions) that are often difficult to obtain retrospectively.

8. At its seventy-first session, the Scientific Committee expressed its gratitude for the significant work of the expert group[3] in the preparation of its evaluation of second primary cancer after radiotherapy. In response to numerous constructive comments, the technical evaluation of radiation oncology, biology, dosimetry and epidemiology relevant to the development of second primary cancer after radiotherapy was revised substantially. The Committee discussed and approved for publication the scientific annex on second primary cancer after radiotherapy. In addition, the Committee reviewed and endorsed the three electronic attachments to that annex for presentation on the UNSCEAR website.

---

[3] The expert group comprises 37 experts (including four critical reviewers) from 17 Member States and four observers from IAEA, the International Commission on Radiological Protection (ICRP) and WHO.

### 2. Evaluation of public exposure to ionizing radiation

9. In pursuit of its mandate, the Scientific Committee regularly reviews and evaluates global and regional exposures of the public to radiation. At its sixty-sixth session, the Committee agreed to update annex B to its 2008 report[4] and in 2020 it commenced the evaluation of public exposure to ionizing radiation. At its sixty-ninth session, the Committee supported the application of the methodology for estimating exposures of the public due to radioactive discharges as presented in its 2016 report.[5] At its seventieth session, the Committee endorsed further updates to the methodology and approach regarding quality criteria to be applied in the evaluation as presented in the appendix to annex B to the present report.

10. At its seventy-first session, the Scientific Committee expressed its gratitude for the significant work of the expert group[6] in considering the data on public exposure to ionizing radiation submitted by 61 Member States as at December 2023, the review of relevant literature published since 2007, and relevant data from eight regional and international organizations. The Committee discussed and approved for publication the scientific annex on evaluation of public exposure to ionizing radiation. In addition, the Committee reviewed and endorsed the six electronic attachments to that annex for presentation on the UNSCEAR website.

## B. Present programme of work

### 1. Epidemiological studies of ionizing radiation and cancer

11. At its sixty-sixth session, the Scientific Committee agreed to update annex A to its 2006 report[7] and commenced its evaluation of epidemiological studies of radiation and cancer by establishing an expert group and initiating a comprehensive literature review based on the principles and criteria for ensuring the quality of the Committee's reviews of epidemiological studies of radiation exposure.

12. At its seventy-first session, the Scientific Committee acknowledged the significant progress made by the expert group[8] in drafting the scientific annex on the selected cancer sites and in summarizing the literature review on epidemiological studies of radiation and cancer. In addition, the Committee endorsed the methodology for lifetime cancer risk calculations for four proposed scenarios, which combines approaches used in annex A to its 2006 Report [9] and the risk projections and comparative analysis presented in annex A to its 2019 Report,[10] using extensive demographic data obtained on five geographical regions. The Committee also acknowledged the timely development of the evaluation, which is scheduled for approval at its seventy-second session, in 2025.

---

[4] *Sources, Effects and Risks of Ionizing Radiation: United Nations Scientific Committee on the Effects of Atomic Radiation 2008 Report to the General Assembly*, vol. I (United Nations publication, 2010), annex B.

[5] *Sources, Effects and Risks of Ionizing Radiation: United Nations Scientific Committee on the Effects of Atomic Radiation 2016 Report to the General Assembly with Scientific Annexes* (United Nations publication, 2017).

[6] The expert group comprises 57 experts (including four critical reviewers) from 20 Member States and four observers from the European Commission, IAEA, OECD/NEA and WHO.

[7] *Effects of Ionizing Radiation: United Nations Scientific Committee on the Effects of Atomic Radiation 2006 Report to the General Assembly*, vol. I (United Nations publication, 2008), annex A.

[8] The expert group includes 27 experts (including four critical reviewers) from eight Member States and one observer from the International Agency for Research on Cancer.

[9] *Effects of Ionizing Radiation: UNSCEAR 2006 Report*, vol. I, annex A.

[10] *Sources, Effects and Risks of Ionizing Radiation: United Nations Scientific Committee on the Effects of Atomic Radiation 2019 Report to the General Assembly with Scientific Annexes,* (United Nations publication, 2020, and corrigendum), annex A.

## 2. Effects of ionizing radiation on the circulatory system

13. At its sixty-seventh session, in 2020, the Scientific Committee agreed to commence an evaluation of diseases of the circulatory system resulting from radiation exposure. At its sixty-eighth session, it endorsed the project plan for commencement in 2021. At its seventieth session, the Committee welcomed the considerable work carried out by the expert group[11] and discussed the first results of the literature review. It provided feedback on the scope of topics to be included in the review and endorsed the proposed structure of the evaluation.

14. At its seventy-first session, the Scientific Committee welcomed the progress made by the expert group and reviewed the first draft annex. The Committee accepted that the expert group would not attempt to make lifetime risk projections, given the marked heterogeneity of the available data. The Committee requested the expert group to prepare a progress report and finalize the draft of the scientific annex, including the results of the review of new literature published up to 2023 and the group's conclusions on diseases of the circulatory system caused by radiation exposure, and to present the annex for approval by the Committee in 2025.

## 3. Effects of ionizing radiation on the nervous system

15. At its sixty-seventh session, the Scientific Committee agreed to commence an evaluation of the effects of ionizing radiation on the nervous system, when the necessary resources became available. At its seventieth session, the Committee noted the establishment of the expert group and endorsed the revised timelines for the implementation of the evaluation, with approval of the report expected in 2027. At its seventy-first session, the Committee took note of the updated content of the report and noted the progress reported with regard to expanding the expert group[12] and developing literature searches, also noting that the literature review was in progress, with the first draft of the annex expected to be available for the seventy-second session, in 2025.

## 4. Strategy to improve collection, analysis and dissemination of data on radiation exposure, including consideration of the Committee's ad hoc working group on sources and exposure

16. At its seventy-first session, the Scientific Committee endorsed the proposed actions by the ad hoc working group on sources and exposure and agreed to: (a) endorse the proposal for a new UNSCEAR Global Survey on Medical Exposure, to be initiated in 2026 and carried out in two phases; (b) continue outreach efforts with the network of national contact persons to promote dialogue with a view to addressing common issues and concerns among Member States and provide related training and tools; and (c) request the secretariat to review and update the UNSCEAR online global survey platform, including with regard to the availability of questionnaires in languages other than English.

17. The Scientific Committee expressed appreciation for the work performed by the ad hoc working group on sources and exposure and approved the extension of its work until the Committee's seventy-second session, in 2025. The ad hoc working group currently consists of two small groups of experts, one on medical exposure and one on occupational exposure, and it was agreed that a new small group of experts would be created on public exposure. The purposes of these groups are to monitor the implementation of the recommendations outlined in the 2022 strategy to improve the collection, analysis and dissemination of data on radiation exposure and to advise the Committee on future data collection and exposure assessments.

---

[11] The expert group comprises 20 experts (including two critical reviewers) from 12 Member States and one observer from ICRP.

[12] The expert group comprises 36 experts (including five critical reviewers) from 13 Member States and two observers from ICRP and OECD/NEA.

18. The Scientific Committee re-emphasized the need for Member States to provide data on public, occupational and medical exposures to ionizing radiation in order to ensure that future assessments are as comprehensive as possible. Furthermore, the Committee welcomed the growing number of national contact persons (currently 112) and alternate national contact persons and expressed appreciation for the participation of national contact persons in the webinars, surveys and meetings organized by the secretariat and the ad hoc working group on sources and exposure and conducted since the seventieth session.

### 5. Implementation of the public information and outreach strategy for the period 2020–2024 and adoption of the public information and outreach strategy for the period 2025–2029

19. At its seventy-first session, the Scientific Committee noted the secretariat's progress report on public information and outreach activities in the period 2020–2024 and endorsed the public information and outreach strategy for the period 2025–2029. The strategy is aimed at disseminating, in particular among the scientific community, decision makers, the general public, young professionals, students and the media, the Committee's findings on levels, effects and risks of exposure to ionizing radiation. The main elements of the strategy are:

(a) Strengthening the Committee's input to the science-policy interface within and outside the extensive United Nations network;

(b) Enhancing engagement with scientific, diplomatic, academic and professional communities;

(c) Encouraging the involvement of young professionals in the work of the Committee.

20. The Scientific Committee welcomed the online publication of the UNEP booklet entitled *Radiation: Effects and Sources* in Italian, bringing the total number of language editions of the booklet to 16,[13] and also welcomed the planned publication of the booklet in Bulgarian in 2024 and in Urdu in 2025. The Committee urged the secretariat to update the booklet on the basis of its reports and annexes published since the most recent update, issued in 2016 (including those referred to in chapter III of the present report), with the goal of publishing the updated booklet in order to mark the seventieth anniversary of the Committee. It also encouraged the secretariat to further translate and promote the booklet.

21. The Scientific Committee welcomed the publication, in all six official languages of the United Nations, of the summary of its findings as contained in annex B to its 2020/2021 report. All the Committee's publications are available on the UNSCEAR website.[14] The Committee welcomed the ongoing work of the secretariat to publish the website in all six official languages of the United Nations in 2024.

22. The Scientific Committee invited member States to support the implementation of the public information and outreach strategy for the period 2025–2029 by undertaking outreach activities at the national and regional levels (e.g. dedicated UNSCEAR-related events and days).

## C. Update on the Committee's long-term strategic directions

23. At its sixty-sixth session, the Scientific Committee approved its long-term strategic directions and plan for the period 2020–2024. An update on progress in that regard is set out below.

---

[13] The 16 languages include the six official languages of the United Nations (Arabic, Chinese, English, French, Russian, Spanish) and 10 other languages (Czech, Dutch, German, Hindi, Indonesian, Italian, Japanese, Korean, Persian and Portuguese) (see www.unscear.org/unscear/en/publications/radiation-effects-and-sources.html).

[14] www.unscear.org/unscear/en/publications/index.html.

### 1. Establishing working groups on sources and exposure and on effects and mechanisms

24. At its seventy-first session, the Scientific Committee prolonged the mandate of the ad hoc working group on effects and mechanisms to continue its activities until the Committee's seventy-second session, in 2025. The prolongation would enable the ad hoc working group to: (a) continue monitoring the progress of the ongoing scientific evaluations of effects and mechanisms; (b) support the timely finalization of those evaluations, as needed; (c) support and monitor progress in the implementation of both the present and future programme of work; and (d) evaluate new scientific developments.

25. Also at its seventy-first session, the Scientific Committee prolonged the mandate of the ad hoc working group on sources and exposure with a focus on the implementation of the updated strategy for improving the collection of data on radiation exposure and a focus on the establishment of a subgroup on public exposure. The Committee noted that the ad hoc working group, together with small groups of experts on, respectively, occupational exposure, medical exposure and public exposure, would continue to: (a) monitor the literature and provide advice to the Bureau and Committee for ongoing data collection; and (b) evaluate available and new data sources relevant to the Committee's exposure evaluation, in order to work with the secretariat in preparation for the Committee's future evaluation on medical exposure to ionizing radiation, in 2026, and subsequently, on occupational and public exposure to ionizing radiation.

### 2. Inviting, on an ad hoc basis, scientists from other States Members of the United Nations to participate in the Committee's evaluations

26. The Scientific Committee noted that the secretariat and the Bureau had taken steps to involve 30 additional scientists since its seventieth session, including scientists from other States Members of the United Nations,[15] in supporting the Committee in conducting ongoing evaluations. That was particularly relevant for the ongoing evaluations of the effects of ionizing radiation on the nervous system and of the effects of ionizing radiation on diseases of the circulatory system, and the approved evaluations of public exposure to ionizing radiation and of second primary cancer after radiotherapy.

### 3. Increasing the Committee's efforts to present its evaluations and summaries thereof in a manner that attracts readers without compromising scientific rigour and integrity

27. The Scientific Committee referred to the outreach activities reported in paragraphs 19 to 22 above.

### 4. While maintaining its lead in providing authoritative scientific evaluations to the General Assembly, liaising closely with other relevant international bodies to avoid duplication of efforts

28. The importance of the Scientific Committee's evaluations in providing the scientific basis to relevant entities of the United Nations system and other international bodies for establishing international safety standards has continued to be demonstrated in the period since the seventieth session of the Committee. The Committee noted that its secretariat was a member of the Inter-Agency Committee on Radiation Safety, and that the current scientific basis provided by the Scientific Committee for the international safety standards remained valid. The Committee also noted that the secretariat continued to collaborate with IAEA, participating as an observer to the IAEA Commission on Safety Standards and Radiation Safety

[15] Austria, Italy, Netherlands (Kingdom of the) and Switzerland.

Standards Committee, and that the secretariat cooperated with several other international organizations[16] and regional organizations.[17]

29. The Scientific Committee welcomed and supported the continued cooperation of the secretariat with entities within the United Nations system and with other intergovernmental organizations with a view to promoting the Committee's work and exploring synergies and joint activities that would contribute to that work and support the collection and analysis of scientific data. The Committee specifically acknowledged the research framework agreement signed with the European Commission in June 2023, the memorandums of understanding signed with the International Commission on Radiological Protection (ICRP) in May 2024 and the ongoing dialogue with the Nuclear Energy Agency of the Organisation for Economic Co-operation and Development and the International Labour Organization, and requested the secretariat to report on cooperation with other entities at the seventy-second session.

## D. Future programme of work

30. Established by the Scientific Committee at its sixty-fifth session, in 2018, the ad hoc working group on effects and mechanisms supports the Bureau and the secretariat in monitoring the progress of the ongoing scientific evaluations and in evaluating new scientific developments between the sessions for consideration by the Committee.

31. At its seventy-first session, the Scientific Committee noted that, owing to considerable delays caused by the limited availability of resources in the secretariat, three scientific evaluations that had been planned to begin in the period 2020–2024 had not yet started. Among those was the new evaluation of the effects of ionizing radiation on the eye, which is now planned to begin in early 2025. The Committee confirmed the importance of the planned evaluation of the effects of ionizing radiation on the immune system and the overview on non-cancer health effects of exposure to ionizing radiation, the starting dates of which will be set in accordance with the availability of resources in the period 2025–2029.

32. Also at its seventy-first session, the Scientific Committee identified priorities for its future programme of work for the period 2025–2029 by discussing the assessment reports provided by the ad hoc working group on effects and mechanisms on three potential topics identified at the seventieth session. Of those, the evaluation of levels of radiation exposure to and effects on wildlife populations and ecosystems was considered to be urgent and should, if possible, be initiated early in the programme period 2025–2029. The ad hoc working group on effects and mechanisms was asked to assess to what extent the other two topics, the effects of prenatal exposure to ionizing radiation and the effects of ionizing radiation on ageing and lifespan, could be included in the planned overview on non-cancer health effects of exposure to ionizing radiation. In addition, as requested by the ad hoc working group on sources and exposure, the commencement of a new evaluation of medical exposure to ionizing radiation is planned for the period 2025–2029. Lastly, the Committee considered three topics for potential white papers – namely: (a) biomarkers and signatures of radiogenic diseases; (b) dose-effect relationships; and (c) survey of biophysical models of radiation action currently in use and their biological relevance – which may be initiated by the secretariat as resources allow.

33. However, given the current budgetary constraints and liquidity issues faced by the United Nations, it was not possible to establish a detailed programme timeline.

[16] Including the Preparatory Commission for the Comprehensive Nuclear-Test-Ban Treaty Organization, the International Civil Aviation Organization, ILO, UNEP, the Office for Outer Space Affairs, WHO, the International Agency for Research on Cancer, OECD/NEA, ICRP, the International Radiation Protection Association, the International Organization of Medical Physics and the Scientific Advisory Group for the Treaty on the Prohibition of Nuclear Weapons.

[17] Including the European Commission and the Ibero-American Forum of Radiological and Nuclear Regulatory Agencies.

34. Bearing in mind the high-quality and important scientific work conducted by the ad hoc working group on effects and mechanisms in monitoring the Scientific Committee's programme of work, the Committee extended the mandate of the ad hoc working group for one year to support and monitor progress in the implementation of the present and future programme of work and to evaluate new scientific developments.

## E. Administrative issues

35. The Scientific Committee welcomed the increased regular budget allocation for 2024 approved by the General Assembly and the Assembly's approval of two new staff positions in the secretariat.[18] However, the Committee expressed concern with respect to the reduction in the regular budget allotment in 2024, which had been affected by the liquidity challenges faced by the United Nations. The 2024 allotment of 60 per cent of the regular budget allocation for non-staff implementation costs had allowed the minimum resources necessary to prepare, organize and hold the seventy-first session of the Committee in accordance with its mandate. The Committee welcomed the voluntary financial support and in-kind contributions of 16 member States who had funded their travel to attend the seventy-first session in those exceptional circumstances.

36. While the Scientific Committee acknowledged that the contributions made to the general trust fund by six Member States[19] had allowed work to progress in a number of areas, such a method of funding continued to be neither predictable nor sustainable. The Committee welcomed General Assembly resolution 78/71, in particular paragraph 23, in which the Assembly requested the Secretary-General to strengthen support for the secretariat of the Committee. The availability of the increased regular budget allocation for 2024, and of increased allocations in the future, is essential for enabling the secretariat to provide adequate and efficient long-term services to the Committee in a predictable, sustainable and independent manner. The resource requirements for the Committee's work have increased owing to the ever-increasing amount of literature to be covered and the more formal methods adopted for evaluation.

37. The Scientific Committee agreed to hold its seventy-second session in Vienna from 16 to 20 June 2025.

[18] See General Assembly resolutions 77/119, 78/71 and 78/254 A.

[19] Australia, Austria, Belgium, Germany, Norway and Spain.

# Chapter III

## Scientific reports

38. The following two scientific annexes were approved by the Scientific Committee at its seventy-first session: (a) second primary cancer after radiotherapy; and (b) evaluation of public exposure to ionizing radiation.

## A. Second primary cancer after radiotherapy

39. At its seventy-first session, the Scientific Committee reviewed and approved for publication the annex on second primary cancer after radiotherapy. The conclusions reached by the Committee in that annex are summarized as follows:

(a) Ionizing radiation has been used to treat cancer for over a century. Treatment using radiation, which is primarily delivered through external beam technologies, comprises approximately 50 per cent of all cancer treatments. In general, 40 per cent of all cured cancer patients received radiotherapy as part of their treatment. As a result of improvements in cancer screening, treatment options and treatment efficacy, global cancer survival rates have increased, leading to an ever-expanding world population of cancer survivors;

(b) The evolution of radiotherapy and imaging has allowed better targeting of tumouricidal doses to the tumour and reductions in doses to surrounding tissues. Nevertheless, some undesired radiation doses are still delivered to surrounding tissues and organs. Efforts to characterize and quantify radiation doses close to and at a distance from the primary tumour are necessary to improve treatment and facilitate future epidemiological analyses of second primary cancer risk. Although it is generally understood that, for a given radiation dose, children are generally at more risk of tumour induction than are adults, more study is required to fully understand the combined and independent effects of age at exposure and attained age;

(c) As cancer survival rates improve, a growing number of individuals are now living well beyond the completion of their primary cancer treatments. While a primary cancer relapse or metastatic spread is a main concern in the years immediately following treatment, a second primary cancer becomes an important issue in the long-term. Cancer survivors have a higher risk of developing a new primary cancer than does the general population. Contributors to this higher risk include genetic susceptibility to cancer development, behavioural and environmental factors, and treatment components such as chemotherapy and radiotherapy;

(d) It is important to note that second primary cancers causally linked to radiation exposure represent only a fraction of all second primary cancers, even in patients treated with radiotherapy. Only rarely, for a specific tumour type, is radiation an attributable cause with high probability (e.g. in the case of sarcomas arising in areas exposed to high radiation doses);

(e) To obtain information on risks of second primary cancer after radiotherapy, an extensive literature search was conducted, and a meta-analysis was performed on haematopoietic tissues, connective tissues, female breasts, lungs, gastrointestinal organs, thyroid and brain. Information on radiation-related cumulative second primary cancer rates and associated latency periods was also reviewed;

(f) The meta-analysis yielded risks per unit dose, which were then compared with cancer risks per unit dose from other radiation epidemiological studies, such as those based on data from Japanese atomic bomb survivors. The calculated excess relative risks per unit dose, derived from the meta-analyses of the seven second primary cancer sites, were generally lower than the risks reported in other types of radiation epidemiological studies. For sarcoma (connective tissues), such comparisons are statistically compatible. For thyroid cancer, statistical compatibility depends on the choice of the comparison study. For other sites (haematopoietic tissues, female breasts, lungs, gastrointestinal organs and brain), the pooled estimates of risks from the radiotherapy cohorts are in general statistically significantly smaller than the corresponding risks from radiation studies of cohorts not subject to radiotherapy;

(g) On the basis of the relevant literature, the Committee is aware that between 5 and 15 per cent of cancer survivors may develop a second primary cancer. However, the Committee considered that only a small proportion of the total second primary cancers were likely to be attributable to radiotherapy. The annex on second primary cancer after radiotherapy has refined the Committee's general understanding of how many of the total second primary cancers can be attributed to radiotherapy. The absolute numbers depend upon the specific tissues at risk and the radiation doses received during radiotherapy treatment. In view of the significant benefit of radiotherapy, cancer patients should not be dissuaded from undergoing radiotherapy solely on the basis of concerns regarding the possible development of a second primary cancer. Nevertheless, the future design and development of radiotherapy should involve dedicated efforts to reduce second primary cancer induction.

## B. Evaluation of public exposure to ionizing radiation

40. The Scientific Committee has conducted an evaluation of worldwide levels of, and trends in, public exposure to natural and human-made sources of radiation based on three sources: (a) data from the UNSCEAR global survey on public exposure; (b) reviews and analyses published in peer-reviewed literature since 2007; and (c) data from regional and international organizations. The evaluation was critically dependent on the availability of reliable national data. The Committee expressed its gratitude to the national contact persons in 61 Member States and the other national experts who were involved in collecting, submitting and checking national data. The Committee emphasized that Member States' efforts were needed in the future to maintain and further extend the Committee's network of national contact persons and improve reporting of public exposure data for enhanced quality and reliability of future evaluations of sources and levels of exposure to ionizing radiation, as well as to improve geographical representation in the datasets.

41. Internal exposure of the public to natural radiation sources arises from inhalation of radon and thoron and their decay products, and ingestion of uranium and thorium series radionuclides and potassium-40. External exposures arise from cosmic radiation and terrestrial radionuclides. On the basis of a large number of measurements of radionuclide concentrations in soil, air and food samples taken in many countries of the world, global average annual doses from natural radionuclides were estimated. Exposure from industries involving naturally occurring radioactive materials was assessed either on the basis of data on radionuclide discharges from facilities, provided through the global survey on public exposure, or on the basis of published papers and reports.

42. While radiation protection is not in the mandate of the Scientific Committee, the Committee uses, for practical and pragmatic reasons, the radiation protection quantities introduced by ICRP and the International Commission on Radiation Units and Measurements. These quantities are used and reported by Member States and have been used in previous UNSCEAR reports. The quantities have changed over time.

43. Exposures from radon and thoron were calculated using the same dose coefficients provided in previous UNSCEAR reports,[20] in accordance with the

[20] *Sources, Effects and Risks of Ionizing Radiation: United Nations Scientific Committee on the Effects of Atomic Radiation 2000 Report to the General Assembly with Scientific Annexes*, vol. I (United Nations publication, 2000), annexes A and B; and *Sources, Effects and Risks of Ionizing Radiation: UNSCEAR 2008 Report*, vol. I, annex B.

Scientific Committee's recent review.[21] Other components of public exposure were evaluated using dose coefficients currently used by ICRP.[22]

44. The range of the average annual effective dose to the public from natural radiation sources was estimated to be 1–14 millisieverts (mSv). That range remained consistent with previous estimates of 1–13 mSv.[23] In the scientific annex on evaluation of public exposure to ionizing radiation, the global annual average effective dose from natural sources was estimated to be about 3.0 mSv, of which the major contribution (60 per cent) was from inhalation of radon, thoron and their decay products, which represented about 1.8 mSv (see the figure below). The ingestion of uranium and thorium series radionuclides and potassium-40 represented a smaller contribution, of about 0.5 mSv. External exposure to terrestrial radionuclides and cosmic radiation contributed 0.40 mSv and 0.30 mSv, respectively.

45. While the estimate of the global annual average effective dose has changed from 2.4 mSv, provided in previous reports,[24] this change does not necessarily reflect an actual change in public exposure. It reflects methodological improvements and a greater diversity in the data available from different locations and regions, which were not available for previous Committee estimates. This has had the resultant effect of a changed global average value. For example, the updated estimates of exposure to radon and its decay products are now more robust, covering over 60 per cent of the world's population, whereas the previous estimate reflected less than 40 per cent of the world's population.

**Global average annual effective dose from natural sources**

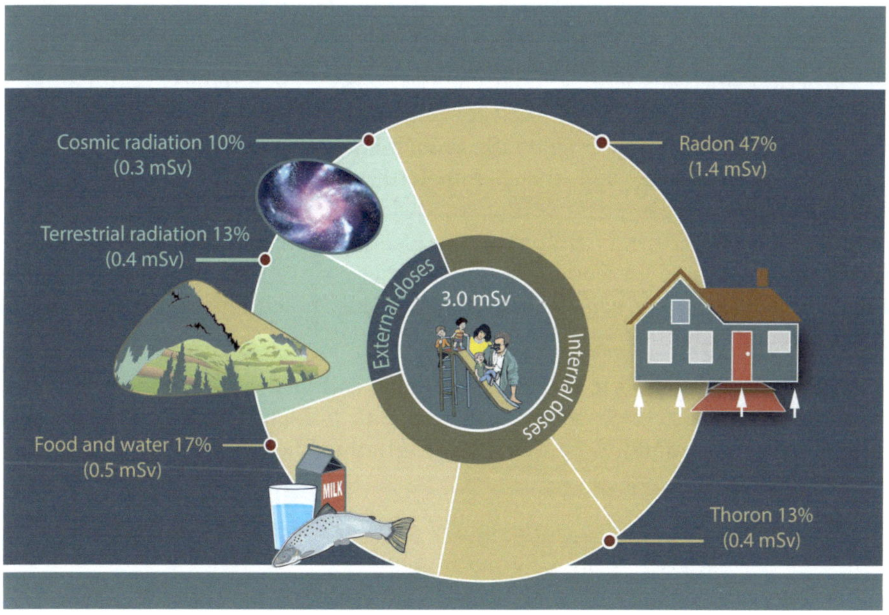


*Source:* Adapted from UNEP, *Radiation: Effects and Sources.* Copyright 2016 by UNEP.

46. Compared with natural sources, exposure of the public from human-made sources is generally lower, except in the rare situation of major accidents.

47. Exposures to the public as a consequence of nuclear power production were evaluated on the basis of discharge information and by means of the assessment

---

[21] *Sources, Effects and Risks of Ionizing Radiation: UNSCEAR 2019 Report*, annex B.

[22] N. Petoussi-Henss and others, "ICRP Publication 144: dose coefficients for external exposures to environmental sources", *Annals of the ICRP*, vol. 49, No. 2 (October 2020) and K. Eckerman and others, "ICRP Publication 119: compendium of dose coefficients based on ICRP Publication 60", *Annals of the ICRP*, vol. 42, No. 4 (August 2013).

[23] *Sources, Effects and Risks of Ionizing Radiation: UNSCEAR 2000 Report*, vol. I, annexes A and B; and *Sources, Effects and Risks of Ionizing Radiation: UNSCEAR 2008 Report*, vol. I, annex B.

[24] *Sources, Effects and Risks of Ionizing Radiation: UNSCEAR 2000 Report*, vol. I, annexes A and B; and *Sources, Effects and Risks of Ionizing Radiation: UNSCEAR 2008 Report*, vol. I, annex B.

methodology from the UNSCEAR 2016 report,[25] with some minor modifications. While the Scientific Committee has updated its estimates of public exposure from electricity generated from nuclear power sources, it has not updated its evaluation of exposures from other forms of electricity-generating technologies, including those based on the combustion of coal, natural gas, oil and biofuels and on geothermal, wind and solar power, as reported in the UNSCEAR 2016 report. In that report, it was noted that, when considering the amount of electricity generated in the year 2010 by each technology, the coal cycle had resulted in the largest collective dose to the global public and workers combined, followed by the nuclear fuel cycle.

48. Estimated annual doses to members of the public from nuclear power production facilities generally did not exceed a few tens of microsieverts. The results were slightly higher than those presented in the UNSCEAR 2016 report,[26] owing to the use of updated discharge data and population distribution information, and the application of an approach that took into account differences in the age-related dose coefficients and the proportion of different age groups in the population.

49. The worldwide collective effective dose per unit of electricity generated from nuclear power plants, and from uranium mining and milling, was estimated and is presented in the annex on evaluation of public exposure to ionizing radiation.

50. On the basis of an analysis of data on public exposures from other applications of sources of ionizing radiation, including incidental exposures resulting from medical, industrial and research applications, and from the use of consumer products and other goods, it was concluded that average worldwide annual doses from those sources ranged from a few microsieverts to several hundred microsieverts. Such exposures, however, have affected a large proportion of the world population.

51. In the period since the publication of the UNSCEAR 2008 report,[27] estimates of past and current public exposure at nuclear weapons test sites have been updated at Marshall Islands; Mururoa and Fangataufa, French Polynesia (France); New Mexico, United States of America; and the Semipalatinsk region of Kazakhstan. Past exposures at many of those sites were estimated to have been well above natural background levels, immediately following the tests, and there remains the potential for significant exposures in certain circumstances. However, continuing public exposures at those sites have been found to be generally much lower than those from natural background radiation.

52. Public exposure relating to the legacy of other military applications of nuclear and radioactive material, such as nuclear weapons production, maintenance and decommissioning, was found to be generally negligible, except for the radiological consequences of major accidents.

53. The Scientific Committee has previously published detailed reports relating to the accidents that occurred at the Chornobyl nuclear power plant in 1986 and the Fukushima Daiichi nuclear power plant in 2011. In the affected areas, there has been a continuous reduction of radionuclide levels in soil, air, water bodies, vegetation and foodstuffs, owing to radionuclide decay and migration within ecosystems and to environmental countermeasures in those areas. Doses to members of the public residing in those areas have decreased accordingly, and current annual doses to members of the public residing in the areas surrounding the Chornobyl nuclear power plant, in areas of Belarus, the Russian Federation and Ukraine, range from tens of microsieverts to a few millisieverts. In the area of the Fukushima Daiichi nuclear power plant, current annual doses to residents in non-evacuated municipalities range from a few microsieverts to 0.3 mSv. Since late 2023, treated water has been routinely discharged from the Fukushima Daiichi nuclear power plant site. Those releases commenced after the submission period for data to be considered in this current evaluation had closed. The Committee has acknowledged that data from environmental

[25] *Sources, Effects and Risks of Ionizing Radiation: UNSCEAR 2016 Report.*

[26] Ibid.

[27] *Sources, Effects and Risks of Ionizing Radiation: UNSCEAR 2008 Report,* vol. I, annex B.

monitoring and assessments are now available. Those data and any subsequent publications in the scientific literature will be monitored under the Committee's ongoing programme of work and considered for inclusion in a future evaluation.

54. A comprehensive uncertainty analysis of public exposures could not be included in this evaluation of public exposure, owing to the nature of the available data. Descriptive statistics were used to quantify the distributions of the environmental activity concentrations and public exposures, where possible. However, an updated methodology to quantify uncertainties and variabilities in dose assessments has been proposed to guide future evaluations.

55. Databases of relevant data have been created by various international organizations, including the Scientific Committee. The success of such efforts depends on the voluntary submission by Member States of annual and other reports and relevant data. Ensuring wide geographical representation in the Committee's future reports will depend on Member States' greater involvement in such activities.

# Appendix I

## Members of national delegations attending the sixty-sixth to seventy-first sessions of the United Nations Scientific Committee on the Effects of Atomic Radiation in the preparation of its scientific reports for 2024

| | |
|---|---|
| Algeria | S. Chelbani (Representative), Z. Lounis Mokrani (Representative), M. Ait-Ziane, D. T. Errahmani, A. Merad, M. Mezaguer |
| Argentina | A. J. González (Representative), D. Álvarez, A. Cánoba, M. Di Giorgio, M. G. Ermácora, A. E. Rossini |
| Australia | G. Hirth (Representative), C. Lawrence, S. Solomon, P. Thomas, I. Williams |
| Belarus | S. Sychyk (Representative), A. Stazharau (Representative), A. Ashurkevich, A. Avetisov, V. Drobyshevskaya, A. Nikalayenka, A. Rozhko, L. Sheuchuk, A. Yaumenenka |
| Belgium | S. Baatout (Representative), H. Vanmarcke (Representative), J. Vives i Batlle, G. Biermans, H. Bosmans, F. Dekkers, H. Engels, F. Jamar, M. Locquet, L. Mullenders, H. Slaper, P. Smeesters, P. Sonveaux, P. Willems |
| Brazil | L. Vasconcellos de Sá (Representative), D. de Souza Santos, P. Rocha Ferreira |
| Canada | J. Chen (Representative), D. Bracken Chambers, J. Burtt, P. Demers, J. Gaskin, R. Lane, K. Sauvé, B. Thériault, R. Wilkins |
| China | S. Liu (Representative), L. Chen, L. Dong, Y. Fa, Y. Gu, H. Guan, M. Huang, Y. Li, X. Lin, L. Liu, Q. Liu, Sh. Liu, J. Luo, L. Ma, G. Song, Q. Sun, J. Wang, Y. Wang, Q. Wu, X. Xia, S. Xu, D. Yang, L. Yuan, P. Zhou |
| Egypt | M. R. Ezz El-Din (Representative), M.A.M. Gomaa (Representative), A. A. Abdelaal, T. M. Morsi |
| Finland | A. Auvinen (Representative), S. Salomaa (Representative), R. Bly, E. Salminen, T. Siiskonen |
| France | D. Laurier (Representative), L. Lebaron-Jacobs (Representative), Y. Billarand, V. Blideanu, J.-M. Bordy, S. Candéias, J. Guillevic, C. Huet, A. Isambert, S. Jacob, J.-R. Jourdain, D. Klokov, K. Leuraud, F. Ménétrier, G. Pina, S. Roch-Lefevre, M. Simon-Cornu, R. Tamarat, J. Thariat |
| Germany | A. Friedl (Representative), P. Jacob (Representative), S. Baechler, L. Brualla, C. Engelhardt, C. Fournier, F. Gering, U. Gerstmann, T. Jung, J. C. Kaiser, K. Kammerlander, M. Kreuzer, R. Michel, W.-U. Müller, S. Neveling, W. Rühm, U. Schneider, S. Tapio, L. Walsh, W. Weiss, D. Wollschlaeger, H. Zeeb |
| India | A. Ghosh (Representative), S. K. Jha (Representative), A. Vinod Kumar (Representative), B. Das |
| Indonesia | N. R. Hidayati (Representative), E. Hiswara (Representative), T. Handayani, E. Kunarsih, E. D. Nugraha, D. H. Nugroho, T.B.M. Permata, H. Prasetio, N. Rahajeng, I. Untara |
| Iran (Islamic Republic of) | M. R. Kardan (Representative), K. Akbarzadeh, A. Rahimi Khoshmakani |

| | |
|---|---|
| Japan | M. Akashi (Representative), R. Kanda (Representative), T. Nakano (Representative), K. Akahane, K. Furukawa, T. Iwasaki, I. Kawaguchi, M. Kowatari, K. Ozasa, K. Tani, S. Yoshinaga |
| Mexico | J. Aguirre Gómez (Representative), G. Molina (Representative), M. E. Cuecuecha Juárez, R. F. Ortega, M. B. Robles |
| Norway | C. Robinson (Representative), P. Strand (Representative), K. Gulliksrud, L. K. Juvet, D. Oughton |
| Pakistan | R. A. Khan (Representative), M. Usman (Representative) |
| Peru | P. Fuentes Rivera Carmelo (Representative), A. Lachos Dávila (Representative), V. A. Muñante |
| Poland | M. Waligórski (Representative), L. Dobrzyński, K. Fornalski, M. Janiak, D. Kluszczyński, M. Kruszewski, P. Olko, J. Welsh |
| Republic of Korea | K.-W. Jang (Representative), H. S. Kim (Representative), B. S. Lee (Representative), S. H. Park (Representative), C. Choi, J. Jang, J. H. Jang, J. Jeong, S. Ji, U. Jung, S. Kang, B. S. Kim, H. Kim, J.-I. Kim, M. Kim, H. Lee, J. K. Lee, J. Lee, R. Lee, W. J. Lee, E K. Paik, J. Park, S. Seo, S. W. Seo, K. M. Seong, M. C. Song, J. Yoo, H. Yu |
| Russian Federation | A. Akleev (Representative), T. Azizova, S. Fesenko, S. Ivanov, V. Ivanov, L. Karpikova, S. Kiselev, D. Kononenko, E. Melikhova, S. Mikheenko, S. Romanov, V. Romanov, S. Shinkarev, R. Takhauov, V. Usoltsev, P. Volkova |
| Slovakia | L. Auxtová (Representative), D. Galanda (Representative), M. Berčíková A. Ďurecová, P. Fojtik , A. Froňka, P. Papirnik, K. Petrová, L. Tomášek |
| Spain | A. M. Hernández Álvarez (Representative), C. Álvarez García, J. M. Fernández Soto, M. T. Macías Domínguez, J. C. Mora Cañadas, D. Pérez-Sánchez, B. Robles Atienza, M. Sánchez Sánchez, F. J. Usera Mena, E. Vañó Carruana |
| Sudan | E.H.O. Bashier (Representative), A. M. Elamin Hassan (Representative), N. M. Hassan Suliman |
| Sweden | A. Almén (Representative), E. Forssell-Aronsson (Representative), I. Lund (Representative), A. Hägg, P. Hofvander, A. Wojcik |
| Ukraine | D. Bazyka (Representative) |
| United Arab Emirates | A. Al Shehhi (Representative), U. Al Bastaki, S. Al Mansoori, T. M. Almansoori |
| United Kingdom of Great Britain and Northern Ireland | S. Bouffler (Representative), A. Bexon, R. Haylock, R. Wakeford, W. Zhang |
| United States of America | V. Holahan, Jr. (Representative), A. Ansari, W. Bolch, H. Grogan, N. Harley, B. Napier, D. Pawel, G. Woloschak |

# Appendix II

## Scientific staff and consultants cooperating with the United Nations Scientific Committee on the Effects of Atomic Radiation in the preparation of its scientific reports for 2024

T. Anderson
M. Balonov
V. Berkovskyy
W. Bolch
H. Grogan
B. Napier
U. Schneider
K. Thiessen
L. Walsh

## Members of the Committee's ad hoc working group on the effects of radiation exposure and the mechanisms by which they occur at the sixty-sixth to seventy-first sessions

A. Friedl, Chair (Germany)
A. Auvinen, Rapporteur (Finland)
Z. Lounis Mokrani (Algeria)
J. Vives i Batlle (Belgium)
J.-R. Jourdain (France)
N. Hidayati (Indonesia)
K. M. Seong (Republic of Korea)
V. Ivanov (Russian Federation)
D. Pérez-Sanchez (Spain)
D. Pawel (United States)
P. Jacob, Chair (Germany)
L. Lebaron-Jacobs, Rapporteur (France)
M. Di Giorgio (Argentina)
R. Wilkins (Canada)
H. Zeeb (Germany)
K. Ozasa (Japan)
A. Akleev (Russian Federation)
A. Hernández Álvarez (Spain)
S. Bouffler (United Kingdom)
G. Woloschak (United States)

## Members of the Committee's ad hoc working group on sources and exposure at the sixty-sixth to seventy-first sessions

L. Vasconcellos de Sá, Chair (Brazil)
A. Ansari, Rapporteur (United States)
D. de Souza Santos (Brazil)
U. Gerstmann (Germany)
S. Romanov (Russian Federation)
P. Hofvander (Sweden)
J. Alsuwaidi (United Arab Emirates)
V. Holahan (United States)
J. Chen, Chair (Canada)
P. Thomas (Australia)
S. Liu (China)
A. Kryshev (Russian Federation)
A. Almén (Sweden)
A. Al Shehhi (United Arab Emirates)
A. Bexon (United Kingdom)

## Secretariat of the United Nations Scientific Committee on the Effects of Atomic Radiation

B. Batandjieva-Metcalf (sixty-sixth to seventy-first sessions)

F. Shannoun (sixty-sixth to seventieth sessions)

M. Zimmermann (sixty-seventh to seventy-first sessions)

L. Beaton (sixty-ninth and seventy-first sessions, seconded)

A. Gaw (sixty-eighth session, seconded)

N. Bushra (seventieth and seventy-first sessions, UNV)

# ANNEX A

# SECOND PRIMARY CANCER AFTER RADIOTHERAPY

## CONTENTS

*The attachments cited in this annex are electronically available for download from*
*https://www.unscear.org/unscear/en/publications/2024_1_Attachments.html*

# LIST OF ABBREVIATIONS

| | |
|---|---|
| AAA | Analytical anisotropic algorithm |
| ABM | Active bone marrow |
| ALL | Acute lymphoblastic leukaemia |
| AML | Acute myeloid leukaemia |
| AS | Assigned share |
| BCC | Basal cell carcinoma |
| CACO | Case control |
| CBCT | Cone-beam computed tomography |
| CCCS | Collapse cone convolution-superposition |
| CCSS | Childhood Cancer Survivor Study |
| CI | Confidence interval |
| CLL | Chronic lymphocytic leukaemia |
| CNS | Central nervous system |
| CT | Computed tomography |
| EAR | Excess absolute risk |
| EBRT | External beam radiotherapy |
| EOR | Excess odds ratio |
| ERR | Excess relative risk |
| FGFR2 | Fibroblast growth factor receptor |
| FISH | Fluorescence in situ hybridization |
| HDI | Human development index |
| HR | Hazard ratio |
| IAEA | International Atomic Energy Agency |
| IARC | International Agency for Research on Cancer |
| ICRP | International Commission on Radiological Protection |
| IGRT | Image-guided radiotherapy |
| IMRT | Intensity-modulated radiotherapy |
| IRR | Incidence rate ratios |
| LBTE | Linear Boltzmann transport equation |
| LET | Linear energy transfer |
| LSS | Life Span Study |
| MDS | Myelodysplastic syndrome |
| MR | Magnetic resonance |

| | |
|---|---|
| NCRP | National Council on Radiation Protection and Measurements |
| OAR | Organ at risk |
| OR | Odds ratio |
| PBC | Pencil beam convolution |
| PBS | Pencil beam scanning |
| PTV | Planning target volume |
| RBE | Relative biological effectiveness |
| RR | Relative risk |
| SD | Standard deviation |
| SIRs | Standardized incidence ratios |
| SEER | Surveillance, Epidemiology and End Results Program |
| SPC | Second primary cancer |
| TPS | Treatment planning systems |
| t-AML | Therapy-related acute myelogenous leukaemia |
| t-MDS | Therapy-related myelodysplastic syndrome |
| UNSCEAR | United Nations Scientific Committee on the Effects of Atomic Radiation |
| UPS | Undifferentiated pleomorphic sarcoma |
| VMAT | Volumetric modulated arc therapy |
| WHO | World Health Organization |
| 3D-CRT | 3D conformal radiotherapy |

# I. INTRODUCTION

1. In this annex the United Nations Scientific Committee on the Effects of Atomic Radiation (UNSCEAR) addresses the risk of second primary cancer following radiotherapy of a first primary cancer. It is important to note that the estimated risk of second primary cancers associated with radiotherapy is rather low. Therefore, this risk need not be routinely taken into account in the decision as to whether to include radiotherapy in multimodal cancer treatments. Radiation is one of many tools in the treatment and potential eradication of cancer, with approximately 30% to 50% of all cancer patients receiving some form of radiotherapy. Most importantly, it is estimated that approximately 40% of patients that survived cancer had received radiotherapy as part of their treatment regime [M15]. Furthermore, radiotherapy can be included as a critical component of palliative treatment for some cancer patients. As with nearly all forms of therapeutic medicine, there are side effects of treatment. For radiotherapy, these fall within two broad categories: normal organ toxicities (tissue reactions, previously called deterministic effects); and the induction of a second primary cancer (a stochastic radiation effect). During the radiotherapy treatment planning process, only the former side effect is currently explicitly included in the optimization process of the delivery of the tumour dose.

2. Owing to improvements in both treatment options and radiation delivery technologies, the survival rates of cancer patients have steadily increased. With longer post-therapy life expectancies comes an increasing occurrence of second primary cancers. Cancer survivors have a higher risk of developing a new primary cancer than the risk expected for the general population. The risk for developing a second primary cancer varies according to the organ in which the first primary cancer is located, and various known risk factors, such as patient ethnicity, sex, age at first cancer, treatment type of first cancer, genetic susceptibility for cancer development, as well as lifestyle and environmental factors [M12, S55, V2, W8]. For example, several recent studies on cancer survivors have concluded that survivors of cancers associated with smoking, alcohol, obesity, diet high in red or processed meat are at increased risk of developing a second primary cancer with the same shared lifestyle risk factors, and that these cases constitute a major part of all second primary cancer cases [F2, G2, S55].

3. Increased longer-term survival expectancy for patients treated as children, adolescents, or younger adults increases the time during which health outcomes for these patients can be observed. This is particularly important for younger cancer patients, given their higher overall susceptibility to radiation compared to older adults. It is important to note that the attribution of a second primary cancer to the use of radiation in the treatment of the first primary cancer is not absolute, and that the appearance of the second primary cancer could be attributed to other risk factors, such as patient genetic predisposition to cancer, environment factors, behavioural factors, or the prior use of chemotherapy. The risks of second primary cancer should be quantified to the extent that epidemiological evidence is available. These risks can then be used as the basis for:

*(a)* Selecting the most appropriate form or modality of radiotherapy for a cancer patient;

*(b)* Expanding the optimization process of the radiation treatment to both avoid normal tissue toxicity and minimize the risk of second primary cancer;

*(c)* Enrolling patients in long-term health surveillance programmes, especially those with high genetic risk of cancer induction; and

*(d)* Considering non-ionizing radiation treatment options if they are medically indicated and justified.

## A. Aims and objectives

4. This annex summarizes the current state of knowledge regarding the health risks of second primary cancer following radiotherapy. It is important for establishing the basis for long-term medical care and preventive interventions that will benefit the growing population of cancer survivors. The specific objectives of the annex are:

*(a)* To evaluate the risks of second primary cancers following radiotherapy through a detailed review of published radiation epidemiological studies from which a meta-analysis of risk is performed;

*(b)* To evaluate the contribution of ionizing radiation independently from the contribution of chemotherapy and their combined effects;

*(c)* To promote a rigorous dosimetric framework for the assignment and/or reconstruction of organ doses in patients that can contribute usefully to both retrospective as well as prospective epidemiological studies;

*(d)* To address future research needs to both assemble suitable patient cohorts for risk studies, and to explore how genetics and radiotherapy interact, as well as the link between radiotherapy and other carcinogenic agents.

5. The annex comprises seven chapters, supported by three electronic attachments. Chapter I introduces the aim, objectives, background for the evaluation, and the scope of the literature review conducted. Chapters II to IV contain summaries of relevant information regarding oncology, radiobiology, and dosimetry. Chapter V deals with the available information on epidemiology and sets the basis for the risk estimates presented in the annex, while chapter VI summarizes future research needs. Finally, chapter VII sets out general conclusions. A glossary of terms is presented at the end of the annex.

## B. Background

### 1. The global cancer burden

6. Cancer is a leading cause of death globally and is anticipated to surpass cardiovascular disease as the leading cause of death in most countries over the course of this century [B41]. At present, cancer is the leading cause over cardiovascular disease in 57 countries, including China, while cardiovascular disease leads over cancer in 70 countries, including Brazil and India.

7. The Cancer Atlas [A2], estimated that there were 18.1 million cases and 9.6 million cancer deaths in 2018 (including non-melanoma skin cancers) worldwide, with 1 in 4 men and 1 in 5 women developing the disease, and 1 in 8 men and 1 in 11 women dying from it. In addition, there were 43.8 million persons living with cancer who were diagnosed within the previous five years. These figures are in agreement with the estimates of the International Agency for Research on Cancer (IARC) and the Global Cancer Observatory of the World Health Organization (WHO), according to which there were 20 million new cases of cancer and 9.7 million deaths from cancer worldwide in 2022 [I3]. The most commonly diagnosed cancers worldwide were those of the lungs (2.48 million), female breast (2.30 million), colon/rectum (1.93 million), and prostate (1.47 million). The most common primary sites leading to cancer death were

lung (1.82 million deaths), colon/rectum (904,000 deaths), liver (758,000 deaths), and female breast (666,000 deaths) [I3].

8. It has been estimated that, out of approximately 1.2 million new cancers diagnosed every year in the United States, 95,000 are second primary cancers, although not all of them are associated with radiotherapy, and therefore account for 6–10% of all cancer diagnoses, making second primary cancer the fourth or fifth most common cancer in the United States [B24]. It has also been estimated that between 5% and 15% of childhood cancer survivors develop a second primary cancer within 20 to 30 years of the diagnosis of their first cancer [D9, F16, N9, O2, R5].

9. There is substantial global heterogeneity in both cancer incidence and mortality rates due to differences in exposure to risk factors and in life expectancy [W14]. The Cancer Atlas estimated that in 2018 half of the new cancer cases and cancer deaths in the world occurred in Asia and that approximately one fourth of all new cases (4.2 million) and one fifth of deaths (1.9 million) occurred in Europe, despite the region having less than one tenth of the global population [A2]. As shown in figure I, age-standardized rates for the majority of cancer types are higher in countries with higher values of human development index (HDI), although some cancers such as sarcoma, cervical cancer, and nasopharyngeal cancer are more frequent in countries with low HDI. Differences among countries in the occurrence of most common cancers indicate variations in the epidemiological burden, which can be attributed in part to the lack of robust early screening programmes and the underreporting of cases in countries with more limited access to healthcare services. While in absolute terms, the burden is highest in countries with high and very high HDIs, in relative terms, the increases will be proportionally greater in countries with low and medium HDIs. The anticipated surge in the annual number of new cancer patients over the coming decades, particularly in countries with low or medium HDIs (figure II) are coincident with the lack of cancer therapies access such as radiotherapy systems (figure III) [I1].

Figure I. Age-standardized rates for cancer incidence and mortality according to the human development index[a] as of 2018 [W14]

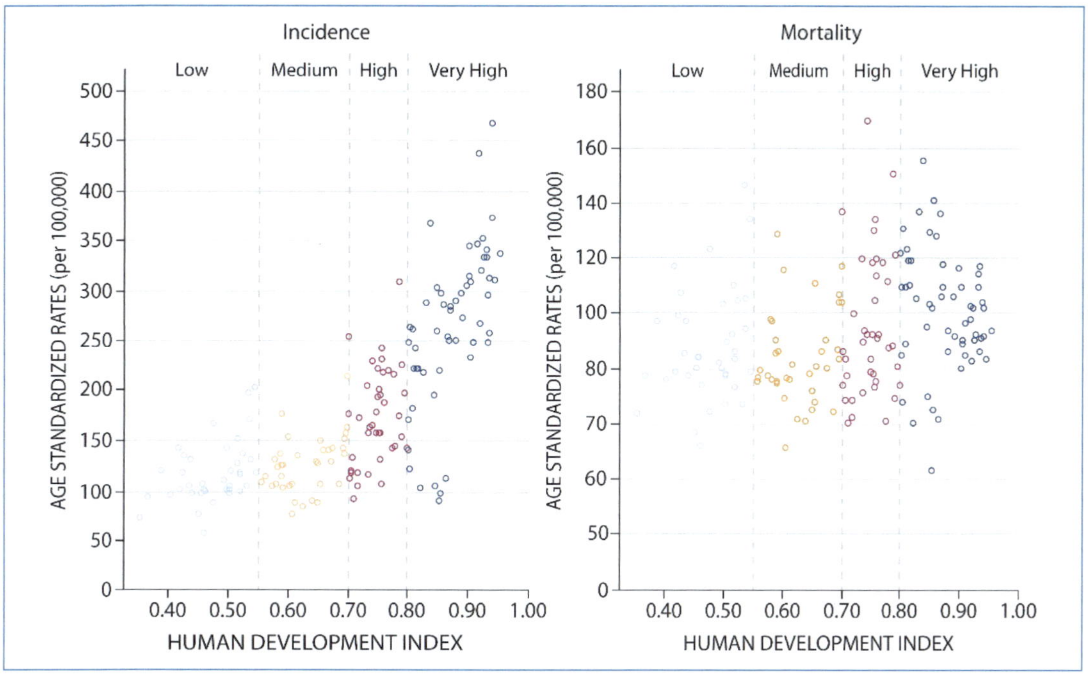


[a] The Human Development Index (HDI) is a summary measure of average achievement in key dimensions of human development: a long and healthy life, being knowledgeable and having a decent standard of living. The HDI is the geometric mean of normalized indices for each of the three dimensions.

Figure II. Projected trends in annual numbers of breast cancer cases in Ethiopia (HDI=049) and in Germany (HDI=095) over the period 2010–2110 [W14]

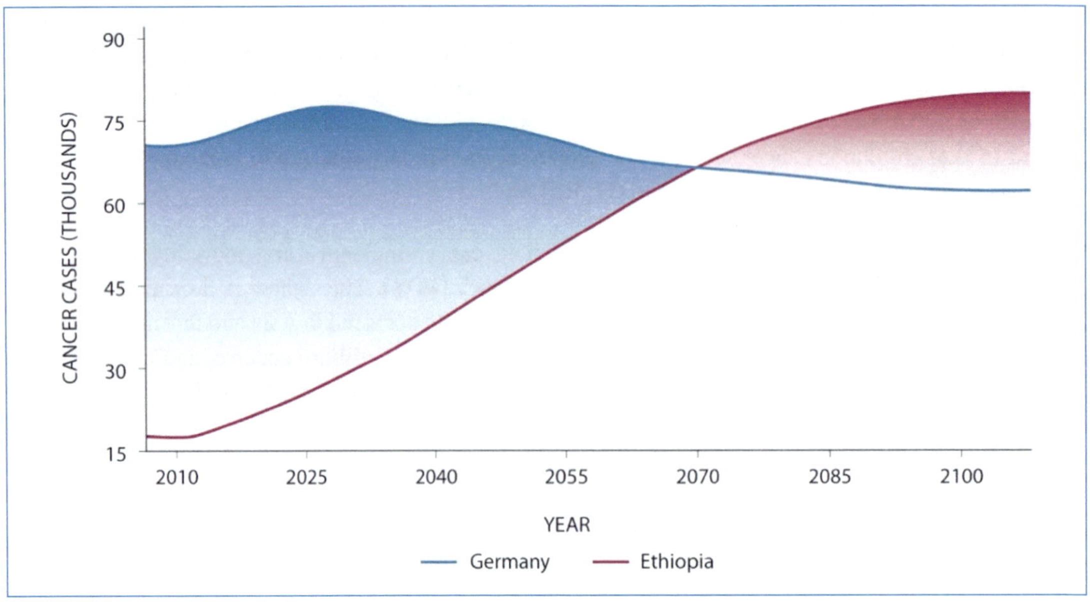


Figure III. Estimates of the number of people served by one radiotherapy unit category for each country [I1]

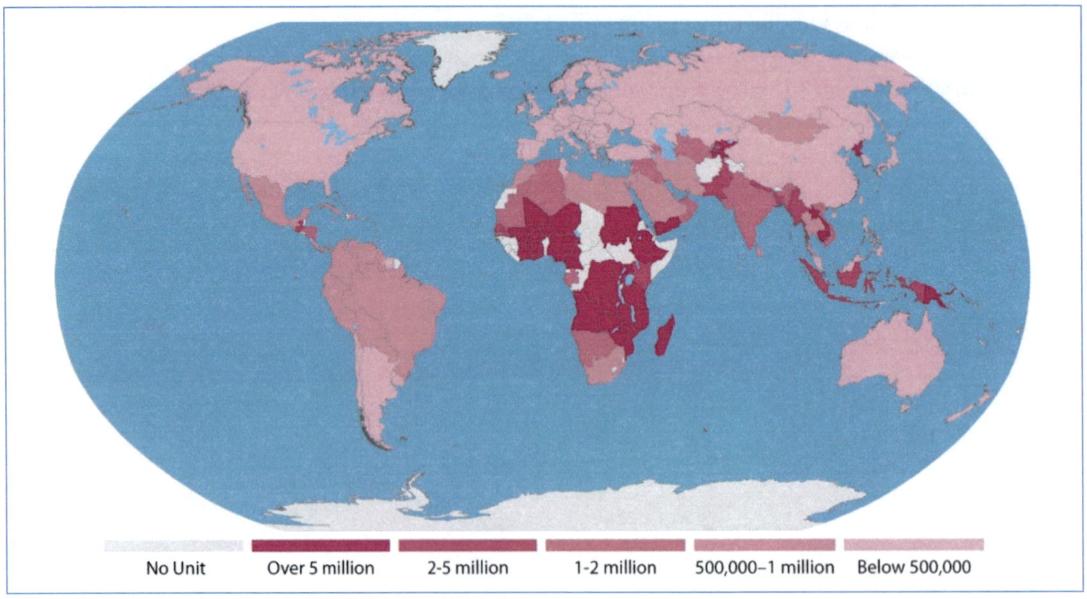


## 2. Global cancer survival rates

10. Advances in the technology of cancer therapy, the adoption and promotion of early screening and detection, and the advancement of supportive post-therapy healthcare have all contributed to a steady rise in the five-year relative survival rates of cancer patients worldwide [B5]. Data collected by IARC through its Global Cancer Observatory [I3], highlight increases in sex- and age-averaged five-year survival rates for various cancers across the seven participating countries: Australia, Canada, Denmark, Ireland, New

Zealand, Norway, and the United Kingdom. In 1995, the net five-year cancer survival rates in these countries averaged 52.5% for colon cancer, 10.5% for lung cancer, 9.1% for liver cancer, and 4.5% for pancreatic cancer. As of 2014, the net five-year survival rates were 64.4%, 14.7%, 13.4%, and 7.9%, respectively, representing increases of between 23% and 76%.

11. The global distribution of five-year cancer survival rates varies significantly by continent and country as reported in CONCORD, a programme for worldwide surveillance of trends in cancer survival. The third cycle of this programme, CONCORD-3, has updated the global surveillance of trends in cancer survival rates to include patients diagnosed up to 2014 in more than 70 countries [A20]. The data reported cover 18 of the most common cancers, and are representative of 75% of all the cancers diagnosed worldwide each year. Figure IV shows the global distribution of five-year net survival rates for seven geographic regions: Africa, Asia, Central America, Europe, North America, Oceania, and South America.

12. In the United States, the overall five-year relative survival rate for all cancers combined that were diagnosed between 2009 and 2015 was 67% [S38]. For all stages combined, survival rates were highest for prostate cancer (98%), melanoma of the skin (92%), and female breast cancer (90%); and lowest for cancers of the pancreas (9%), liver (18%), lung (19%), and oesophagus (20%). Five-year relative survival rates for adolescent and young adult patients aged between 15 and 39 years, diagnosed between 2009 and 2015, were generally similar across age groups (range: 83%–86%) and were comparable to the survival rate in children (84%), but substantially higher than those in adults aged 40 years or more (66%).

13. Cancer is currently treated using a range of modalities, such as radiotherapy, surgery, chemotherapy, immunotherapy, and hormonal therapy. Radiotherapy remains an important component of cancer treatment with approximately 50% of all cancer patients being treated with this modality during the course of their illness [B5, D14]. It is estimated that radiotherapy contributes towards 40% of cancer treatments that are demonstrated to be curative.

14. The burden of second primary cancers associated with radiotherapy depends on a number of factors that vary from country to country, and within a country from decade to decade: the lifetime baseline incidence of cancer, the proportion of cancers treated with radiotherapy and the doses delivered during radiotherapy. For example lifetime baseline risks of incidence of all solid cancers in Germany, based on data between 2013 and 2015, were 40.0% (95% CI: 38.7, 41.3) and 35.1% (95% CI: 33.8, 36.5) for men and women, respectively [W2]. Taking these risks as typical of high-income countries and making the assumption that half of all cancer patients would receive radiotherapy as part of their treatment, it can be estimated that around 20% of people living in a high-income country are likely to receive radiotherapy at some stage during their life.

Figure IV. Global distribution by continent and country of age-standardized five-year net survival for adults (15–99 years) diagnosed during 2010–2014 with *(a)* colon cancer or *(b)* breast cancer (women), and *(c)* children (0–14 years) diagnosed with acute lymphoblastic leukaemia [A20]

*(a)* Colon cancer

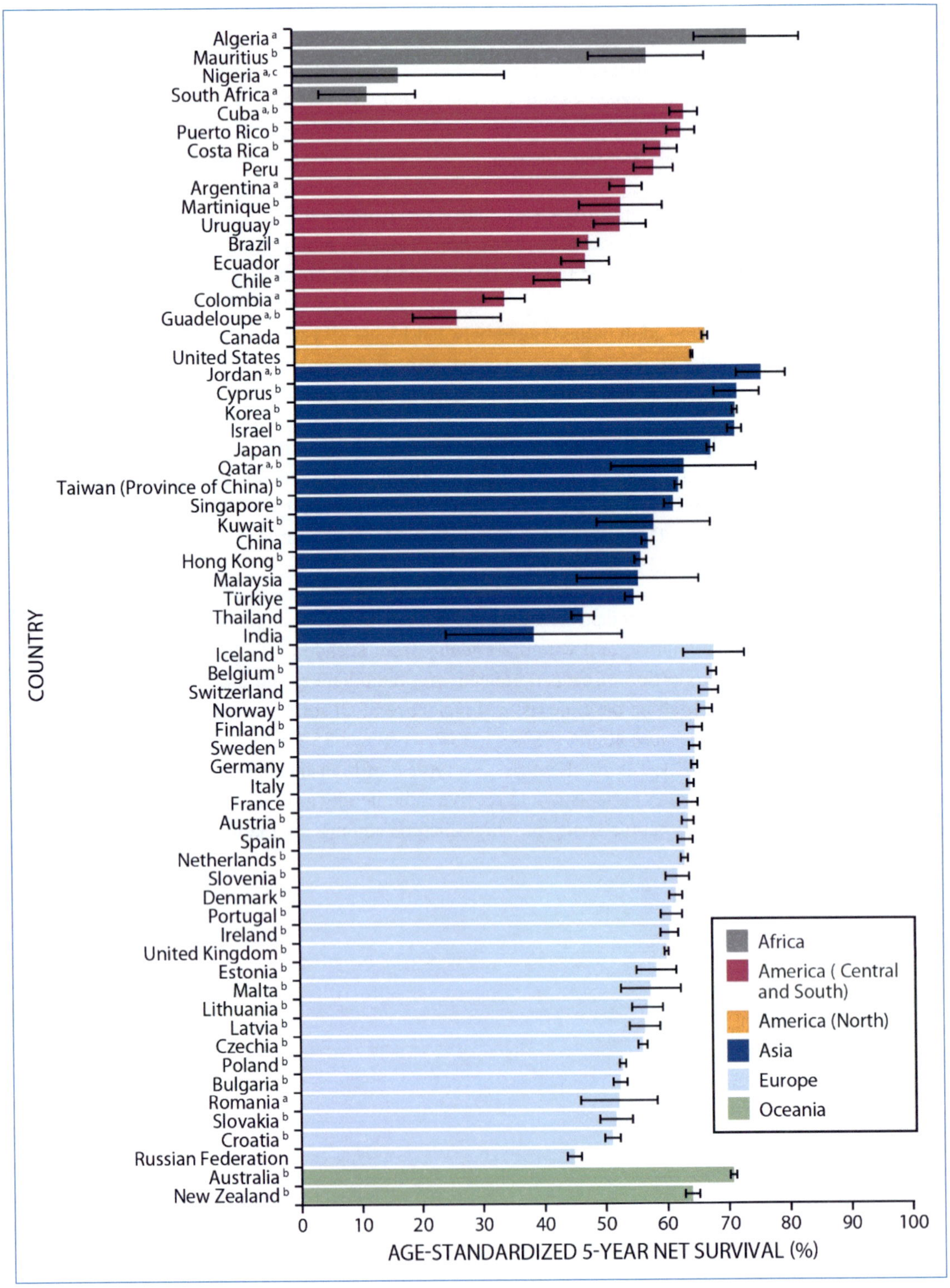

*(b)* Breast cancer (women)

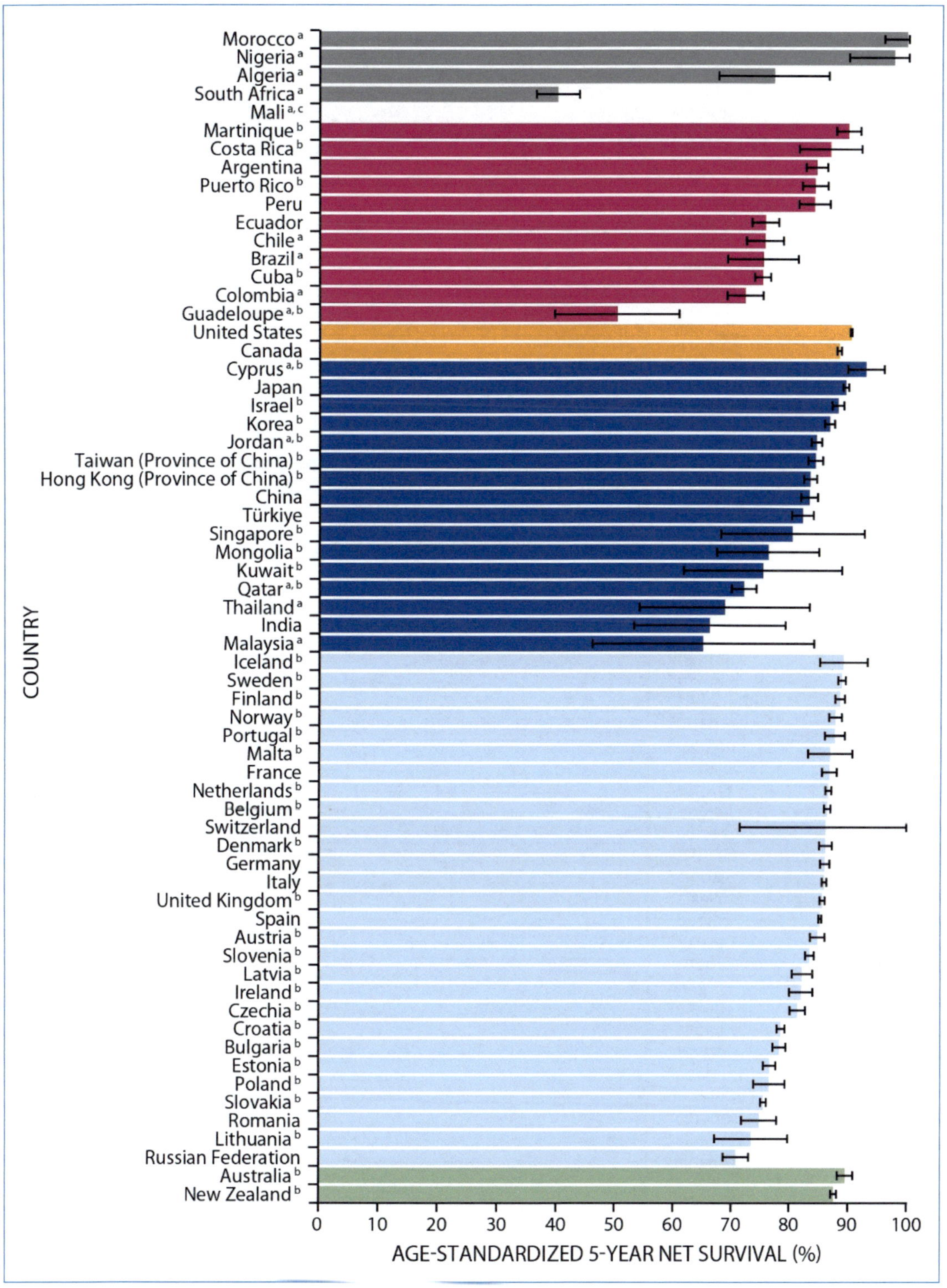

*(c)* Children (0–14 years) diagnosed with acute lymphoblastic leukaemia

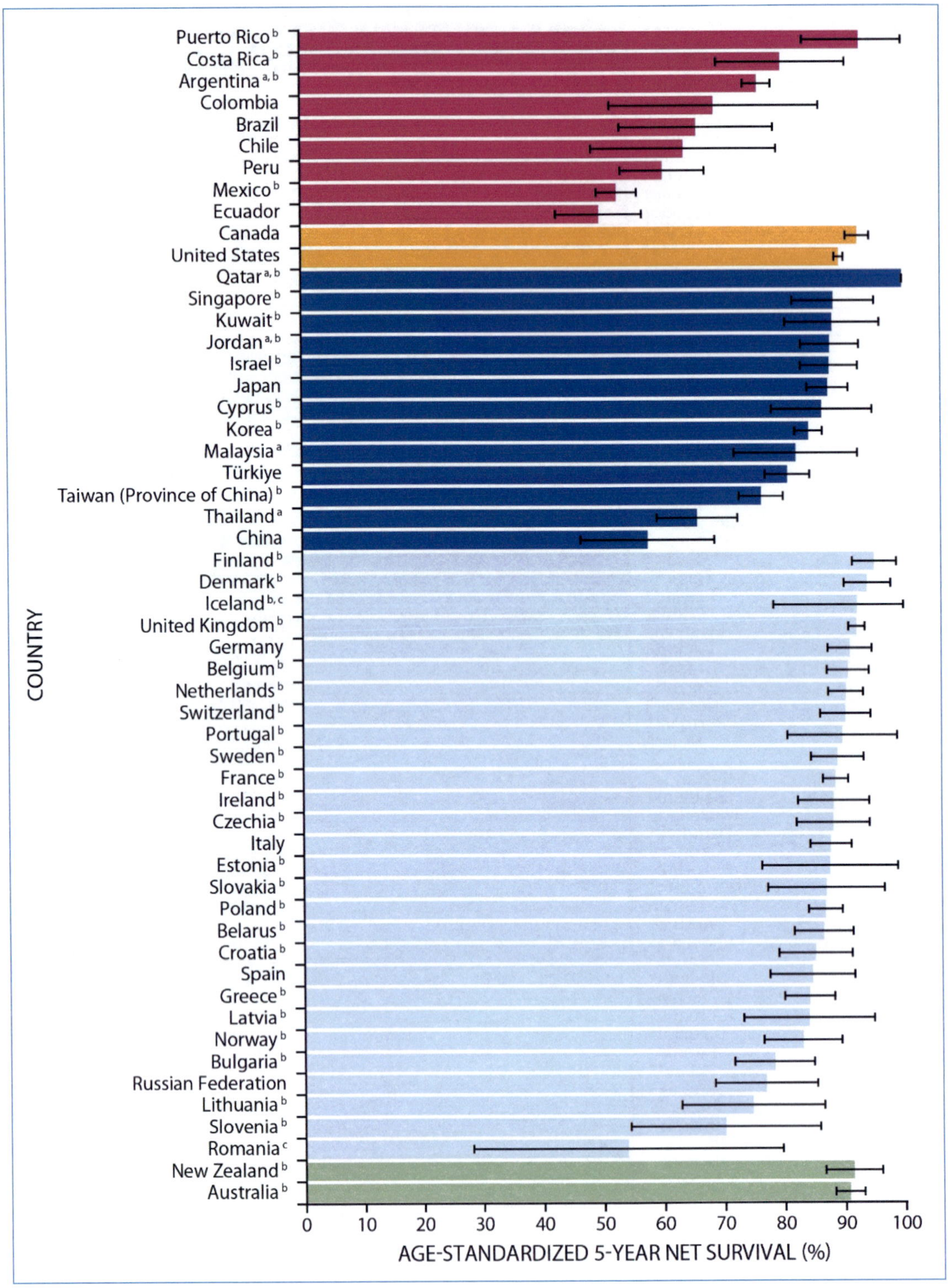


[a] National estimate flagged as less reliable because the only available estimates are from a registry or registries in this category.

[b] Data with 100% coverage of the national population.

[c] National estimate not age-standardized.

15. In 2016, there were an estimated 3.05 million cancer survivors in the United States treated with external beam radiotherapy, accounting for 29% of all cancer survivors. The majority of cancer survivors who had received radiotherapy were breast (40%) and prostate (23%) cancer patients, followed by head and neck cancer (5.8%), lymphoma (5.6%), uterine (3.9%), and rectal cancer (3.8%) patients [B53]. The number of cancer survivors who receive radiotherapy in the United States is projected to reach 4.17 million by 2030. Figure V shows the number of survivors of the 10 most prevalent cancers in the United States who received or are projected to receive radiotherapy for the period 2000 to 2030 with a breakdown by cancer site.

Figure V. Cancer site-specific distribution of five-year cancer survivors in the United States treated with radiation [B53]

The y-axis changes scale to allow visualization of breast and prostate cancer survivors (top half of the graph) as well as trends for cancers with an overall lower prevalence (bottom half of the graph)

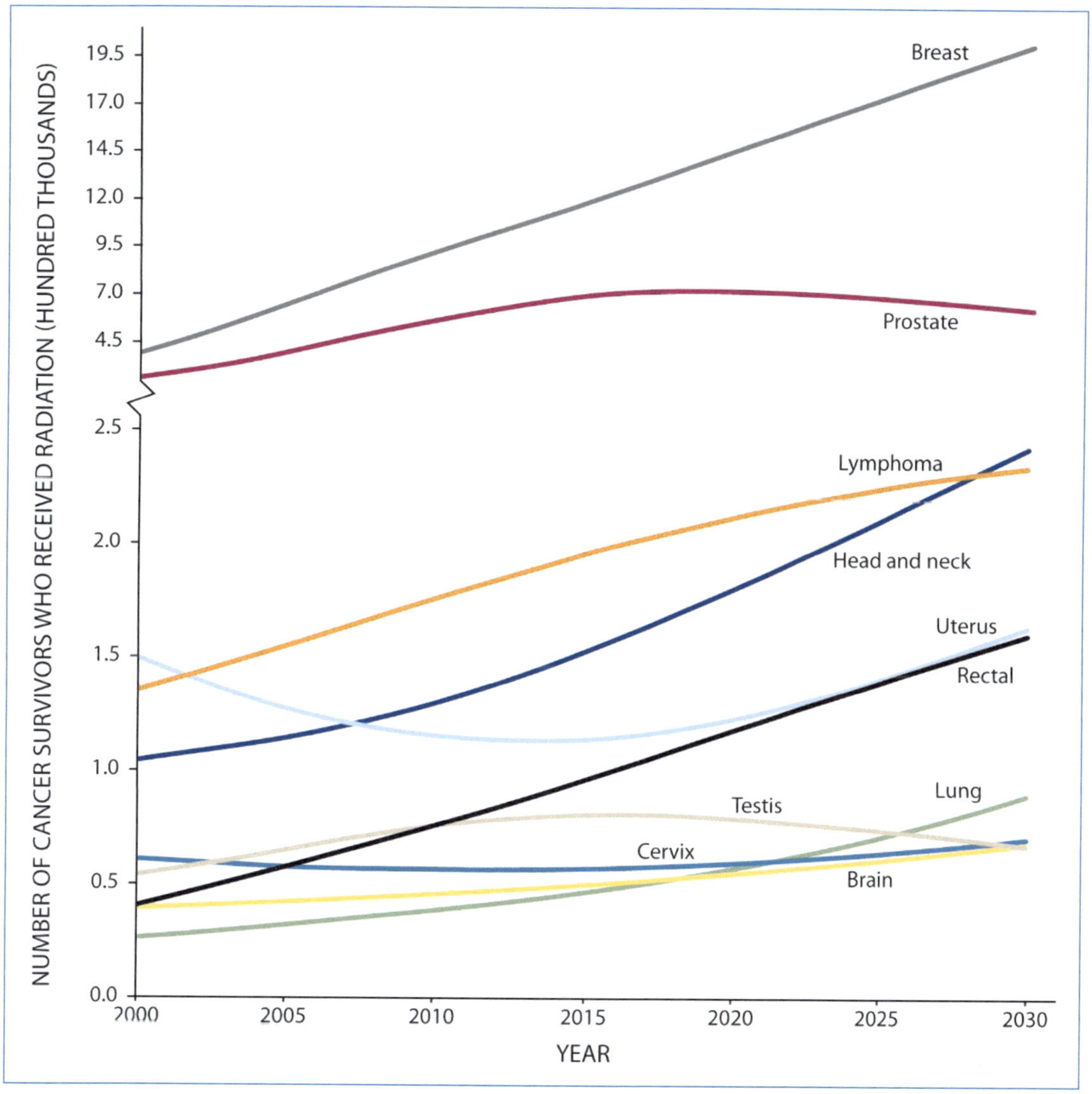

## 3. Definition of a second primary cancer

16. A second primary cancer is a new cancer arising in a person that had previously developed a primary cancer. In contrast to recurrences or metastases, a second primary cancer originates from another progenitor cell than the first primary cancer and can be differentiated from it by histological or molecular methods. A second primary cancer can arise spontaneously or be associated with genetic or environmental risk factors. In addition, some second primary cancers are caused by the treatment of the first cancer with radiotherapy and/or chemotherapy. At the present time, it is not feasible to morphologically distinguish a radiation-associated second primary cancer from those not associated with radiotherapy [T20]. Future technologies may well permit such an identification related to genetic changes specific to radiation-induced cancers, but this diagnostic accomplishment has yet to materialize. Several sets of criteria have been proposed to categorize a second primary cancer as radiation-induced, although more specifically, these conditions increase the likelihood that the second primary cancer is radiation-induced. Examples of these criteria are those based on an original proposal by Cahan et al. [C2] or more recently by Singh et al. [S42]:

*(a)* A radiation-induced cancer must have arisen in an irradiated tissue volume;

*(b)* A sufficient latent period must have elapsed between the initial irradiation and the clinically assessed second primary cancer;

*(c)* The treated tumour and/or the clinically assessed second primary cancer must have been biopsied (the two tumours must represent different histologies);

*(d)* The tissue in which the clinically assessed second primary cancer arose must have been normal (both metabolically and genetically) prior to radiation exposure.

17. Newer technologies are under current development that may assess somatic mutational profiles and clonal relationships that allow distinctions between recurrence or metastatic spread of the first primary cancer and a newly formed second primary cancer.

18. It is also acknowledged that the overall risk of radiation-induced second primary cancers is small, and therefore, large-scale epidemiological studies of irradiated patients followed for prolonged periods (cohort studies) are typically required to provide reliable estimates of risk [N6]. Cancer registry data are now sufficiently robust to provide overall estimates of second primary cancer risks among survivors of specific cancers; in particular, dose-response relationships can be evaluated in epidemiological research such as in case-control studies (chapter V).

## 4. Evolution of techniques for radiation delivery to cancer sites

### (a) *Techniques for external delivery of radiation*

19. Techniques and technologies for delivering radiotherapy have evolved significantly over the past decades [B5, H40]. The earliest forms of external radiotherapy systems used either orthovoltage X-rays (since 1896), or radioactive sources of $^{60}$Co, from 1951, and $^{137}$Cs, from 1955, to deliver external beams of photons to the patient in what was termed “teletherapy” or “therapy at a distance”. With the advent of the medical linear accelerator or LINAC—in 1953 at Hammersmith Hospital in London (United Kingdom), and in 1956 at Stanford University (United States)—it became possible to irradiate tumours with megavoltage external photon beams without relying on radioactive sources. The size and entrance

direction of these external beams were first determined by one or more 2D planar X-ray images of the cancer patient. For each external megavoltage photon beam, regions of dose delivery and regions of dose avoidance could be defined through the construction of patient-specific shaping blocks for conforming the irradiated field.

20. With the arrival of 3D anatomic imaging of the patient by computed tomography (CT) in the mid-1970s, 2D radiotherapy gave way to 3D conformal radiotherapy (3D-CRT) in the mid- to late 1980s. Beam conformality was achieved by the introduction of beam blocking devices in the linear accelerator gantry, called multi-leaf collimators. Further conformality of the beam, which resulted in an improved avoidance of critical organs, became possible in the mid-1990s through the establishment of intensity-modulated radiotherapy (IMRT). This technique uses multi-leaf collimators which move during irradiation to modulate the field size, shape, and dose rate. An IMRT was shown to improve the therapeutic ratio for several tumour sites over 3D-CRT.

21. As irradiated fields became more conformal, the potential to miss the target tumour volume due to organ motion and variations in patient set-up became greater. When critical anatomical structures are close to the tumour, a slight positional error may also lead to the delivery of inadvertent radiation to a normal organ or tissue. Kilovoltage image-guided radiotherapy (IGRT) with improved soft tissue contrast was therefore developed starting in the early 2000s to avoid these problems. Radiotherapy systems were upgraded to include initially two orthogonal kV X-ray imaging devices, later a CT scanner with a cone-like irradiation field (cone-beam computed tomography or CBCT), and more recently magnetic resonance (MR) scanners for image guidance. These imaging devices allow more precise positioning of the patient prior to each irradiation. This improvement in same-day positional accuracy has made it feasible to increase the treatment dose to the tumour, allowing in turn for improvements in the therapeutic ratio. IGRT has also lead to the development of adaptive radiotherapy by which real-time adjustments are made to the radiotherapy treatment plan originally generated on the pre-treatment planning CT image of the cancer patient.

22. Another form of IMRT that has been developed rather recently is the volumetric modulated arc therapy (VMAT) which delivers the dose continuously as the treatment machine rotates around the patient. This technique accurately shapes the absorbed dose to the tumour while reducing the dose to the organs surrounding the tumour. Further technological advances in the early 2010s have enabled stereotactic body radiotherapy, a technique which delivers very high individual doses precisely over only a few treatment fractions to ablate small, well-defined primary and oligometastatic tumours anywhere in the body. Any tissue adjacent to the tumour is likely to be damaged because of the high absorbed dose delivered. However, as the amount of normal tissue in these high-dose regions is small, clinically significant toxicity has been shown to be low.

23. For all forms of external beam radiotherapy (EBRT) which utilize megavoltage photon beams, distributions of absorbed dose have a maximum at tissue depth near the entrance to the patient (build-up region), and then decline roughly exponentially with depth along the beam central axis. Normal organs and tissues can absorb radiation energy through either internal photon scatter and the production of secondary particles, or through further beam delivery following transit through the tumour, called exit dose. Alternative means of EBRT employ the use of charged particles, particularly protons or carbon ions, which have energy-specific ranges. During traversal of body tissues, the rate of energy loss of a charged particle (linear energy transfer or LET) increases with decreasing particle energy, reaching a peak in energy deposition rate (the Bragg peak) at the end of their physical range in tissue. Exploiting the Bragg peak, charged-particle radiotherapy can be delivered close at the tumour location, yielding an exit dose near zero, in sharp contrast to that of megavoltage radiotherapy photon beams. While initially proposed in 1946 by Robert Wilson, proton radiotherapy of cancer patients began in 1954 at the Berkeley Radiation Laboratory in the United States. Starting in the mid-to-late 2000s, proton therapy centres have

greatly expanded worldwide with an increasing focus on the treatment of childhood cancers, owing to the need for reduced non-target dose delivery. Higher values of LET have been achieved through carbon ion radiotherapy, which was first utilized clinically in cancer treatment in 1994 in Chiba, Japan.

### (b) *Techniques for internal delivery of radiation*

24. An alternative radiotherapy modality involves the introduction of a radioactive source within the patient's body. There are two broad classes of internal radiotherapy: brachytherapy, and radiopharmaceutical therapy. Brachytherapy involves the placement of encapsulated radionuclides (typically emitters of beta or gamma radiation). There are two general forms of brachytherapy: high dose-rate and low dose-rate. For high dose-rate brachytherapy, the radionuclide source is temporarily placed in the tumour for a few minutes and then removed. The positions of the source can be precisely adjusted to create a customized distribution of the absorbed dose. High dose-rate brachytherapy is most commonly used for the treatment of prostate, cervical, colorectal, and head and neck cancers. Brachytherapy for cervical cancer was performed as early as 1919. For low dose-rate brachytherapy, the radioactive sources are positioned inside or immediately adjacent to the tumour for a minimum of several days, but they may be left in position permanently. Low dose-rate brachytherapy is most commonly used to treat brain tumours, but also can be applied to the treatment of prostate cancer. Pulsed dose rate brachytherapy is also a therapy option in which higher activity radionuclide sources are used, compared to those used in low dose-rate brachytherapy. Absorbed doses are therefore delivered in short exposures of 10 to 30 minutes every hour to approximately the same total prescription dose as used in low dose-rate brachytherapy.

25. For patients with cancer in its advanced stages, characterized by tens to hundreds of metastatic lesions, EBRT is not an option, and therefore, a systemic cancer treatment is needed. Options include chemotherapy or various forms of immunotherapy. Radiopharmaceutical therapy was first applied to the treatment of thyroid cancer in 1941 using the radionuclide $^{131}$I. Promising forms of systemic cancer therapy include the use of radiopharmaceuticals labelled with beta-emitting radionuclides, but there is an increasing clinical interest in the use of radionuclides that emit higher LET alpha particles [I7]. Targeting of the radiopharmaceutical to the cancerous lesions can include a variety of mechanisms, such as cell sequestration, passive diffusion, active transport, chemotaxis, antibody-antigen complexation, receptor binding, or physiochemical adsorption. The majority of therapeutic radiopharmaceuticals are delivered intravenously, and therefore, prior to tumour uptake and body excretion, a variety of normal organs and tissues receive a radiation dose through blood transit. If the mechanism of tumour uptake is not unique to body tissues, other source organs may take up the radiopharmaceutical delivering an off-target radiation dose. Late effects, such as induction of second primary cancers in off-target tissues, have to be considered because of the continuing increase of patients treated with radiopharmaceutical therapies. Radioembolization is another form of internal radiotherapy in which encapsulated radionuclides such as $^{90}$Y are encased in glass or resin-based particles for injection into the arterial blood supply of the host organ—most notably the liver—where they are captured within the capillary bed of the treated tumour for localized irradiation.

## 5. Treatment planning and dose delivery to normal tissues in radiotherapy

26. In radiotherapy, radiation is delivered to achieve a therapeutic benefit within the gross tumour volume, with a margin for microscopic tumour extensions called the clinical target volume. In addition, uncertainties in positioning and delivery require the use of a planning target volume (PTV) and corresponding irradiation of healthy tissue in this volume. Related to the target volume is the treated

volume, which is the volume of tissue planned to receive a prescribed dose which might be much larger than the PTV for conventional radiotherapy [I6]. Tissues outside the PTV do not benefit from the delivery of radiation, but their irradiation is unavoidable. Non-target dose within the treated volume is the dose within the primary field borders and includes entrance and exit dose along the beam path. Non-target dose outside the treatment volume is delivered from:

*(a)* Leakage radiation from the treatment head of the linear accelerator;

*(b)* Radiation scatter and other interactions within the collimators and other beam modifiers;

*(c)* Radiation scatter within the patient's body (e.g., Compton collisions of photons);

*(d)* Secondary radiations (e.g., neutrons generated by nuclear absorption reactions from high-energy photons or protons), which may be produced within the patient's body or within the treatment head of the radiotherapy device.

27. Non-target doses can be classified according to three approximate dose levels. High doses (>30 Gy or >50% of the prescription dose to the tumour) are typically optimized directly during the planning process. Intermediate doses (3 to 30 Gy or 5–50% of the prescription dose) are also often addressed during treatment optimization but are not generally the focus of the treatment plan. Low doses (<3 Gy or <5% of the prescription dose) are not generally considered during treatment planning. However, even if these lower doses are considered, they are difficult to measure, characterize, or model in the planning system [K30]. It is noted that the level of doses defined for these three categories are higher than those commonly defined for dosimetry related to radiological protection.

28. In a modern treatment planning system (TPS), high-dose regions and areas within the primary beam path are typically well described. However, the accuracy of the dose estimates beyond a few centimetres outside the treatment volume is usually poor [H47, H52]. In these cases, alternative methods are required to assess the organ doses to the patient. Low radiation doses outside the treated volume are a concern for the induction of second primary cancers.

## C. Scope of the literature search

29. The Committee has conducted a comprehensive literature review to identify published studies and papers providing information on oncological, biological, dosimetric or epidemiological aspects related to second primary cancer after radiotherapy. The scope of the review included results of peer-reviewed studies published in the scientific literature and publications of relevant international organizations. The primary focus of the review was on previously published epidemiological studies dealing with second primary cancer risk due to radiotherapy, which provided estimates of organ dose in their patient cohorts.

30. All papers were reviewed using previously agreed screening criteria and quality questions that followed the Committee's evaluation criteria published in the UNSCEAR 2017 Report, annex A [U6]. Epidemiological studies that did not meet the screening criteria, such as studies in non-original publications were excluded from the review. Publications meeting the requisite screening criteria were further assessed with respect to their quality.

31. A total of 3,095 articles were identified for review, of which 1,445 were assessed as meeting the criteria for a detailed review and included in this evaluation. Of these 264 articles were reviewed as part of two topics or more. Owing to the wide scope of the literature review, the search was performed using

the PubMed database and repeated, as closely as feasible, within the EMBASE and Web of Science databases. The literature search was conducted with a combination of query search strings that included terms associated with "second primary cancer" and "radiotherapy" as well as the four specialty areas: oncology, radiobiology, dosimetry, and epidemiology.

32. A detailed description of the epidemiological studies included in this evaluation for the purpose of the risk meta-analysis is presented in chapter V. The publication years of these studies ranged from 1979 to 2020.

# II. ONCOLOGY

## A. Introduction

33. It is often very difficult to differentiate the carcinogenic effect of radiotherapy in isolation from the effects of other cytotoxic therapies used in current medical practice. Therefore, the approach adopted in this chapter was to review all available literature that included radiotherapy, either alone or in combination with other various types of therapeutic modalities. This approach was considered more realistic than reviewing the very limited number of papers that summarize cases of patients treated with radiotherapy alone.

34. From a clinical point of view, there is a lack of concerted clinical research that provide insightful guidance to either cancer survivors or healthcare providers in managing survivors after their remission. Most clinical programmes for cancer survivors currently focus on the detection of relapse of the prior cancer as early as possible. The aim of this chapter is to provide a summary of current clinical data and information, and to make suggestions on the requirements necessary to reach consensus on methods that can be used to diagnose and treat second primary cancers after radiotherapy.

## B. Radiotherapy in cancer therapeutics

35. Radiotherapy is one of the main therapeutic modalities in the treatment of cancer and used both for the cure and palliation of cancer patients. Since the development of radiotherapy technology, significant progress has been made to improve outcomes and to minimize the adverse effects of radiotherapy. Efforts to deliver as much radiation dose as possible to cancer cells and as little radiation dose as possible to normal cells are still ongoing.

### 1. Acute adverse effects

36. Acute adverse effects occur during a course of radiotherapy or shortly after the completion of the treatment in a period which can last up to several weeks following cessation of treatment. These effects are observed in tissues with rapid cellular turnover. The severity of the radiation injury depends on several factors such as the anatomical area treated, the total radiation dose delivered, the dose per fraction and the radiation quality. The extent of cell damage and the time interval of repair play important roles in the

appearance of these clinical features. Symptoms are usually transient and can be recovered completely [B4]. Some of the common sites and symptoms of acute adverse effects are as follows:

*(a) Skin:* erythema, dry and moist desquamation, hair loss and ulceration;

*(b) Gastrointestinal mucosa:* nausea, vomiting, diarrhoea, and xerostomia;

*(c) Bone marrow:* myelosuppression and infection (opportunistic and pathogenic);

*(d) Pharyngeal and oesophageal mucosa:* mucositis and dysphagia.

## 2. Late adverse effects

37. Late adverse effects usually occur in a period of several months to several years after the completion of the course of radiotherapy. These effects are seen in late responding normal tissues, such as central (oligodendroglia) or peripheral (Schwann cells) nervous tissue, kidneys (tubular epithelium), blood vessels (endothelium), dermis (fibroblasts), and skeleton (osteoblasts and chondroblasts). These effects can be severe and recovery from these effects is often limited. Some of the specific late adverse effects are also related to the same factors as acute adverse effects, which include, but are not limited to, organs treated, cumulative absorbed dose, dose per fraction and irradiated volume [B4]. However, the relationship between acute and late adverse effects is still an area of active research. Examples of late adverse effects include:

*(a) Infertility*: Ionizing radiation can cause direct DNA damage to ovarian follicles which leads to ovarian dysfunction. Cranio-spinal irradiation can also elicit significant hormonal changes in women by disrupting the hypothalamic-pituitary-ovarian axis and affect their ability to become pregnant later in life [W17];

*(b) Cardiac toxicity*: Cardiac toxicity can be a late adverse effect in various cancers for which chest radiotherapy is recommended, for example breast cancer, Hodgkin lymphoma, lung cancer, and other cancers in which the heart receives some radiation dose. Radiation causes fibrosis of all components of the heart and increases the risk of coronary arterial disease, cardiomyopathy, valvulopathy, arrhythmias and pericardial disease [B13];

*(c)* Second primary cancers: Some cases of second primary cancers have an established relationship with prior radiotherapy. This adverse effect is critical especially in young cancer survivors since it may take 15 to 20 years or longer until a second primary cancer is diagnosed. Furthermore, second primary cancers can be life-threatening. The distribution of the incidence of second primary cancer after radiation is bimodal. The first peak is within several years of radiation exposure, usually driven by haematological malignancies such as acute leukaemia. The second peak is seen over 10 years following radiotherapy and is driven by the appearance of various solid cancers [H43, M3].

# C. Diagnosis of second primary cancers

## 1. Diagnostic criteria

### (a) *Histological type*

38. It is crucial to assess the histopathology of a tumour in a cancer patient to determine if the tumour is a newly and independently developed cancer or a metastasis of the original cancer. A second primary cancer needs to be compared histologically in parallel with the prior cancer because, by definition, it must have a different set of histological characteristics from the prior cancer.

39. Since the histological classification of tumours is evolving and tumour cells are heterogeneous even within the same tumour, it is sometimes difficult to decide whether the two histologic findings are the same or similar, or are different by microscopic assessment only. Therefore, genomic and mutational profiling to assess the clonal relationship between multiple tumours from a patient can also be necessary for the purpose of distinguishing multiple primary cancers from metastases derived from the original cancer site. Recent studies which distinguish a second primary cancer from clonally identical tumour pairs (i.e. a metastasis) have used several advanced technologies, such as loss of heterozygosity analysis of selected chromosome regions [G5], genome-wide copy number profiling [V18], array copy number variation [O4], genomic and transcriptomic approach [M20], or other comprehensive genetic analyses for the somatic mutational landscape [I15]. New technologies are still being developed to assess somatic mutational profile and clonal relationships between tumours.

### (b) *Past medical history of exposure to radiotherapy*

40. The term "second primary cancer after radiotherapy" in this annex is based on a patient's past medical history of exposure to radiotherapy for therapeutic purposes [B15]. If a patient has a history of prior cytotoxic therapies including radiotherapy, it may mean that some of the subsequent second primary cancer has arisen because of carcinogenic effects associated with the cytotoxic therapies, although it does not necessarily follow that the previous therapy has been the cause.

### (c) *Latency period*

41. Many studies have reported various values for the latency period between the time when the first cancer was diagnosed or treated with cytotoxic therapies and the time when the second primary cancer was diagnosed. The latency period for second solid cancers has been reported to be 10 years or longer, up to several decades in some cases [A3, A7, A9, A14, A25, B2, F2]. Compared to second primary solid cancers, therapy-related myelodysplastic syndrome (t-MDS) and acute myelogenous leukaemia (t-AML) generally have a shorter latency period, which can be several months or several years since treatment of the first primary cancer [A13, F3]. Hiraoka et al. have reported a case of t-AML developed within two months after chemoradiotherapy which included delivery of a dose of 60 Gy in 30 fractions in addition to cisplatin and 5-fluorouracil for oesophageal cancer [H35].

42. The specific length of interval between the first and second primary cancer diagnosis, or the length of interval between the therapy for the first primary cancer and the diagnosis of the second primary cancer, has not yet been fully clarified. However, it has been reported that the latency period of second

primary solid cancers is generally 10 years or more, which is a longer period than that of second primary haematologic malignancies, which typically ranges from several months to several years.

### (d) *Bone marrow findings for the diagnosis of therapy-related myeloid neoplasms*

43. Both t-MDS and t-AML are distinct categories of bone marrow neoplasms that develop following cytotoxic therapy for prior non-myeloid neoplasms. According to the WHO classification of myeloid neoplasms and acute leukaemia, t-MDS together with t-AML constitute the group of therapy related myeloid neoplasms [A26]. The WHO revised their hierarchical classification structure of various tumours in the 5th edition of the WHO report on classification of haematolymphoid tumours, published in 2022. The term "therapy-related" has now been changed to "post-cytotoxic therapy" in the new edition [K17]. The de novo myelodysplastic syndrome (MDS) and acute myeloid leukaemia (AML) are a group of clonal bone marrow neoplasms and can be diagnosed by morphological examination, which includes dysplastic features in unilineage or multilineage cells, and the assessment of the percentage of blasts in the peripheral blood or bone marrow (blasts constitute 5–19% of the cells in the peripheral blood or bone marrow). Recurrent chromosomal abnormalities such as 5q or 7q deletions are significant in diagnosing MDS or AML, either in de novo or therapy-related cases. However, the frequency and complexity of the chromosomal abnormalities are much higher in the t-MDS and t-AML than in their de novo counterparts. A recent study reported that recurrent chromosomal abnormalities are present in 40–70% of patients with de novo MDS and in 90–95% of patients with t-MDS at diagnosis [L10].

## 2. Genetic tests for second primary cancers

44. The detection of both somatic and germline variants is helpful for patient care. Germline testing is recommended for cancer patients who have a suspected hereditary disease based on their presentation and family history. These tests can help determine if there are specific genetic alterations that could provide information regarding the risk of developing other cancers. Detecting somatic variants (i.e., mutations in the tumour), on the other hand, can provide valuable insights for patient care, for example in the selection of therapeutic agents predicted to be effective. As the number of druggable mutations and cancer specific driver mutations increases, precision medicine is becoming a general practice in the field of oncology. Tumour sequencing usually detects both somatic and germline variations, which means pathogenic germline mutations can also be detected in tumour somatic testing. However, recent studies recommend performing a germline analysis separately using a paired normal tissue when performing tumour somatic sequencing [L18]. The use of massively parallel sequencing technology, which is also called next-generation sequencing, in clinical diagnosis is now widely accepted owing to its high throughput and ability to analyse multiple genes simultaneously. As next-generation sequencing has become affordable, both germline and somatic genetic variation testing have been rapidly incorporated into the clinical decision-making process in cancer treatment. At this stage, standardizing the interpretation and reporting variants data arising from various pipelines of sequencing platforms are important and beneficial considerations for clinicians in the treatment decision-making process.

### (a) *Germline mutations for genetic susceptibility*

45. It is known that some individuals with certain hereditary disorders such as ataxia telangiectasia or Fanconi anaemia experience an increased sensitivity to DNA-damaging agents such as radiation [Y4]. Multiple genetic pathways which contain DNA damage repair, oxidative stress, cell cycle control and

complex gene-gene or gene-environment interactions contribute to the development of radiotherapy and/or chemotherapy related neoplasms [B3].

46. Cancer predisposition genes have been tested on a single gene basis or on multiple gene panels selected according to the cancer phenotype in a patient or in a patient's family [D28, E8, F5]. However, none of the genetic factors that have been reported so far have absolute sensitivity or specificity as a risk predictor of second primary cancer after radiotherapy. Therefore, a genome-wide analysis through whole exome sequencing or whole genome sequencing are currently recommended to detect pathogenic or likely pathogenic variants that have clinical impact in the patient [H53]. The American College of Medical Genetics and Genomics and the Association for Molecular Pathology jointly provide consensus guidelines for the interpretation of sequence variants of genetic disorders [R6].

### (b) *Somatic mutations for precision medicine*

47. Some mutations that have been reported in a sporadic cancer have also been known to be observed in a second primary cancer after cytotoxic therapies. For example, somatic alteration of the *PIK3CA* gene, which is one of the most prevalent driver genes in sporadic breast cancer, is occasionally observed in the radiation-associated breast cancer [B12]. Behjati et al. also found that ionizing radiation transforms adjacent cells in the radiotherapy field into cancer cells, and the process of transformation requires additional driver mutations [B12]. In a study of radiation-induced sarcomas, the prevalence of *TP53* somatic mutations at approximately 58% was significantly higher than the prevalence of 16.8% seen in radiation-naive sporadic sarcomas [G15]. The *AML1* gene also was reported as a target of gene alteration by ionizing radiation and/or anticancer drugs in an experiment where a type of leukaemia harbouring *AML1* translocations have been reported after radiation and/or chemotherapy [H14].

48. Despite the mutational signatures reported in various cancer types, it should be noted that there are no currently available biomarkers for a specific type of second primary cancer after cytotoxic radiotherapy. Therefore, a diagnosis of a radiation-induced cancers cannot be definitively made by testing somatic mutations in tumour tissues. The main clinical use of detecting somatic mutations in cancer tissues is to identify driver mutations which may be helpful for selecting systemic targeted therapies (e.g., vemurafenib in BRAF-mutated melanoma, erlotinib in EGFR-mutated lung cancer, or PARP inhibitors in *BRCA1/2*-mutated breast and ovarian cancers) [B16].

## D. Examples of cases of second primary cancer after radiotherapy

49. As part of the review of oncological information on second primary cancer after radiotherapy it is also useful to give examples of clinical cases that can be encountered in the real practice of cancer care. Three cases are described in this section. Case 1 is typical of second primary solid cancer, where breast angiosarcoma was developed in the field of prior radiotherapy. In this instance, the patient had not received any other form of cytotoxic therapy beyond radiotherapy. Case 2 is more complicated and ambiguous than Case 1 regarding the attribution of radiotherapy to second primary cancer induction. The patient in Case 2 developed lung cancer in the field of prior radiotherapy for breast cancer. However, the patient had also received cytotoxic chemotherapy for the first primary cancer, and therefore it is less clear whether the second cancer was related to the patient's radiation treatment. In addition, alternate scenarios regarding causation could be conceivably taken in consideration because of the histological type of the newly developed lung cancer. Case 3 is an example of haematological malignancy after chemoradiotherapies. The patient described in this case had experienced three different cancers before they were finally

diagnosed with MDS. The three cases in this section were provided by the Korea Cancer Center Hospital of the Korea Institute of Radiological and Medical Sciences in Korea and the Cancer Center Baclesse in France after obtaining approval by their internal review boards for inclusion in this evaluation.

**Case 1: Second primary angiosarcoma in the area of radiation for the prior breast cancer**

50. In 2011, a 60-year-old female patient was discovered to have microcalcifications of American College of Radiology Appropriateness Criteria 4 at the junction of the inferior quadrants of the right breast during a screening mammography. The patient underwent tumourectomy and sentinel node biopsy. The histology report showed a 7-mm invasive grade 2 ductal adenocarcinoma with 80% oestrogen and 10% progesteron receptor positivity, HER2 negative, with in situ ductal carcinoma, safe resection margins, and a pN0/2 cancer staging.

51. The patient underwent whole breast postoperative radiotherapy of a prescribed dose of 48 Gy given in 20 fractions of 2.4 Gy per fraction using two tangential field-in-field beams of 6 MV photons (figure VI) with a linear accelerator (the biologically equivalent dose was 52 Gy). The patient did not receive a boost dose to the tumour bed per institutional strategy for patients over 60 years of age at the time of treatment and experienced whole breast grade 1 dermatitis and desquamative grade 2 in the inframammary fold. Hormonal therapy using anti-aromatase was then delivered for five years. The patient had persistent radiation-induced breast erythema and oedema one year after treatment.

52. Ten years later, during annual follow-up visits the patient presented with an atypical slow-growing, poorly limited angiomatous nodule of the inferomedial quadrant of the right breast and telangiectasia. The lesion was within the 40 Gy isodose line, consistent with dose levels associated with a risk of radiation-induced angiosarcoma (figure VI). The MR imaging showed that there was a gadolinium-enhanced lesion 12×8×11 mm in size with subcutaneous extent anterior to the lesion (22 mm). Fiducials—visible in the inferomedial mammary quadrant on follow-up CT in addition to fiducials for a laser based in-room setup during radiotherapy—and a well-limited scar from previous breast surgery could be seen next to the enhanced lesion. The biopsy showed that the epidermis was hyperplastic and included a pigmented lesion with malignant vascular proliferation and haemorrhages. The results of the immuno-histochemistry test showed that malignant cells were positive for ERG and CD34, negative for cytokeratin AE1-AE3 and Melan-A. They also showed strong and diffuse positivity for CD31 and C-MYC.

## Figure VI. Radiotherapy treatment plan and follow-up PET-CT

(A′) 48 Gy isodose on initial 3D radiotherapy plan from A. (A) initial 3D radiotherapy plan for whole right breast irradiation using 2 tangential opposed 6 MV photon beams targeted to breast ductal adenocarcinoma of the inferomedial quadrant of the right breast. (B) Follow-up PET-CT shows a poorly limited subcutaneous angiosarcoma in the inferomedial quadrant of right breast. The lesion is located in the 40 Gy-isodose line of the previous irradiation

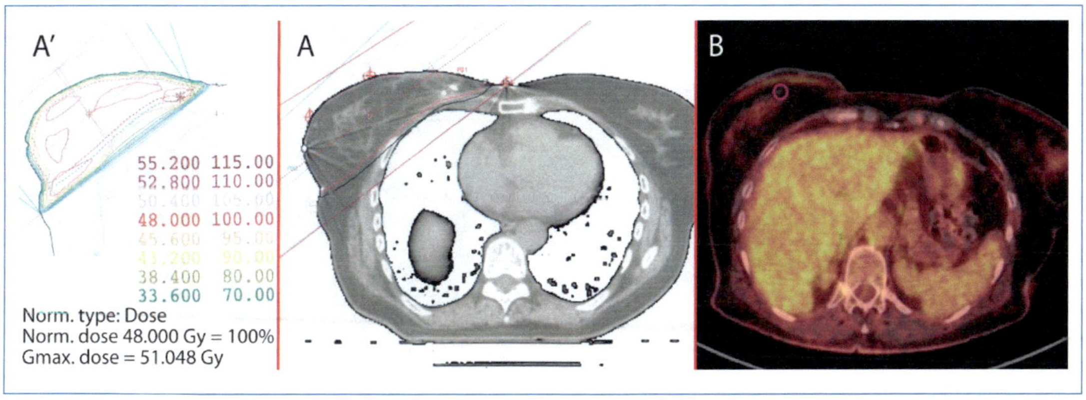

53. The patient underwent three months of paclitaxel chemotherapy, mastectomy (with safe 1 cm margins) and another three months of paclitaxel chemotherapy. During the subsequent follow-ups, a multifocal dermis/infradermis relapse on the inframammary fold of the same breast was observed. As of 2023, two years after being diagnosed with radiation-induced angiosarcoma, the patient was alive with multiple cutaneous lesions from angiosarcoma under continuing paclitaxel chemotherapy.

54. This case presents a typical example of second primary cancer angiosarcoma occurring in an irradiated area following radiotherapy for breast adenocarcinoma 10 years earlier. The radiation-induced causality may be considered unambiguously in that there was no known genetic cancer-prone individuals or familial predisposition (according to Amsterdam's criteria), no other genotoxic treatment, clear correlation between sites of the irradiated primary and second cancer, and different histologies between the first primary cancer and the second primary cancer therefore ruling out a possible relapse of the primary cancer. Sarcomas typically occur in areas of intermediate to high doses. This case is typical for radiation-induced sarcoma but not radiation-induced carcinoma, which may occur in areas that receive lower doses.

**Case 2: Second primary lung cancer in the area of radiotherapy for the prior breast cancer**

55. In March 2003, a 39-year-old woman underwent a health check-up and the mammogram showed a 2.5 cm irregular mass with microcalcification in the right upper quadrant of the right breast. The patient was subsequently referred to the surgery department for a breast biopsy which resulted in the diagnosis of an invasive ductal carcinoma. After the patient underwent breast surgery, a histological study confirmed that the carcinoma cells showed positivity for oestrogen receptor, progesterone receptor, and human epidermal growth factor receptor 2.

56. Following breast conserving surgery with axillary dissection, the patient was then treated with chemotherapeutic agents (four cycles of cyclophosphamide and doxorubicine) and with external beam radiation (figure VII, left). Adjuvant radiotherapy was delivered to a total dose of 60–62 Gy in 2 Gy fractions over seven weeks. Furthermore, a radiation dose of 50 Gy was delivered in 25 fractions to the entire right breast and right supraclavicular lymph node areas. The breast was treated using opposing tangential fields with 6 MV photons and the supraclavicular field with $^{60}Co$ external beams. Following whole-breast irradiation, a tumour bed boost dose of 10 Gy in 5 fractions was delivered using 11 MeV electrons, and a posterior axillary boost dose of 12 Gy in 6 fractions was delivered using $^{60}Co$ beams. After completion of the treatment, the patient was under state of no evidence of disease for 13 years.

Figure VII. (Left) Radiotherapy treatment plan to treat invasive ductal carcinoma in the right breast: field configurations and isodose distribution in the right breast and axillary region. (Right) CT scan of the newly visible mass in the right middle lobe of lung and chest wall after being in disease-free state for 13 years

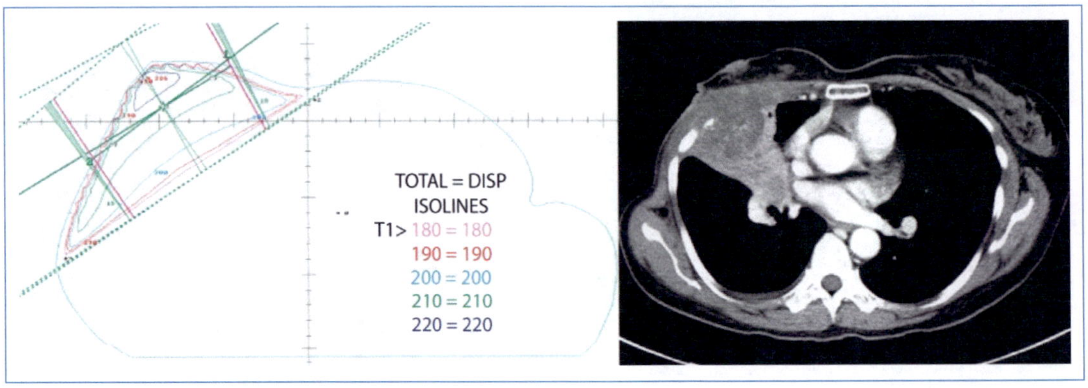

57. In November 2016, the same patient visited the pulmonology clinic for a persistent cough over a period of more than six months. A chest radiograph showed haziness in the right lung. A subsequent CT scan showed a 2.9 cm mass obstructing the right middle lobe bronchus and a 4.7 cm mass in the right middle lobe and chest wall with erosion of the right third to fifth ribs (figure VII, right). The location of the newly developed right chest wall mass overlapped the area previously treated with radiotherapy.

58. A histological study via bronchoscopic biopsy indicated that the patient had a squamous cell carcinoma. The patient was diagnosed with non-small-cell lung cancer, squamous cell carcinoma in the right lung with *EGFR* gene exon 19 deletion. In spite of treatment with neoadjuvant chemotherapy and tyrosine kinase inhibitor (Gefitinib) for two months, the cancer progressed and metastasized to the opposite lung. The patient had received palliative radiotherapy a couple of times, including treatment to the right lung and chest wall mass due to pain in March 2018. The physicians switched to another regimen of chemotherapy over the period October 2017 to October 2020. Nevertheless, the patient's general condition deteriorated, and she passed away in December 2020.

59. This case may represent an example of second primary lung cancer which developed at the site of radiotherapy received 13 years earlier to treat breast cancer. However, other alternate scenarios are also possible in which the second primary cancer could have been a de novo squamous cell lung cancer even though it was coincidentally developed at the site (in-field) of the prior radiotherapy. Development of de novo squamous cell carcinoma of lung is common, particularly among smokers, and it typically originates in the central part of the lungs and extends into the surrounding tissue or nearby structures. Given the frequency and the pattern of growth of the squamous cell carcinoma, even though the smoking history of the patient was not available from her medical record, there is a reasonable possibility that this patient developed a de novo lung cancer. However, a possible alternative explanation is that the second primary lung cancer was caused by the carcinogenic effects of chemotherapeutic agents previously used for the treatment of breast cancer.

**Case 3. Therapy-related myelodysplastic syndrome after three prior cancers treated with cytotoxic chemoradiotherapies**

60. In June of 2009, a 56-year-old woman presented herself at the gynaecology unit complaining of a painful vulvar lesion. A pelvic examination by colposcopy revealed an ulcerative lesion localized on the left labium majus and an incisional biopsy was performed. The biopsy was histologically diagnosed as squamous cell carcinoma. After the patient underwent a wide local excision with inguinal lymph nodes dissection, she was also treated with concurrent chemoradiotherapy with a prescription dose of 50.4 Gy in 28 fractions with 5 cycles of weekly cisplatin chemotherapy. Following completion of the treatment, she had been under the state of no evidence of disease.

61. In June 2010, the patient was seen in the gastrointestinal clinic following reports of epigastric discomfort. She was subsequently diagnosed with low-grade mucosa-associated lymphoid tissue lymphoma after gastro-fiberscopy and endoscopic submucosal dissection.

62. In 2012, the patient was again referred to the surgery department at age 59 years due to a painful mass in her right breast. A core needle biopsy revealed the tumour to be an invasive ductal carcinoma, which was oestrogen receptor negative, progesterone receptor negative, human epidermal growth factor receptor 2 negative, P-AKT negative, and Ki-67 positive. She subsequently underwent breast conserving surgery, adjuvant radiotherapy at a prescription dose of 50 Gy with 5 cycles of cyclophosphamide, methotrexate and fluorouracil chemotherapy.

63. In September 2013, during the follow-up after the last chemotherapy set/session/course for breast cancer, she developed symptoms of dizziness and weakness. Blood test revealed pancytopenia in her peripheral blood, and she was referred to the haemato-oncology department. In the bone marrow study, hypocellular marrow

with 16% of myeloblasts were observed (figure VIII-A). A chromosomal study of the bone marrow blasts with G-banding technology showed complex karyotype containing 5q deletion and monosomy 7 (figure VIII-B), which were also identified by fluorescence in situ hybridization. The fluorescence in situ hybridization finding included deletion 5q (27%), monosomy 7 (33%), trisomy 8 (4%) and deletion 20q (26%).

Figure VIII. Bone marrow morphology and chromosome analysis

(A) The bone marrow aspiration (Wright staining, x1000) showing 16% of myeloblasts among with scanty haematopoietic cells with dysplasia on erythroid and granulocytic lineages. (B) Chromosome study by G-banding technique of the bone marrow blasts showing a complex karyotype

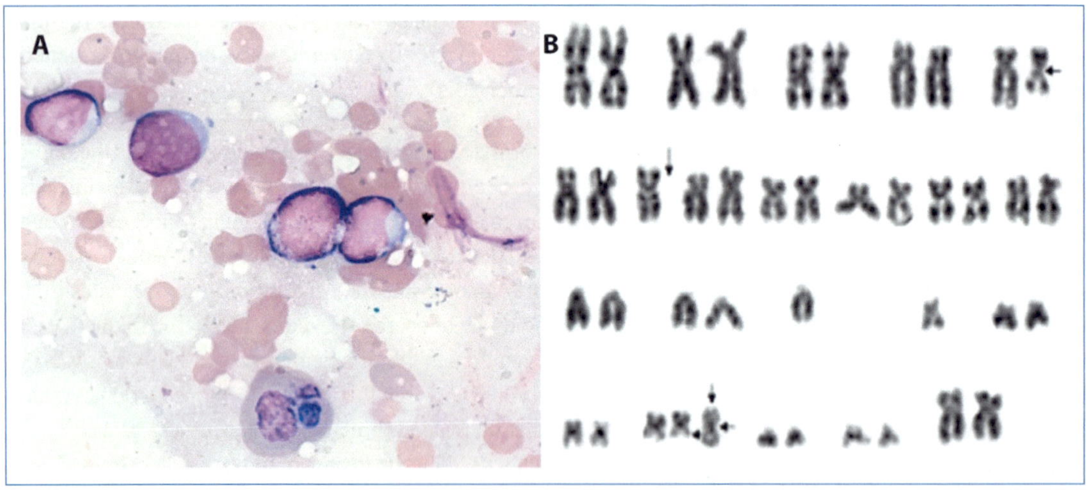


64. Based on the bone marrow findings and her treatment history with cytotoxic therapies for the previous three malignancies, the patient was diagnosed with t-MDS. She was treated with seven cycles of azacitidine and two cycles of decitabine. Massively parallel sequencing with a targeted gene panel of clonal haematopoiesis mutation was performed with the patient's bone marrow aspirates at the time of diagnosis of t-MDS. Somatic mutation in the *TP53* gene: NM_001126112.2 (*TP53*): c.659A>C, p.(Tyr220Ser) was detected with the variant allele frequency of 18.5%. The mutation had not been detected in July 2012 when the patient was in the midst of being treated for breast cancer before being diagnosed with t-MDS. Despite the full range of therapies given, the patient died in February 2015 due to a urinary tract infection.

65. The patient in this case had experienced three other solid cancers before being diagnosed with t-MDS. She had been given both chemotherapy and radiotherapy for those prior cancers. Although the sequential relations between the prior cancers were complex, the last diagnosis of t-MDS was obvious because the bone marrow finding, and the chromosomal abnormalities met the diagnostic criteria of the t-MDS/t-AML group of the WHO classification. Complex karyotype including monosomy 7 as well as *TP53* mutations are known to be frequently identified in therapy-related myeloid neoplasms [K11, N16].

## E. Incidence of second primary cancers

66. The incidence of second primary cancers varies with the tissue or organ where the first primary cancer originated, the age of the patient at treatment, individual susceptibility, with further correlations to the dose distribution in the tissue, the size of the irradiated volume, the total dose delivered and the

dose rate. In some cases, it is difficult to follow up for a long period after completion of the treatment, and therefore, the incidence of second primary cancers might be underestimated especially for those with long latency periods [T23].

67. There are still few registries available globally from which the exact number of second primary cancer cases diagnosed with the same diagnostic criteria can be extracted. In many studies on the incidence of second primary cancers, the standardized incidence ratio (SIR) has been calculated according to the site of cancer [B10]. In this section, the SIR of each second primary cancer is used to compare its incidence with the one in the general population by cancer site. The SIRs vary among cancer sites and types of treatment [B54, F16]. According to Burt et al., the incidence of second primary cancers is higher in breast cancer patients who received radiotherapy than in patients who did not receive radiotherapy as shown in table 1 [B54]. SIR values given in table 1 for both radiotherapy and non-radiotherapy patients may include contributions from patients who received combination therapy with chemotherapeutic agents.

Table 1. Standardized incidence ratios for second primary cancers in breast cancer patients compared to the general female population [B54]

| *Second cancers after the first breast cancer* | *Standardized incidence ratio (95% CI)* | | *p-value*[a] |
|---|---|---|---|
| | *Non-radiotherapy patients* | *Radiotherapy patients* | |
| Bones and joints | 1.14 (0.77, 1.62) | 1.73 (1.14, 2.52) | 0.18 |
| Brain and other nervous system | 0.79 (0.69, 0.9) | 0.88 (0.74, 1.03) | 0.32 |
| Cervix uteri | 0.71 (0.62, 0.81) | 0.69 (0.57, 0.82) | 0.84 |
| Colon excluding rectum | 0.95 (0.92, 0.99) | 0.97 (0.92, 1.02) | 0.59 |
| Contralateral breast | 2.62 (2.57, 2.66) | 2.68 (2.61, 2.74) | 0.15 |
| Corpus and uterus | 1.28 (1.23, 1.34) | 1.42 (1.34, 1.49) | 0.004 |
| Kidney | 0.81 (0.74, 0.89) | 0.98 (0.88, 1.09) | 0.010 |
| Leukaemia[b] | 0.92 (0.85, 0.99) | 1.4 (1.29, 1.52) | <0.001 |
| Lung, bronchus, and other respiratory organs | 0.87 (0.85, 0.9) | 1.03 (0.99, 1.07) | <0.001 |
| Lymphoma[b] | 0.84 (0.79, 0.89) | 0.94 (0.87, 1.01) | 0.03 |
| Melanoma of the skin | 0.97 (0.9, 1.05) | 1.15 (1.06, 1.26) | 0.003 |
| Myeloma[b] | 0.9 (0.82, 1.0) | 0.87 (0.75, 1.01) | 0.71 |
| Oesophagus | 1.06 (0.92, 1.22) | 1.57 (1.32, 1.84) | <0.001 |
| Ovary | 1.15 (1.08, 1.22) | 1.22 (1.12, 1.31) | 0.27 |
| Pleura | 0.62 (0.01, 3.43) | 4.53 (1.22, 11.59) | 0.09 |
| Soft tissue including heart | 1.03 (0.87, 1.22) | 2.12 (1.81, 2.47) | <0.001 |
| Thyroid | 1.16 (1.05, 1.28) | 1.09 (0.97, 1.22) | 0.42 |
| All cancers | 1.2 (1.19, 1.22) | 1.33 (1.31, 1.35) | <0.001 |
| All solid cancers | 1.27 (1.25, 1.28) | 1.38 (1.36, 1.4) | <0.001 |
| All haematologic cancers | 0.88 (0.84, 0.91) | 1.06 (1.01, 1.12) | <0.001 |

[a] P-value compares the SIR with and without radiotherapy.

[b] Three cancers (leukaemia, lymphoma, myeloma) are categorized as haematologic cancers.

## 1. Therapy-related haematologic malignancies (t-MDS and t-AML)

68. Therapy-related haematologic malignancies, such as t-MDS and t-AML have been reported as a late complication of chemoradiotherapy for various cancers, including breast cancer [B54], lung cancer [W4], Hodgkin [B36] and non-Hodgkin lymphoma [T13], acute lymphoblastic leukaemia [B50, N8, P14], sarcomas [B25, B50], ovarian [T15] and testicular cancers [T16], and prostate cancer [W7].

69. Patients with childhood cancers, such as acute lymphocytic leukaemia, lymphomas, Ewing tumour, medulloblastoma and rhabdomyosarcoma, treated with chemotherapy and/or radiotherapy had a greatly increased risk of subsequently being diagnosed with t-AML as compared to the general population (SIR=45.6 (95% CI: 35.3, 59.0)). However, about 98% of t-AML patients received treatment that included chemotherapeutic agents for their first primary cancers and 34% of t-AML patients received combination therapy including radiotherapy. Only a small percentage of t-AML patients received radiotherapy as the only treatment for their prior cancers. The data available did not allow to distinguish between the effect of radiotherapy and that of chemotherapy [B50].

70. In a retrospective cohort study of elderly prostate cancer patients, the incidence of subsequent t-MDS/t-AML was reported at 0.5% (157 cases out of 32,112). Among patients who received surgery, 0.3% developed t-MDS/t-AML with a median latency period of 3.43 years (range: 0.13–9.01). In patients who underwent radiotherapy, however, the incidence of developing t-MDS/t-AML was 0.7%, with a median latency period of 2.95 years (range: 0.27–9.48). Compared with the group of patients who had only surgery, the group who received radiotherapy was more likely to develop t-MDS/t-AML (HR=1.51 (95% CI: 1.07, 2.13)) with a shorter latency period [W7].

## 2. Sarcomas

71. The incidence of second primary sarcoma varies with the type of first primary cancer. Bazire et al. estimated the cumulative incidence rate of second primary sarcoma after radiotherapy for non-metastatic breast cancer over a period of 10 to 15 years compared to the expected incidence rate in the general population. The SIR for sarcoma was 7.79 (95% CI: 5.39, 11.24), which indicates that the incidence of sarcoma was significantly higher in breast cancer survivors who received radiotherapy than in the general female population ($p<0.001$) [B9]. On the other hand, in a population-based cohort study from Canada, consisting mostly of men (72.1%, 125,080 out of 173,580 patients) the SIR for sarcoma among patients with prior abdominopelvic cancers (non-metastatic cancers of the prostate, bladder, colon, rectum or anus, cervix, uterus or testis) treated with radiation compared with the general population was estimated to be 2.41 (95% CI: 1.57, 3.69) [H36], much lower than the one in Bazire's study.

72. Among all types of sarcomas, angiosarcoma contributes to up to 40% of all radiation-induced sarcomas [I12, L1]. The incidence rate of radiation-induced angiosarcomas has been reported from 0.09% to 0.16% following radiotherapy for a primary breast cancer [F1, G10, L8]. Another study reported a pooled incidence of radiation-induced angiosarcomas at about 2.6 per 10,000 [D22]. Radiation-induced sarcomas can also originate in the irradiated bone with osteosarcomas, or in soft tissue with malignant fibrous histiocytomas [B15].

## 3. Breast cancer

73. An increased incidence of second primary breast cancer after radiotherapy has been reported in patients who received irradiation on the chest area for a primary breast cancer [B17, X1], Hodgkin lymphoma [N3], osteosarcoma and Ewing sarcoma [E4].

74. Among second primary cancers after adjuvant radiotherapy of breast cancer, contralateral breast cancer is one of the mainly reported prior cancers along with lung cancer, although the cumulative risk of second primary cancer was low, at approximately 6 and 10 cases per 1,000 women with irradiated breasts at latency periods of 5 and 10 years after breast radiotherapy, respectively [X1]. Compared with the group of breast cancer patients who did not receive radiotherapy, the radiotherapy group had significantly increased risks of contralateral breast cancer (hazard ratio (HR)=1.268 (95% CI: 1.112, 1.445)) and lung cancer (HR=1.218 (95% CI: 1.049, 1.565)) [X1].

75. Radiotherapy has usually been used for lymphoma patients as either a single modality or in combination with other systemic therapies [Y1]. Second primary breast cancer is the most frequently diagnosed solid cancer in female survivors of Hodgkin lymphoma treated with radiotherapy in the chest area. Data show a 2.4-fold increase in the risk of developing a second primary breast cancer among survivors of Hodgkin lymphoma compared to the baseline risk of developing a primary breast cancer in the general population [W10]. Among Hodgkin lymphoma survivors, the excess absolute risk for second primary breast cancer strongly increases with attained age, which is an indication of the increasing burden of excess cancers with advancing age of Hodgkin lymphoma survivors [V7]. On the other hand, a lower incidence of second primary breast cancer was reported in the cohort of Hodgkin lymphoma survivors if the patients were treated with a reduced irradiation field size and delivered doses [N3, V7].

## 4. Lung cancer

76. Second primary lung cancer has been reported among patients who previously had contracted malignancies and were treated with radiation in the chest area, such as the breast [B17, B54, X1] and the larynx or oral cavity [F2], or for Hodgkin lymphoma [B15, N3].

77. An analysis of the cohort of breast cancer patients showed a significant increase in second primary lung cancer in the group of patients who received radiotherapy compared to those who did not ($p<0.001$). In the United States there was an increase in the incidence rate of second primary lung cancer in breast cancer patients receiving radiotherapy compared to the baseline rate of the general population; the reported SIR was 1.03 (95% CI: 0.99, 1.07). However, the incidence rate of second primary lung cancer among patients who did not receive radiotherapy was lower than the baseline rate of the general population, with a reported SIR of 0.87 (95% CI: 0.85, 0.9) [B54] (table 1).

## 5. Gastrointestinal cancer

78. Whether radiotherapy increases the incidence of subsequent gastrointestinal cancers is still an unresolved question. Some data from clinical studies show an increase in incidence of rectal cancer on prostate cancer patients after radiotherapy [H27, J3]. This increase can be associated with the location of the rectum which is in the direct radiation area. However, other studies do not support this finding [H34].

79. According to the population-based cohort from the Korean Central Cancer Registry, patients with prostate cancer had a lower incidence of second primary cancers as compared to men in the general

population (SIR=0.75 (95% CI: 0.72, 0.78)), which was significant for second primary cancers of the oesophagus, stomach, liver, gallbladder, bile duct, and pancreas. In contrast, patients in this cohort who received radiotherapy for the prior prostate cancer had an increased risk of second primary rectal or recto-sigmoidal cancers (SIR=1.03), colon cancer (SIR=1.11), and cancer of the small intestine (SIR=1.15) although these risks were not statistically significant [J3].

## 6. Thyroid cancer

80. In radiotherapy for head and neck cancers, the thyroid is one of the most radiosensitive organs at risk. Although thyroid cancer following radiotherapy is less common than benign conditions such as hypothyroidism [M22], it is still worrisome especially for childhood cancer survivors. Friedman et al. reported an increased incidence of second thyroid cancer occurring five years or more after radiation treatment of childhood cancer compared to the expected incidence rate in the general population [F16].

81. Bhatti et al. quantified the long-term risk of thyroid cancer associated with radiotherapy among five-year survivors of childhood cancer (leukaemia, Hodgkin lymphoma and non-Hodgkin lymphoma, central nervous system (CNS) cancer, soft-tissue sarcoma, kidney cancer, bone cancer, neuroblastoma) diagnosed between 1970 and 1986 from the Childhood Cancer Survivor Study (CCSS) using the follow-up of the cohort up to 2005 [B27]. The overall ratio of observed to expected numbers (SIR) of thyroid cancer cases was 14.0 (95% CI: 11.7, 16.8). When patients are grouped by previous treatment type, the SIR was 17.9 (95% CI: 14.8, 21.8) in the group of patients who received radiotherapy, and 4.9 (95% CI: 2.8, 8.6) in the group of patients who did not ($p<0.001$). A higher radiation risk was also observed in patients exposed at younger age ($p<0.001$), in the female population ($p=0.008$), and in patients exposed for longer periods ($p<0.001$) [B27].

## 7. Brain tumour

82. Cranial radiotherapy is a mainstay modality for treatment of some intra- and extracranial tumours. Along with a variety of neurological complications of radiotherapy, intracranial tumours can occur within the brain, meninges, bones and other connective tissue elements as a late complication of cranial radiotherapy. Gliomas and meningiomas are the most frequently reported second primary brain tumours [R3, Y3]. The term “radiotherapy-induced” used in this subsection, was taken from the original text in the papers reviewed and should not be interpreted to mean that causality between radiotherapy and second primary cancer had been proven.

83. Taylor et al. [T6] analysed childhood cancer patients of meningiomas, gliomas, and other CNS neoplasms included in the British Childhood Cancer Survivor Study who received radiotherapy and intrathecal methotrexate. The study estimated a SIR for gliomas of 10.8 (95% CI: 8.5, 13.6). Leary et al. [L6] showed that the cumulative incidence rate of developing radiotherapy-induced glioma was 0.48% at 10 years, 0.87% at 15 years, 1.13% at 20 years, and 2.58% at 35 years after the cranial radiotherapy was used to treat the primary cancer. In total, 1–4% of patients undergoing cranial radiotherapy for paediatric cancers developed radiotherapy-induced glioma [L6]. According to a systemic review of 296 radiation-induced glioma cases by Yamanaka et al. [Y3], the most frequent locations are the frontal lobe (32.7%), temporal lobe (26.9%), parietal lobe (22.8%), cerebellum (17.8%) and spinal cord (4.5%). These results are outlined in figure IX. By contrast, sporadic high-grade gliomas in the cerebellum and spinal cord are quite rare [Y3].

Figure IX. Reported locations for second primary gliomas after radiotherapy [Y3]

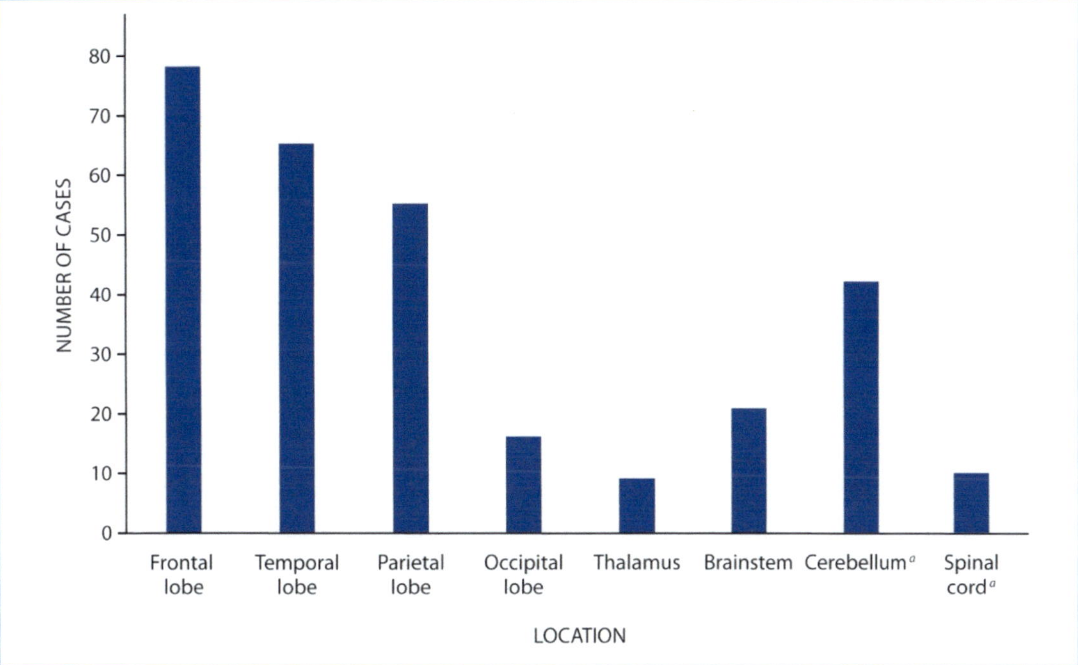


[a] Cerebellum and spinal cord are the locations where radiotherapy-induced gliomas are frequently reported, whereas, sporadic high-grade gliomas have been rarely reported [Y3].

84. Radiation-induced meningioma is one of the most common types of brain tumour to occur following cranio-spinal radiotherapy [G8, R3, Y3]. A retrospective study of 54 patients with 110 radiation-induced meningiomas found that the most common prior cancers treated with radiotherapy were medulloblastoma, (18.5% of patients), pilocytic astrocytoma (13%), and leukaemia (26%). They also reported the locations of the second primary meningioma as defined by the International Consortium on Meningioma (figure X) [G8].

Figure X. Reported locations for second primary meningiomas after radiotherapy [G8]

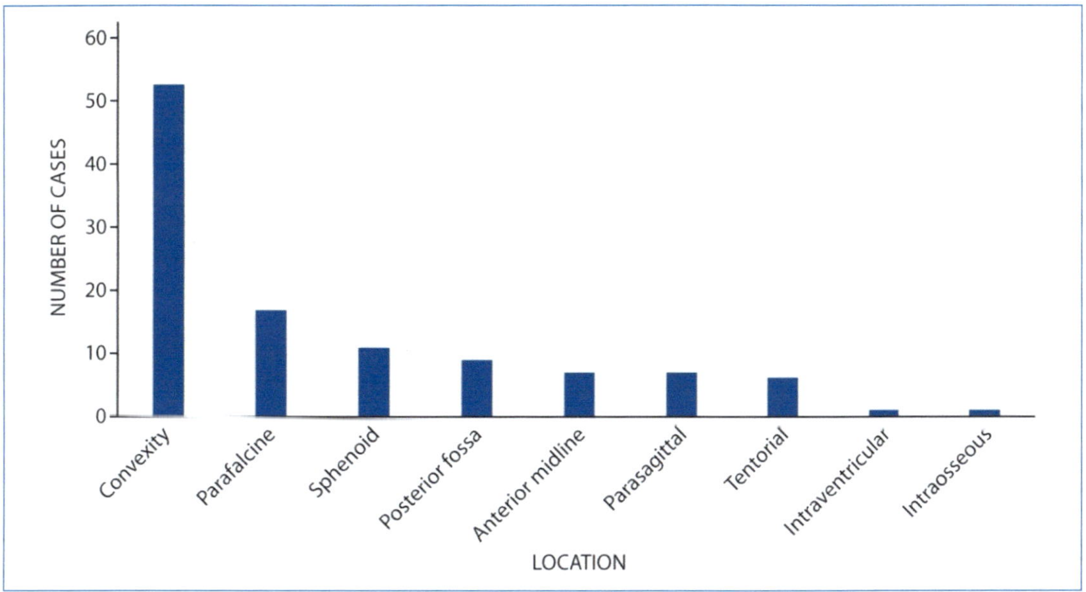

## F. Prognosis of second primary cancers

85. There are few publications that provide comprehensive information on survival rates or treatment outcomes after treating second primary cancer induced by radiotherapy [T21]. Various papers in the open literature indicate that the prognosis of second primary cancer is poorer than the one for the first primary cancer when more than one cancer has developed in a patient [B7, D20, D22, G3, L16, V8]. However, a few studies reported no difference in prognosis between the first and second primary cancers [N2, T10].

86. A recent study by Ye et al. classified the oesophageal cancer patients recorded in the Surveillance, Epidemiology and End Results (SEER) database of United States' National Cancer Institute (NCI) into four groups according to whether the oesophageal cancer was the only one (EC-only), the first of two or more primary cancers (1st EC), the second of two or more primary cancers (2nd EC-SPC), or the third of three or more primary cancers (3rd EC-SPC) [Y6]. They reported that the median and the mean survival times of the 1st EC group were 47 months (95% CI: 43.87, 50.13) and 74.67 months (95% CI: 72.12, 77.22), respectively, and the five-year survival rate was 39% ($p$<0.01). The median and the mean survival times of the EC-only group were 9 months (95% CI: 8.86, 9.14) and 32.16 months (95% CI: 31.58, 32.74), respectively, and the five-year survival rate was 13% ($p$<0.01). The survival rates of the 1st EC and EC-only groups were significantly better than those of the other groups ($p$<0.01). Since the combined EC-only and 1st EC group can be considered a sporadic oesophageal cancer group, while the others (2nd EC-SPC and 3rd EC-SPC groups) can be regarded as the second primary oesophageal cancer group, the results of this study seem to indicate a poorer prognosis of second primary oesophageal cancer as compared to the sporadic one. An interesting point is that the survival rate of the 1st EC group was better than that of the EC-only group. The authors explained that more aggressive treatment (such as multiple surgeries and repeated radiotherapy or chemotherapy) might have caused the better prognosis in the 1st EC group than in the EC-only group [Y6].

87. Radiation-induced sarcoma is a rare complication of radiotherapy to the breast or chest wall in the treatment of breast cancer. The prognosis of female patients with radiation-induced breast angiosarcoma, which is known to be more common than spontaneous angiosarcoma of breast, is poor with a mean recurrence-free survival of 15.9 months and mean overall survival of 27.4 months [D22]. In a study on clinical prognostic factors for improved outcome, primary breast angiosarcoma had improved local control and overall survival compared to radiation-induced breast angiosarcoma [G3]. Two-year local control for primary versus radiation-induced was 89% versus 51% ($p$=0.064), respectively. Two-year overall survival for primary versus radiation-induced was 89% versus 61% ($p$=0.016), respectively [G3].

88. Undifferentiated pleomorphic sarcoma (UPS) is also a commonly diagnosed subtype of radiation-induced sarcoma. Dineen et al. conducted a survival and recurrence analysis for radiation-induced and sporadic UPS in a large cohort of 1,068 UPS patients [D20]. Patients with radiation-induced UPS demonstrated worse clinical outcomes compared to patients with sporadic UPS. Radiation-induced UPS represented 5.1% of the UPS population. The five-year disease-specific survival was 52.2% for radiation-induced UPS patients (n=55) compared with 76.4% for patients with sporadic UPS (n=1,013) ($p$<0.001). A matched-cohort analysis also demonstrated that the five-year disease specific survival rate was significantly worse for radiation-induced UPS than for the sporadic UPS (52.2% versus 73.3%, $p$=0.002). Higher local recurrence rates were observed for patients with radiation-induced UPS than for patients with sporadic UPS (54.5% versus 23.5%, $p$<0.001) [D20].

89. Radiation-induced fibrosarcoma in patients with a history of prior radiotherapy for nasopharyngeal carcinoma was also associated with a poorer prognosis compared to sporadic fibrosarcoma of the head and neck [Z3]. Forty-two patients with radiation-induced fibrosarcoma had a poorer five-year disease specific survival rate compared to 124 patients with sporadic fibrosarcoma patients (36.2% versus 50.4%, $p$=0.026). However, treatment modality was significantly different between the two groups, with only

64.3% of patients in the radiation-induced fibrosarcoma group receiving surgery with or without adjuvant treatment versus 91.1% of patients in the sporadic fibrosarcoma group ($p<0.001$). Multivariate analysis of the combined group indicated that patient group ($p=0.032$), tumour, node, stage in the metastasis classification system ($p=0.019$), histologic grade ($p=0.046$), and treatment modality ($p<0.001$) were independent variables affecting disease specific survival [Z3].

90. Radiation-induced osteosarcoma of the maxilla and mandible is rare but has been shown to be an aggressive second primary cancer with a very poor prognosis. In the study by Liao et al., the total incidence rate of radiation-induced osteosarcoma between 1964 and 2012 was 0.084%, with 45 out of 53,760 cases of nasopharyngeal cancer receiving radiotherapy. The overall survival rates at 1, 2 and 3 years for the cohort of 45 patients who received radiotherapy for their radiation-induced osteosarcoma were 53.3%, 35.6%, and 13.5%, respectively. All patients died within five years after diagnosis of radiation-induced osteosarcoma [L16].

91. In several studies on comparing the prognosis of MDS or AML and their therapy related counterparts (t-MDS and t-AML), chromosomal karyotype has been known as an independent prognostic factor [A15, K13, S22]. Schoch et al. [S22] compared the pattern of karyotype abnormalities of t-AML (n=93) with de novo AML (n=1,091) and reported that the favourable, intermediate, and unfavourable cytogenetics were observed in 25.8, 28.0, and 46.2% of t-AML cases, and in 22.2, 57.3, and 20.4% of de novo AML. The frequency of each karyotype was compared between therapy-related and de novo AML in table 2. The median overall survival rate was shorter in the group of t-AML patients than in the group of de novo AML patients (10 versus 15 months, $p=0.0007$). The favourable and unfavourable cytogenetics had a prognostic impact with respect to overall survival in both the t-AML ($p=0.001$ and 0.0001) and de novo AML ($p<0.0001$ and <0.0001) patient groups [S22]. The WHO classification which groups t-MDS and t-AML together and classifies them as therapy related myeloid neoplasm was largely based on the study which compared t-MDS and t-AML patients and showed no significant differences in survival, with a uniformly poor outcome, if they harbour complex karyotypic abnormalities [E7].

Table 2. Frequency of karyotypes in t-AML compared to de novo AML [S22]

Karyotype by nomenclature of the International System for Human Cytogenomic Nomenclature (ISCN).
t: translocation; inv: inversion; NS: not significant

| *Karyotype* | *Cytogenetic group* | *t-AML (n=93)* | *de novo AML (n=1 091)* | *p-value* |
|---|---|---|---|---|
| t(8;21) | Favourable | 6.4% (n=6) | 7.5% (n=82) | NS |
| inv(16) | | 14.0% (n=13) | 6.7% (n=73) | 0.009 |
| t(15;17) | | 5.4% (n=5) | 8.0% (n=87) | NS |
| Normal | Intermediate | 14.0% (n=13) | 42.4% (n=463) | <0.001 |
| Other abnormal | | 14.0% (n=13) | 14.9% (n=163) | NS |
| 11q23 | Unfavourable | 12.9% (n=12) | 3.7% (n=40) | <0.001 |
| Complex aberrant | | 26.9% (n=25) | 11.3% (n=123) | <0.001 |
| Other unfavourable | | 6.4% (n=6) | 5.5% (n=60) | NS |

92. A recent study on the clinicopathologic features of outcomes of t-MDS and t-AML patients with complex chromosomal karyotypes showed that the presence of *TP53* mutation indicated poorer outcome, in particular, multihit *TP53* mutation indicated poorer outcome than monoallelic mutations. Multivariable analysis identified the presence of multihit *TP53* mutation as the strongest predictor of worse outcome, whereas, neither a diagnosis of AML versus MDS nor therapy-relatedness independently influenced overall survival (figure XI) [W13].

Figure XI. Overall survival of patients with MDS and AML based on the status of their prior therapy history and *TP53* mutation [W13]

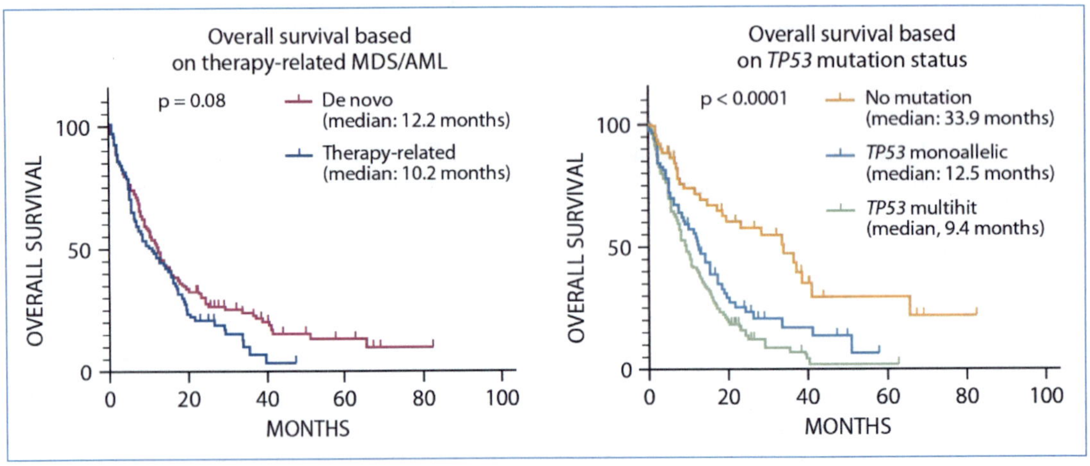


## G. Summary and discussion

93. In this chapter, the current knowledge of diagnosis, incidence, and prognosis of second primary cancer after radiotherapy was described and clinical example cases were presented. The incidence of second primary cancers is very low and attribution to radiotherapy as the sole cause can be difficult to establish. Additionally, because second primary cancers are histologically distinct and arise in different organs from the initial primary cancer, their diagnosis and treatment often fall under different medical specialties.

94. Because of the carcinogenicity of ionizing radiation, a risk-benefit analysis is generally performed by a multidisciplinary team before planning a course of radiotherapy for cases where it is considered necessary. This risk-benefit analysis generally considers that the benefit of active treatment should exceed the risk of adverse effects, such as development of a second primary cancer years later. Radiotherapy is based on delivering the maximum absorbed dose to the target volume while sparing normal tissues. The available radiotherapy guidelines are generally based on the characteristics of a population rather than those of the individual receiving the treatment. Therefore, radiotherapy should be tailored according to the patient's individual factors in the future [M6]. Various factors including the type of prior cancers, individual susceptibility to cancer, genomic instability, radiation susceptibility, and the patient's lifestyle are integrated in the medical decision to perform personalized radiotherapy [B38, M24, T11]. For example, in adult patients of in situ breast carcinoma, the standard treatment generally consists of lumpectomy followed by adjuvant radiotherapy. However, for those patients with a genetical predisposition to develop cancers, such as *BRCA1/2* or *TP53* germline mutation carriers, total mastectomy with immediate reconstruction can be recommended to avoid adjuvant radiotherapy [E9]. In spite of the adverse effects, radiotherapy cannot always be omitted even in advanced cancer cases.

Radiotherapy may still be advocated for genetically predisposed patients with locally advanced breast cancer, if they have a high probability of locoregional relapse. In such cases, radiotherapy techniques may be adapted to minimize the risk of adverse effects [E10, F15, H44, L22, M21, P4, P9, W16].

95. Radiotherapy currently relies mainly on the use of photons. Although good conformality can be achieved by advanced technologies such as IMRT or stereotactic body radiotherapy, photons induce non-target dose spillage in adjacent tissue areas at risk [K30]. More spatially accurate targeting with hadron therapy which includes proton, helium, and carbon ion therapy is expected to reduce the stochastic risk of second primary cancers by reducing the volumes of normal tissues irradiated [C12]. However, while doses can be limited outside the treatment field, it is necessary to obtain long-term in-field data on second primary cancer.

96. Given the rarity and the unpredictable onset times of second primary cancers, studies in this field should be well-structured, long-term investigations employing multidisciplinary approaches. In most prospective clinical trials, primary endpoints typically focus on overall patient survival and locoregional recurrence of the tumour. A key limitation of current research on second primary cancers following radiotherapy is their predominantly retrospective and observational design. As awareness of second primary cancer risk grows in treatment decision-making, there is an urgent need for well-designed clinical trials. These studies should specifically consider second primary cancer as a crucial endpoint and systematically collect categorized therapy data from prospectively enrolled patients.

# III. RADIOBIOLOGY

## A. Introduction

97. This chapter discusses advances in knowledge of molecular mechanisms leading to second primary cancer after radiotherapy. It is important to note that second primary cancers causally linked to radiation exposure represent only a fraction of all second primary cancers, even in patients treated with radiotherapy. Only rarely for a specific tumour that radiation aetiology can be assumed with high probability, for example in the case of sarcomas arising in areas exposed to high radiation doses. The generation of a radiotherapy-induced second primary cancer is a multihit, multistage process where not all hits (i.e., genetic mutations as key drivers of cancer) are linked to the radiation exposure. Factors such as lifestyle, tumour environment, and genetics, among others, play a role in determining a patient's susceptibility to developing a second primary cancer after radiotherapy, and understanding the role of these factors may improve the ability to predict the adverse effects that patients are likely to experience following treatment of the first primary cancer. Additionally, this information would allow more personalized cancer treatment and that constitutes an important step towards personalized radiation oncology and precision medicine. Figure XII shows the pathophysiology and causes of second primary cancers after radiation-induced DNA damage. The constant production of reactive oxygen and nitrogen species, along with the progression of oxidative stress effects, leads to the formation of mutated cells. These mutated cells eventually develop into radiation-induced second primary cancer after a long period of latency, influenced by environmental factors, lifestyle choices, and genetic predispositions [K16].

Figure XII. Schematic of the processes, causes, and pathogenesis of second primary cancers after radiotherapy [K16]

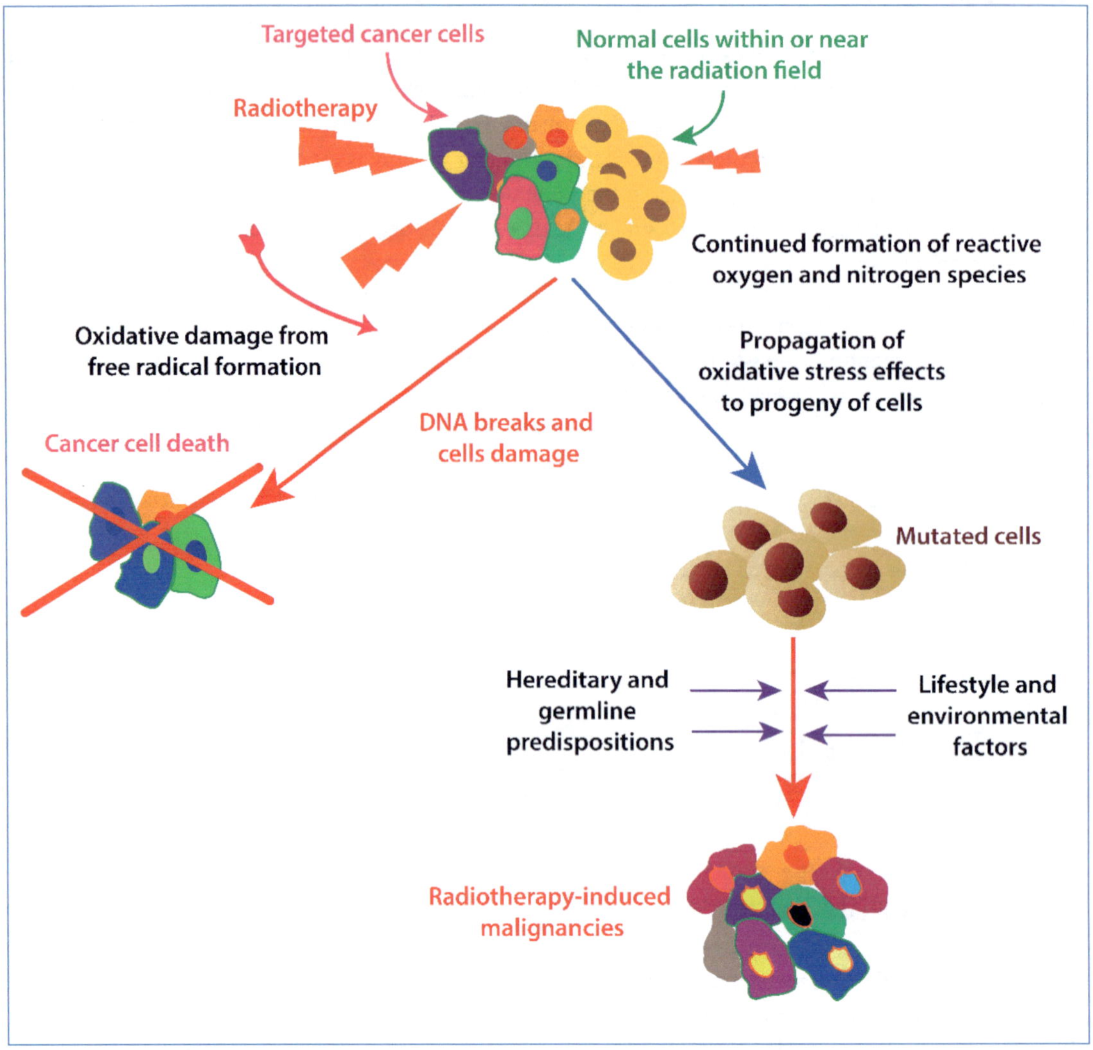


## B. Molecular biology of radiation-induced cancers

98. Cancer is most accurately characterized as a multifaceted progression originating from a single cell that carries a mutation in a cancer-relevant gene, such as the one caused by misrepair of DNA damage. These cells, either directly or subsequent to the accrual of additional mutations or epigenetic factors, acquire growth advantages, progressing towards a proliferative state and ultimately culminating in the transformation of normal human cells into cancerous cells. Multiple important pathways of oncogenesis have been identified, with Hanahan and Weinberg proposing carcinogenesis models with several different hallmarks associated with disease development and/or progression, shared by most and perhaps all types of human cancer, including second primary cancers [H10, H11].

99. Currently accepted models of cancer progression include 10 recognized core hallmarks: sustaining proliferative signalling, evading growth suppressors, resisting cell death, enabling replicative immortality, inducing or accessing vasculature, activating invasion and metastasis, deregulating cellular

metabolism, avoiding immune destruction, genomic instability and mutation, and tumour-promoting inflammation. Recently, the model has incorporated four additional emerging hallmarks and enabling characteristics: "unlocking phenotypic plasticity", "nonmutational epigenetic reprogramming", "polymorphic microbiomes" and "senescent cells" [H12]. These hallmarks are illustrated in figure XIII and provide the current framework for considering the oncogenic process for cancers induced by all mechanisms including radiation.

Figure XIII. The hallmarks of cancer represent a conceptual framework that succinctly summarizes the capabilities of cancer cells and the intricate mechanisms contributing to carcinogenesis (adapted from Hanahan [H12])

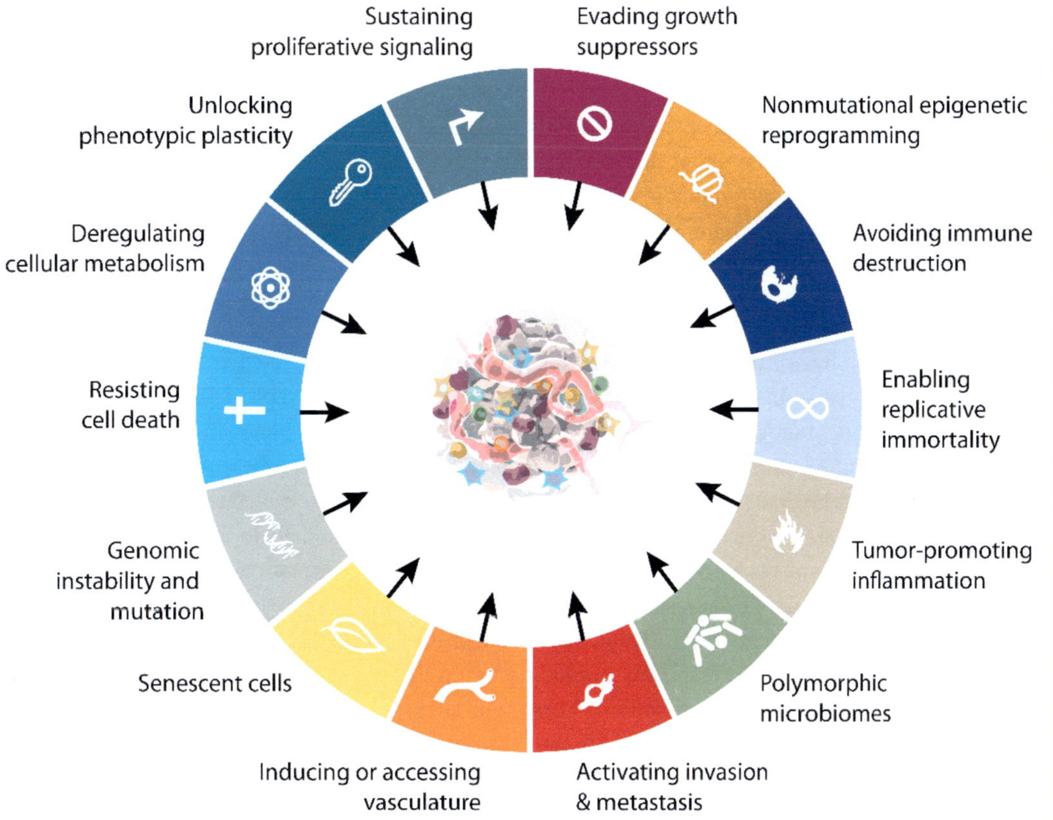


100. These mechanisms have been well-researched in vivo during the past decades with the identification of over 700 dominant oncogenes, fusion partner genes, and tumour suppressor genes associated with cancer [S45]. This list is being updated continuously and includes information on which genes are linked to the hallmarks of cancer [C18].

101. Ionizing radiation has long been known to induce cellular transformation. The Committee has recently reviewed the biological mechanisms relevant for the inference of cancer risks from low-dose and low dose-rate radiation [U8]. A central role in the carcinogenic process is played by radiation-induced direct and indirect DNA damage, where the latter can be caused by radiation-induced radicals and by reactive oxygen and nitrate species produced by cells in response to the irradiation. Missing or incomplete repair and misrepair of the DNA damage can result in mutations, including chromosome aberrations. These mutations can affect function, regulation and quantity of gene products such as proteins and regulatory RNAs, and therefore disturb and dysregulate cellular pathways, leading to the cancer hallmarks described above. Current estimates based on high-throughput sequence analysis of tumour

DNA suggest that clinically detectable tumours carry on average 4 to 5 relevant driver mutations which confer selective advantage to these cells, in addition to many thousands of passenger mutations, which do not confer a selection advantage [A16, G9, I13, M8]. In response to radiation-induced DNA damage, cells initiate complex DNA damage response pathways that affect damage repair mechanisms, chromatin remodelling and epigenetic mechanisms, gene and protein expression, cell cycle regulation, programmed cell death mechanisms, and other mechanisms. All these mechanisms can also contribute to the carcinogenic process if not fully reverting to the original state.

102. The main aim of radiotherapy is to be genotoxic to the tumour, and this therapy achieves cancer cure predominantly through lethal DNA damage in tumour cells. Currently most modalities of radiotherapy are unable to deliver a full tumoricidal dose to the tumour while avoiding delivering a dose to normal tissues. This results in a large spectrum of absorbed doses received by the patients ranging from low doses in areas surrounding the tumour up to very high doses in the tumour target. Therefore, in addition to mechanisms of formation of radiation-induced cancer associated with low doses, as discussed in the UNSCEAR 2020/2021 Report, annex C [U8], mechanisms occurring preferentially at higher doses are also relevant. These include major tissue reorganizations due to a high proportion of cell killing and induction of senescence or differentiation, associated with repopulation by surviving tissue stem cells, and oncogenic stimuli due to inflammation and altered cellular microenvironment.

103. Tissue reconstitution can enhance cancer promotion by several mechanisms [O5]. If the replenishing cells survive prior genotoxic stress, they may carry relevant mutations. This is especially relevant if quiescent cells, which have a higher probability to survive genotoxic insult, re-enter the cell cycle in order to act as replenishing cells. In addition, increased proliferation rates of cells replenishing the stem cell pool increase their mutation burden. Stem cells carrying mutations that make them more resistant towards radiation may outcompete normal stem cells in the replenishing process, and these mutations may also be relevant for cancer induction (e.g., loss of p53 function). For example, in a transgenic mouse model, Fernandez-Antoran et al. [F4] investigated how normal and p53-mutated cells in the proliferative layer of the oesophagus epithelium respond to a radiation dose of 50 mGy. Normal cells stop proliferating and differentiate into non-dividing cells resulting in the loss of any mutations present from the tissue, but the more radioresistant p53 mutant cells exhibit increased proliferation, and their population size in the epithelium increases. It is known that many normal tissues in the elderly population contain premalignant cells carrying cancer-relevant driver mutations. These cells are controlled by surrounding cells in a manner inhibiting their outgrowth, even if the mutations confer growth advantages (cf. review by van Neerven and Vermeulen [V9]). Therefore, the process described by Fernandez-Antoran et al. can explain promotion of pre-malignant cells toward malignant cells after irradiation.

104. A variety of internal and external stressors, including high DNA damage load caused by high radiation dose, can induce programmed cell death pathways. Several of these pathways activate nucleases, such as caspases, that can damage the DNA. In their review, Hawkins and Miles suggested that cells that survive the induction of cell death pathway may carry additional mutations caused by erroneous repair of the nuclease-induced DNA lesions [H24]. It is, however, not clear to what extent cells survive the activation of cell death pathway after high damage load.

105. In vitro studies are limited by the fact that cells are taken out of the context of the entire organism and therefore, some mechanisms cannot be studied well. For example, immune and inflammatory mechanisms, vascular changes, and other complex tissue responses cannot be studied adequately in cells in culture and require whole organisms to be understood accurately. For that reason, most of numerous carcinogenesis studies of cells in culture have limited relevance to the situation for in vivo cancer. Organoid models have progressed to the point where some of these issues have been addressed but study limitations remain [N1].

106. A large volume of work has been performed with transgenic animals engineered to bear germ-line mutations of particular oncogenes or tumour suppressor genes. These studies have helped pinpoint molecular processes in oncogenesis and the enhancement of tumour development by DNA damaging agents as well as other agents. Animal studies are especially relevant to clarify mechanisms of carcinogenesis associated with inflammation and altered cellular microenvironment.

107. A particular feature of second primary cancers after radiotherapy is its ability to develop in the distal organs (low-dose region) or in the irradiation field (high-dose region). It is unclear if distinct biological pathways could be responsible for carcinogenesis induced by both high and low radiation doses. By expanding the size of the susceptible target from the irradiated cell to the entire tissue or a portion of the tissue that has not been irradiated, non-targeted effects may raise the risk of second primary cancer in tissues exposed to low doses of radiation [N12]. Low-dose-mediated activation of glycolysis, mitochondrial function, and redox sensitive pathways suggest a significant role for reactive oxygen species in cellular responses. In contrast, high radiation doses trigger DNA damage response pathways through double strand breaks induction as the primary cellular response. The impact of these molecular and cellular responses on the risk of second primary cancer development at low and high dose remains under active investigation [S4].

108. Mancuso et al. demonstrated the contribution of non-targeted radiation effects to carcinogenesis in bystander tissues on mice. They showed that genetic damage caused by bystander responses contributes to cancer risk in the mouse central nervous system, with drastic acceleration of medulloblastoma in Ptch1+/− mice irradiated with shielded brains. These studies highlighted the importance of the cellular proliferative and/or differentiation state for expression of bystander damage [M5].

## 1. Genetic considerations

109. As discussed by Morton et al. [M30], while the genetic susceptibility to certain cancers is well established, the understanding of genetic susceptibility to treatment-related second primary cancers is limited. Genome-wide association studies [C11], have identified genomic regions linked to treatment related leukaemia in patients treated with chemotherapy, radiotherapy, or combined therapy [K24] and to the occurrence of second primary cancers in Hodgkin lymphoma survivors initially treated with radiotherapy [B20]. These findings substantiate the notion of genetic susceptibility to treatment-related second primary cancers.

110. Some studies on second primary cancer risks have focused on specific genes related to drug metabolism and carcinogenesis. These investigations have identified associations for variants in genes associated with oxidative stress, DNA detoxification (reflecting individual capacities for genotoxic damage repair), and DNA repair, especially in cases of leukaemia associated with radiation treatment [A18, B26, L3, S26, S27, W21, W22]. Examples include the ataxia telangiectasia gene and mutations in DNA damage repair pathway genes, showing associations with contralateral breast cancer [B48, B49]. Additionally, the Fibroblast Growth Factor Receptor (FGFR2) gene has been linked to the development of breast cancer after radiotherapy for Hodgkin lymphoma, indicating that either breast cancer is developed through activation of FGFR2 pathways or that individuals with specific genetic variants in FGFR2 might be more vulnerable to these effects [M1].

111. Familial predisposition to cancer may be of monogenic or polygenic character with different levels of penetrance of the responsible genes [D12]. Rare monogenic cancer predisposition syndromes exist that are associated with extremely high sensitivity to radiation resulting from modifications of autosomal recessive genes with high penetrance. Prominent examples are the ataxia telangiectasia gene, involving

the ataxia telangiectasia mutated gene, Nijmegen Breakage syndrome, involving the NBS gene, Fanconi Anaemia, involving the FA/BRCA pathways, and Bloom Syndrome involving, the BLM gene [P8]. No data exists regarding the risk of second primary cancer after radiotherapy in these patients because the diseases are very rare, and radiotherapy is avoided due to the high risk of side effects [D27].

112. However, the role of genetic background in susceptibility towards radiation-induced second primary cancer is not clear. In breast cancer patients carrying missense or truncated variants of the ataxia telangiectasia mutated (ATM) gene, the excess 10-year absolute risk of developing a second contralateral breast cancer after radiation is low. Bergom et al. estimated an absolute risk of 6.8% (95% CI: 4.9%, 9.6%) in the absence of radiotherapy exposure, compared with 8.9% (95% CI: 6.1%, 13.2%) in analogous patients who were treated with radiation [B16]. However, others found that breast cancer patients under the age of 45 years who carry certain rare missense ATM classified as "variants of unknown significance" exhibit a significantly increased risk of developing contralateral breast cancer as compared to non-carriers after treatment with radiotherapy but was not observed in those who did not receive radiotherapy [R2]. As reviewed by McDuff et al. [M11], ATM genetic variant, exposure to radiation during treatment and age may all influence the risk for contralateral breast cancer. There are also conflicting reports on the sensitizing impact of BRCA mutations that are known to enhance the sporadic risk of breast cancer (also of ovarian cancer for BRCA2). Some studies demonstrate an enhanced risk of radiotherapy-induced second primary cancer in patients with BRCA mutations [K19, P6, T30], while other studies do not report the same findings [M17, P5].

113. It is not clear how many genetic defects that are associated with the first primary cancer can predispose the patient to a second primary radiation-induced cancer. The Radiogenomics Consortium was established in 2009 by the NCI with the objective to expand knowledge of the genetic basis for differences in radiosensitivity and has been actively engaged in conducting exhaustive genome-wide studies with the primary objective of identifying genes linked to late effects resulting from radiation exposure. This initiative seeks to unravel genetic defects associated with the onset of the initial cancer, potentially predisposing individuals to a subsequent, radiation-induced cancer [K14]. This undertaking is ambitious, given that one single nucleotide polymorphism could potentially account for heightened sensitivity. As an example of this complexity, patients with bilateral retinoblastoma exhibit radiotherapy-induced osteosarcoma as the predominant second primary cancer. This observation suggests a potential predisposition linked to heterozygosity for the tumour suppressor gene retinoblastoma 1 [K21]. Genetic mapping in mice further supports this observation, revealing multiple loci conferring genetic susceptibility to alpha-particle-induced osteosarcoma, with retinoblastoma 1 heterozygous mice exhibiting an increased risk of alpha-particle-induced osteosarcoma [R12]. However, mice with heterozygous INK4A—an upstream regulator of retinoblastoma—did not show increased susceptibility but did experience a shorter tumour latency after radiation exposure [G16]. Deletion of INK4-ARF, often by codeletion, is a hallmark of cancer, contributing to aberrant self-renewal capacity of tumour cells [S36].

114. Female patients of Hodgkin lymphoma treated with chest radiotherapy face a notably elevated risk of developing breast cancer. The extent of genetic contribution to this heightened risk remains unclear. Opstal-van Winden et al. [O3] conducted an in-depth examination of 211,155 germline single nucleotide polymorphisms to assess the interaction between genes and radiation in breast cancer risk. Their case-only analysis encompassed 327 breast cancer patients who had undergone chest radiotherapy for Hodgkin lymphoma, in addition to 4,671 first primary breast cancer patients. Nine single nucleotide polymorphisms exhibited statistically significant interactions with radiotherapy concerning breast cancer risk. A comprehensive evaluation was performed, incorporating a polygenic risk score derived from these single nucleotide polymorphisms, referred to as the radiotherapy interaction-polygenic risk score, and an established breast cancer polygenic risk score derived from the general population. The case-control analysis involved 327 chest-irradiated Hodgkin lymphoma patients with breast cancer and 491 chest-irradiated Hodgkin lymphoma patients without breast cancer. Notably, patients in the highest tertile

of the radiotherapy interaction-polygenic risk score demonstrated a 1.6-fold higher breast cancer risk compared to those in the lowest tertile. This study underscores the influence of genetic factors on breast cancer risk subsequent to chest radiotherapy for Hodgkin lymphoma [O3].

115. Best et al. [B20] identified two variants (rs4946728 and rs1040411) of the PRDM1 gene associated with second primary cancer in survivors of Hodgkin lymphoma treated with radiotherapy as children; similar effects were not observed in those treated as adults. The variants are associated with a decreased expression of PRDM1 that is a transcriptional repressor involved in a variety of cellular processes including proliferation, differentiation, and apoptosis [B20].

116. Morton et al. [M31] conducted a genome-wide association study of breast cancer in female survivors of childhood cancer, pooling two cohorts with detailed treatment data and systematic, long-term follow-up: the CCSS and St. Jude Lifetime Cohort. The overall population and subgroups were divided into two classes: patients receiving a radiation dose to the breast of less than 10 Gy and those receiving a dose of more than 10 Gy. Among survivors who received a radiation dose to the breast of more than 10 Gy, a locus on 1q41 was associated with subsequent breast cancer risk. Two rare variants also showed potential associations, indicating a potential genetic susceptibility to radiation-induced breast cancer among this population. On the other hand, Sapkota et al. [S7] conducted a genome-wide association study for subsequent basal cell carcinoma (BCC) among irradiated childhood cancer survivors and identified a risk locus in HTR2A that was specific to survivors with reduced non-genetic risk, suggesting potential genetic factors contributing to BCC development following radiation exposure.

117. These studies highlight the importance of understanding the genetic associations and risks for developing certain types of cancers following radiation exposure among childhood cancer survivors. Further research is needed to confirm these associations and gain deeper insights into the pathophysiology and genetic factors related to subsequent cancers in this vulnerable population.

# C. Contributing factors to second primary cancer risk

118. This section describes the factors that affect the risk of radiation-associated second primary cancers. Although the focus of this section is on the biological mechanisms of action, evidence for their impact comes from epidemiological studies and therefore, some of the studies described in chapter V of this annex are also included in this section which discusses the underlying mechanisms for the development of second primary cancer.

## 1. Radiation dose

119. There is ample evidence that demonstrates that ionizing radiation induces primary cancers in general, whether first or second primary cancer[U4]. Following acute, whole-body exposure to a dose up to 4 Gy, such as after the atomic bombings in Hiroshima and Nagasaki, the risk of all cancers pooled increases with increasing dose [G18]. At doses above 4 Gy, epidemiological analysis of data is truncated due to large dosimetric uncertainties [W1]. However, when these uncertainties are considered by cancer risk models, the incidence of cancer at high doses levels off, and eventually declines with increasing dose [S14]. The decline of cancer incidence is interpreted as the result of the dominance of cell killing over induction of mutations. Bell-shaped dose-response relationships have also been observed in risk models for animals [U1, U9].

120. During a course of radiotherapy, tumour cells receive doses up to 100 Gy delivered at high-dose rate but in a fractionated manner. There is a question of whether the incidence of second primary cancer shows a continuous increase with increasing dose, or if it declines at very high doses, as for first primary cancers induced by whole body exposure [S1]. Epidemiological studies of childhood cancer survivors indicate that the dose-response relationships for most solid second primary cancers in patients treated with external radiation are linear, except for thyroid cancer, for which it is bell-shaped [I11]. However, serious concerns have been raised regarding the precision of dose estimates at the sites of second primary cancers. Second primary cancers arise after a latency period of several years and, due to tissue remodelling and changes in the body mass of a patient, the site of an identified second primary cancer with respect to its position at the time of the therapy is uncertain. Results of studies that were carried out to identify the location of second primary cancer show that most second primary cancers occur at the margin of the planned treatment volume where the dose gradient is very steep [D17, D26]. An experimental study with rats treated with either volumetric modulated arc therapy (VMAT) or a 3D conformal technique with two opposing anterior-posterior and posterior-anterior beams showed that radiation-induced tumours appeared predominantly in regions receiving high doses, with no impact of the low radiation doses absorbed in large body volumes from VMAT [G14].

121. Compared to a single acute dose of radiation, the tolerance of cells exposed to fractionated radiation doses is higher. This is thought to be due to repair of sublethal damages between fractions [E5] and to repopulation events [D15]. Cell and tissue types differ with respect to both repair and repopulation, making them, along with the dose and fractionation scheme, important factors in the dose-response relationship [S47]. This conclusion is supported by results of an in vitro study with two normal cell lines representing elastic and hierarchical tissues [A11].

## 2. Host factors

### (a) *Patient age*

122. The relationship between age at exposure and the risk of radiation-induced cancer is complex. It is generally assumed that the risk of radiogenic cancer is higher in children than in adults because of the longer life expectancy of children and the high potential for development of radiation-induced mutations into a malignancy; and of a high frequency of cell divisions in a growing organism that is associated with a high risk of accumulating mutations that can lead to a malignancy [S23]. However, the results of studies summarized in the UNSCEAR 2013 Report, annex B [U5] demonstrate that the high sensitivity of children to radiation-induced cancer is only observed for brain, skin and thyroid cancer, and leukaemia. The risk for breast and uterine cancer peaks at around the age of puberty, possibly, as a result of the interaction of radiation with sex hormones.

123. Several studies have investigated the age when second primary cancers occur (i.e., attained age) as one of the modifiers of second primary cancer risks. In general, the relative risk (RR) of developing a second primary cancer as compared with the baseline risk in the general population decreases with age, whereas the variation of excess absolute risk (EAR) with increasing age depends on the type of both first and second primary cancer. For example, a study of breast cancer after Hodgkin lymphoma demonstrated that, as the age at diagnosis of Hodgkin lymphoma increased from less than 20 years to between 41 and 50 years, the RR decreased from 17.9 to 1.4 and the EAR decreased from 79 to 11 cases per $10^4$ PY [D3]. However, another study suggested that as the attained age of breast cancer increased, the RR decreased whereas the EAR increased [H38]. In contrast, neither the RR nor the EAR for thyroid cancer after Hodgkin lymphoma varied with attained age [H38]. Since most cancer rates in the general population

increase substantially with increasing age, constant or decreasing RRs with age can be associated with prominently increasing EAR, in the same way as described above for breast cancer after Hodgkin lymphoma. The main factors that lead to variations in second primary cancer risks by age are the baseline cancer incidence in the general population, potential variations in tissue susceptibility to radiation or chemotherapy exposure at different ages, and the roles of other cancer risk variables. More research is required to understand fully the effect of age at exposure and attained age both individually and in combination, in particular on survivors of primary malignancies such as Hodgkin lymphoma survivors.

124. The majority of studies conducted to date have concentrated on age-related effects of radiotherapy exposures. For example, it has been reported that childhood cancer survivors receiving cranial irradiation before the age of 5 years have significantly higher risk of subsequent glioma than children irradiated at ages at or above 5 years [N10]. Similarly, the risk of thyroid cancer has been observed to be highest among children irradiated before the age of 10 years [B27]. The sensitivity to radiation of developing tissues has been identified as the cause of the high risks at younger ages. For second primary breast cancer, the data on the role of the age at exposure are conflicting. While some studies have indicated that the risks are greatest for girls receiving chest-directed radiotherapy during puberty, and that the risks are lowest both before and after puberty, more studies involving a broad sample of people exposed at a variety of ages are required for more definitive conclusions [C17, I10]. Importantly, much less is known about potential age-related effects of chemotherapy exposures.

125. There is growing recognition of the role that lifestyle and medical history factors play in the aetiology of second primary cancers as a result of research into non-treatment risk factors for these cancers. Owing to the cumulative nature of these exposures, older cancer survivors may be more susceptible to developing second primary cancers [M26]; further research is required to determine if interactions between risk factors associated with treatment and other factors are age-dependent.

### (b) *Patient sex*

126. Results from the Life Span Study (LSS) cohort of the Japanese atomic bombing survivors clearly show that the overall excess relative risk (ERR) of radiation-induced solid cancer is higher in women than in men [G18]. The disparity cannot be solely attributed to breast cancer. Due to differences in the shape of the dose-response relationships—linear for women and linear-quadratic for men—the female to male ratio of risk depends on the absorbed dose; the ERR is 6.1 at 0.1 Gy, 3.2 at 1 Gy and a little over 1 at 3 Gy. Based on these results, a disparity in the risk for second primary cancer between male and female patients following high-dose radiotherapy could be expected. The mechanisms that may cause the disparity are not understood, and hormonal differences are usually assumed to be the contributing factor. A study by Brenner et al. [B44] compared solid cancer mortality and incidence, adjusting for the high dose effect to minimize the influence of individuals receiving a dose of over 4 Gy. The findings of this study indicate that there is an upward curvature in the dose-response relationship, which was significant for males in terms of both cancer mortality and incidence, but was not significant among females when considering all solid cancer incidence. The authors suggest that this difference may depend on factors such as the composition of case series, including sex-specific cancers, and the age at exposure or calendar period (1958–1987 versus 1988–2009). Further analysis is needed to confirm this emerging trend.

## 3. Lifestyle and environmental factors

127. A number of modifiable factors are known to influence the risk of cancer [G13] but it is unclear how significantly they affect the risk of radiotherapy-induced second primary cancer. It should be noted

that the ICRP and the Committee on the Biological Effects of Ionizing Radiation (BEIR) assume that, for most cancer sites, radiation increases the background risk of cancer so that multiplicative risk transfer models should be preferred over additive models when the risk is transferred between cohorts [I5, N19]. BEIR's seventh report, commonly referred to as BEIR VII, recommends a 70:30 multiplicative to additive risk transfer model ratio [N19], while the ICRP conservatively recommends a 50:50 ratio [I5]. From a mechanistic point of view, an interaction between two agents is expected if one agent is a cancer initiator and the other is a cancer promoter. Cigarette smoking, obesity, alcohol, sunlight, and infectious agents are the major environmental carcinogens [G13], and they are all both initiators and promoters. Consequently, it is not possible to predict if the combined action of radiation with other factors leads to additive or multiplicative responses, solely based on the mechanism of action. This may vary with cell type and with the cancer type that develops, especially in view of the multistep nature of carcinogenesis [H12].

## (a) *Smoking and alcohol*

128. The influence of smoking on the risk of treatment-related lung cancer has been examined in several studies [K7, N11, P13, T17, V4]. A large international case-control study of Hodgkin lymphoma survivors examined risk of lung cancer in relation to radiation dose, chemotherapy, and smoking [G6, T17]. Together, smoking and treatment had effects that were noticeably stronger than the sum of their individual effects. The elevated risks from radiation and chemotherapy seemed to be multiplied by the increased risks from smoking [T17]. The impact of smoking on the risk of radiation-induced lung cancer in breast cancer survivors has been the subject of numerous studies which found that excess lung cancer risk following post-mastectomy radiotherapy is restricted to smokers, indicating a significant interaction [K7, N11, P13].

129. Alcohol by itself is neither a mutagen nor a carcinogen. However, acetaldehyde, the first metabolite of ethanol oxidation, is carcinogenic [S28], and therefore, alcohol consumption can lead to cancer induction, although not of all cell types. In fact, alcohol consumption was reported to reduce the incidence of several types of blood cancers, triple negative breast cancer and rectal cancer [M14] demonstrating the complexity of alcohol-induced health effects. Whereas the role of alcohol in cancer induction is well-established, the mechanisms by which it prevents cancer are not understood [M14]. With respect to radiotherapy-induced second primary cancer, DiMarzio et al. reported that alcohol consumption did not alter the association between radiotherapy and second primary cancer risk but was associated with lower mortality risk [D19]. The study pointed out that the data regarding alcohol consumption were based on self-reporting without reliable information on the magnitude of the exposure and its duration. A study carried out on a group of cancer survivors following any type of therapy on the relationships between smoking and alcohol consumption revealed that alcohol might increase the second primary cancer risk when the subjects smoke, while non-heavy drinking might not increase the risk when they did not smoke [T1].

## (b) *Patient body mass index*

130. There are numerous emerging biological mechanisms that connect obesity to cancer. They include: alterations in adipokine pathophysiology; factors resulting from ectopic fat deposition; disruption of circadian rhythms; dietary nutrients; factors with potential significance, such as the altered intestinal microbiome; insulin resistance and abnormalities of the IGF-I (insulin-like growth factor-I) system and signalling; sex hormones biosynthesis and pathway; subclinical chronic low-grade inflammation and oxidative stress; and alterations in the microenvironment and cellular perturbations [A28]. It is not known if obesity is a risk modifier for second primary cancers.

### (c) *Reproductive factors*

131. A discussion on reproductive factors has been included in this section, because modifiable cancer risk factors, such as obesity, affect the timing of puberty [L13], while lifestyle factors affect the age of menopause [G12]. Several studies have examined the effect of reproductive factors on the risk of radiation-induced breast cancer. Menopausal age has been shown to modify the risk of breast cancer in survivors of Hodgkin lymphoma and childhood cancer treated with chest radiotherapy [C17, D3, H33, T18, V6]. De Bruin, et al. [D3] reported that 30% of female Hodgkin lymphoma survivors reached menopause before the age of 41 years; the onset of menopause at such an early age was thought to be related to exposure intensive chemotherapy and was associated with a reduced risk of radiotherapy-induced breast cancer of 60% (95% CI: 20, 80). Women with fewer than 10 years of intact ovarian function after radiotherapy had a 70% (95% CI: 40, 80) decreased risk of breast cancer compared with women with 10–20 years of ovarian function after irradiation, while those with more than 20 years of intact ovarian function after radiotherapy had 5.3-fold (95% CI: 2.9, 9.9) increased risk of breast cancer. These risk reductions were noted in women treated before the age of 31 years, but not in those treated between the ages of 31 and 40 years. This difference may have resulted from the closer proximity of the age of the treated women to the menopausal natural course at the time of treatment [D3]. These findings suggest that, after radiotherapy has produced an initiating event, ovarian hormones could play a critical role in promoting tumorigenesis. According to a British study, women who received radiotherapy near menarche had a higher risk of developing breast cancer from radiation, indicating that radiation is more carcinogenic to developing breast tissue [C17]. Similar observations were reported from the cohort of the Japanese atomic bombing survivors [B42]. Other risk factors such as weight, parity, and age at first childbirth have not shown to have any modifying effects, but the small number of women in the published studies does not allow for the detection of smaller interaction effects or risk modification for infrequent exposures. Information regarding hormone replacement therapy was scarce in these studies [C17, D3, H33, K12].

## 4. Modification of treatment-related second cancer risks by chemotherapy

132. Radiotherapy is often combined with chemotherapy. The enhanced efficacy of combining the two therapies (chemoradiotherapy) is attributed to spatial cooperation and enhancement of radiation effects [N17]. Spatial cooperation refers to the action of chemotherapy on distant metastases and radiotherapy on the primary tumour. Enhancement of radiation effect by chemotherapeutic treatment on the primary tumour is considered to be based on:

*(a)* Augmentation of cellular damage by incorporating drugs into cancer cell DNA;

*(b)* Inhibition of radiotherapy-induced DNA damage repair;

*(c)* Redistribution of cell cycle phases leading to accumulation of cancer cells in a radiosensitive phase;

*(d)* Elimination of hypoxic cells; and

*(e)* Inhibition of the cancer cell repopulation.

133. The interaction between radiotherapy and chemotherapy is not restricted to cancer tissue but also applies to normal tissues. Therefore, it is not surprising that chemotherapy increases radiotherapy-induced early and late toxicities [A5, B14]. However, a higher level of toxicity does not necessarily correlate with a higher risk of second primary cancer. The reason for this may be that toxicities, especially

late ones, are mainly caused by cell death while second primary cancer can arise in cells that survive with cancer-related mutations in their DNA.

134. The impact of various chemotherapy modalities on the risk of radiotherapy-induced second primary cancer has been published by several authors. Risk is often given for pooled second primary cancer types because of poor statistical power when particular cancer sites are analysed. D'Angio et al. was one of the first to observe that actinomycin-D reduced the risk of radiotherapy-induced second primary cancer in a cohort of patients with various first primary cancers [D1]. Chemotherapy consisting of cyclophosphamide, vinca alkaloids, and antifolic agents had no effect on the risk of radiotherapy-induced second primary cancer [D1]. A reduced risk of radiotherapy-induced second primary cancer was observed among 20-year or longer survivors of childhood cancers treated by a combination of chemotherapy and radiotherapies compared to radiotherapy alone [T29]. The results and conclusions of these epidemiological studies received support from an experimental study, in which chemotherapy, in combination with radiotherapy, induced a lower level of cytogenetic damage in peripheral blood lymphocytes than for radiotherapy alone. This was true both for damage directly induced by the therapies and for damage induced by a radiation dose given ex vivo to blood samples [W12]. The results negatively correlated with the level of apoptosis, suggesting that the reduced level of cytogenetic damage was caused by a selective elimination of cells damaged by combined chemotherapy and radiation treatment. Therefore, it can be assumed that chemotherapy can increase the killing effect of radiotherapy on normal tissue cells, reducing the risk of second primary cancer.

135. Results of a more recent experimental study on the impact of fractionated cisplatin and radiation treatment on cell growth and accumulation of DNA damage in two normal cell types differing in origin showed that cisplatin reduces the inhibitory effect of radiation on cell proliferation of normal cells and does not strengthen delayed effects resulting from accumulation of DNA damage [A12]. However, studies indicating opposite results also exist. An analysis of breast cancer induced by radiotherapy of childhood cancers revealed that combination of doxorubicin and radiotherapy increases the risk of second primary cancer as compared to radiotherapy alone [V13]. Similarly, it was observed that doxorubicin increases the risk of radiotherapy-induced second primary cancer among Wilm's tumour patients [B46]. Guerin et al. [G24] carried out a pooled analysis of risk factors contributing to second primary cancer following treatment of various first primary cancers and found that chemotherapy generally increased the risk of radiotherapy-induced second primary cancer, with concomitant treatments having a stronger effect than sequential treatments. This result can be interpreted as evidence for a potentiating effect of chemotherapy on radiotherapy-induced DNA damage, leading to an increased level of cancer initiating mutations. A potentiating effect of chemotherapy on radiotherapy-induced leukaemia was also reported by Haddy et al. [H1] who studied survivors of childhood malignancies. The conclusion from all the studies is that there is no single pattern of how chemotherapy modifies the risk of radiotherapy-induced second primary cancer.

136. Studies on combined exposures described in this section only compared the effect of radiotherapy alone with that of radiotherapy plus chemotherapy. The impact of chemotherapy alone was not analysed because the therapeutic procedures did not include such intervention. Without looking at the effect of chemotherapy alone, it is not possible to judge whether chemotherapy interacts with radiotherapy or whether combined effects are simply additive. Results of an analysis of how far chemotherapy alone, applied to treat other cancer entities, induces second primary cancer are inconclusive, to some extent because many studies are small and lacking statistical power. A meta-analysis revealed that treatment with cisplatin is not associated with a significantly increased risk of second primary cancer [L15]. However, high doses of cisplatin, alkylating agents and doxorubicin can lead to second primary cancer among childhood cancer survivors [T29, V13]. These conclusions are supported by a study of the Dutch Childhood Cancer Oncology Group cohort showing that both doxorubicin and cyclophosphamide increase the risk of second primary cancer [T9]. However, other studies show that chemotherapy of adult breast cancer patients leads to a reduction of second primary cancer risk [L14, S9], as does chemotherapy

for Hodgkin lymphoma [S9, T18]. Given these variable results, a conclusion regarding the mode of interaction between radiotherapy and chemotherapy is not possible. It appears, however, that if there is an interaction, it is weak and not readily detectable. This conclusion aligns well with the conclusions from mechanistic studies on combined effects of radiation and chemical toxins [U2, U3, U7].

137. As stated by Wegierek-Ciuk et al. [W12], given the multitude of factors that promote cancer formation [F13, H11], it is not unexpected that, depending on the type of radiation treatment and the dose delivered, the location of the primary cancer, and the response of each patient, the single action of chemotherapy and its combination with radiotherapy can have different effects on the normal tissue of patients. The ability of chemotherapy to cause a second primary cancer can be explained by the mutagenic effect of the drugs used, which trigger cell transformation [B28], and by its inhibitory effect on the immune system, which results in the loss of cancer immunosurveillance [Z5]. The reduction in the risk of second primary cancer associated with chemotherapy can be explained by the elimination of radiation-damaged cells or, as in the case of breast cancer, by ovarian ablation leading to modified hormonal status of the survivors [S10].

## D. Potential biological indicators/biomarkers of second primary cancer risk

138. Current evidence showing that potential biomarkers can be used to effectively predict the risk of second primary cancer is incomplete. Therefore, this is an important area where future studies should be conducted as the prediction of the risk of second primary cancer is crucial, not only to help tailor the initial treatment of first primary cancers to individual patients but also to implement an adequate surveillance plans in the post-treatment period. Ideally, a biomarker of second primary cancer should:

*(a)* Act as a prognostic indicator for the development of a second primary cancer;

*(b)* Allow to compare different radiotherapy techniques and modalities and their effects on healthy tissues, including the evaluation of the risk of second primary cancers after radiotherapy;

*(c)* Act as a biodosimetry tool to be applied to monitor the absorbed dose in the body that can be recorded and associated with the potential development of second primary cancers in prospective studies; and

*(d)* Allow to monitor carefully critical organs such as bone marrow during radiotherapy to reduce the risk of second primary cancers.

139. Prognostic biomarkers for second primary cancers after radiotherapy can include specific genes and proteins that are associated with the risk of developing these cancers. For example, certain genetic mutations or changes in gene expression levels may indicate an increased susceptibility to developing second primary cancers following radiotherapy. Additionally, alterations in the levels of oncoproteins or tumour suppressor proteins could serve as prognostic biomarkers for identifying individuals at higher risk of development of second primary cancer. Evidence suggests that ionizing radiation can cause specific mutation patterns that may contribute to the development of these cancers. Several studies have focused on the mutational signatures associated with radiation exposure in second primary cancers, particularly sarcomas occurring in high-dose radiation fields. These mutational signatures are specific patterns of mutations found in the genomes of these cancers that are indicative of DNA damage caused by ionizing radiation.

140. One prominent mutational signature associated with radiation exposure is characterized by a high frequency of clustered DNA double-strand breaks near indels (insertions or deletions) and subsequent chromosomal rearrangements. Ionizing radiation can cause double-strand breaks, leading to abnormal chromosomal rearrangements and structural variations within the genome, such as translocations, inversions, deletions, and duplications. This signature has been observed in various cancers following radiotherapy or nuclear accidents and it is believed to be caused by errors in DNA repair processes after exposure to ionizing radiation. For instance, Behjati et al. [B12] analysed the genomic profiles of 12 sarcomas that developed as second primary cancers following radiotherapy for an initial malignancy. The researchers identified distinct mutational signatures associated with radiation exposure, including evidence of double-strand break repair mechanisms such as non-homologous end-joining. Kim et al. [K18] used whole-genome sequencing to identify mutational signatures associated with radiation-induced sarcomas and DNA damage repair pathways. Additionally, López et al. [L25] performed genetic profiling on a cohort of 12 gliomas developed after radiotherapy to evaluate their molecular pathogenesis and assess differences in genomic signature as compared to their spontaneous counterparts.

141. In the analysis of 163 medulloblastomas that developed either spontaneously or after X-ray irradiation at doses of 0.05–3 Gy in Ptch1 heterozygous mice, different molecular patterns of loss of the functional Ptch1 locus were observed. In this dose range, S-type tumours, predominantly spontaneous medulloblastomas, broad regions on chromosome 13 were lost, including all consecutive markers distal to Ptch1 locus. In contrast, all tumours that developed after irradiation with a dose of 3 Gy showed losses of limited size around Ptch1, while retaining the distal markers (R-type tumours). The proportion of R-type tumours increased with dose, suggesting that the R-type change is a reliable radiation signature. The authors also observed that 573 genes on other chromosomes were expressed differently between S-type and R-type tumours [I14].

142. Another mutational signature associated with ionizing radiation exposure is a characteristic pattern of base substitutions called the C>T transition signature. This particular mutation type has been linked to oxidative damage induced by reactive oxygen species generated during radiotherapy. Comparing the transcriptome of 12 sarcomas arising in the irradiation field of a primary tumour following radiotherapy with the transcriptome of 12 sporadic sarcomas, a signature of 135 genes discriminating radiation-induced from sporadic sarcomas was identified. The functions of the genes with this signature suggest that radiation-induced sarcomas were subject to chronic oxidative stress probably due to mitochondrial dysfunction [H2].

# E. Summary and discussion

143. During radiotherapy, but irradiation of normal tissues in and around the target is unavoidable. These tissues receive different levels of radiation dose and, therefore, mechanisms of cancer induction for different tissue types must be considered for both low and high doses. Even tissues distal to the field could be affected by non-targeted radiation effects. Although it is recognized that radiation can induce cancer, it is difficult to establish a causal link between a second primary cancer and the radiation treatment of a previous primary tumour.

144. Genetic susceptibility may play a role in the development of second primary cancers from previous radiation treatments, and a few genes have been identified as being linked specifically to an increased risk of the development of second primary cancer after radiotherapy. As it is difficult to differentiate between susceptibility for radiation-induced first and second primary cancers, and second primary

cancers arising from other treatment modalities, the role of genetic background in susceptibility to radiation-induced second primary cancer remains unclear.

145. There are several host factors that can contribute to the risk of second primary cancer after radiotherapy including the age of the patient at the time of treatment of the first primary cancer. In general, it is believed that the risk of second primary cancer decreases with increasing age at first treatment, mostly due to life expectancy following radiation exposure and lower frequency of cell division, although the relationship between age and risk of radiation-induced second primary cancer is complex. Lifestyle factors can also influence the development of second primary cancers, as they can act as cancer initiators or promotors (e.g., smoking) and as such, may interact with radiation.

146. As many cancers are treated with both chemotherapy and radiotherapy, the impact of chemotherapy on the development of second primary cancer must be considered. There is much conflicting information on the mode of interaction between chemotherapy and radiation and without a thorough analysis of the effects of chemotherapy alone on second primary cancer, no conclusion can be made on the influence of chemotherapy on the risk of radiation-induced second primary cancer.

147. The current evidence regarding the applicability of potential biomarkers in predicting the risk of developing a second primary cancer is incomplete. Specific mutational patterns (signatures) related to DNA damage response pathways, cell proliferation, and immune regulation may also be indicative of the likelihood of developing second primary cancers after radiotherapy.

148. Ongoing research in oncoproteomics and gene expression analysis continues to advance our understanding of prognostic biomarkers for second primary cancers after radiotherapy, contributing to improved risk assessment and patient care. The identification of these prognostic biomarkers can enable healthcare professionals to closely monitor high-risk individuals and tailor personalized surveillance and intervention strategies to mitigate the potential impact of second primary cancers. More definitive answers will likely come from the interpretation and analysis of these molecular measures integrated with up-to-date epidemiological investigations through meta-analysis and systematic review processes.

# IV. DOSIMETRY

## A. Introduction

149. Tissue and tumour dosimetry are integral components of the safe and effective use of radiation in the treatment of cancer. The objective of radiation treatment planning is to deliver a tumoricidal radiation dose to the PTV while simultaneously minimizing the dose delivered to surrounding organs, specifically to levels below thresholds for organ toxicity (i.e., tissue reactions). While this objective provide for successful cancer treatment in terms of near-term survival and clinical outcome, it does not address the concurrent longer-term risk of a second primary cancer (i.e., stochastic effects). In this regard, tissue dosimetry—both inside the treated volume, and in regions far from the treated volume—is an essential component of predictive risk models for second primary cancers.

150. In this chapter, different approaches to defining dose regions where second primary cancers may arise are reviewed. The computational techniques for the calculation of the absorbed are presented—both

within, as well as outside the range of validity of the TPS used in radiotherapy. This chapter also considers how biological weighting of the absorbed dose is applied when radiation particles other than photons are used, including those based on radiological protection concepts as well as other methods which more appropriately include effects of dose fractionation. Finally, this chapter provides a concise review of various radiotherapy techniques, with a particular emphasis on the distribution of absorbed doses within the body associated with potential risks for the development of second primary cancers. The review covers historical radiotherapy methods, which account for the majority of treatments in the patient cohorts analysed in the meta-analysis of second primary cancer risk presented in Chapter V. Modern radiotherapy techniques are also addressed, emphasizing the importance of estimates of organ doses in guiding patient surveillance programmes for monitoring of second primary cancer.

151. This chapter includes two electronic attachments that provide further information. Electronic attachment A contains a review of various radiological protection and operational quantities and units used in chapter V. Electronic attachment B provides information on studies describing the various approaches and techniques for tissue dosimetry used in the papers selected in chapter V for the meta-analysis of risks of second primary cancer.

## B. Absorbed dose regions in the patient during radiotherapy

152. In general, the distribution of absorbed dose in a patient who has received EBRT is characterized by a high-dose region inside the PTV, surrounded by a medium-dose region characterized by a large gradient at the boundary with the high-dose region. The rest of the patient's body receives a low dose which mainly originates from the radiation scattered within the patient's tissues but also scattered from the irradiating device and other equipment and objects present in the treatment room. In the high-dose region, tissue reactions caused by the radiation are dominant and are the basis of the optimization process in the treatment planning to limit the risks of tissue complication. In the medium-dose region, tissue reactions and stochastic effects coexist, while in the low-dose region stochastic effects prevail.

153. The extent of these regions has not been clearly established in the radiotherapy and radiobiology communities. Dorr and Herrmann investigated the relationship between the location of a second primary tumour and the absorbed dose at that point from radiotherapy of the first primary tumour [D26]. They found that the majority of second tumours appear from 2.5 cm inside to 5 cm outside the PTV, where the external isodose defines a region receiving less than 6 Gy. Rather than using a dose-based distance scheme, Sánchez-Nieto et al. [S5] defined limits between the three regions in terms of the percentage of the prescribed tumour dose. The medium-dose region was defined as between 50% and 5% of the prescription dose, with these limits concurrently defining the high-dose (≥50%) and low-dose (≤5%) regions, respectively. To avoid limits which depend on the prescribed dose, another approach is to use the absorbed dose thresholds of 2 Gy and 100 mGy as suggested in the analysis of the Japanese atomic bombing survivors [G18].

154. To determine the organ doses in regions peripheral to the region of the radiation treatment, the concept of field borders has been used historically. In conventional 2D-RT, where one field or two opposed fields are used, the concepts of in-field and out-of-field are clearly marked by the field border defined by the collimating devices. In this nomenclature, organs that are partially irradiated by the treatment field would be classed in the near-field dose region. However, in modern irradiation techniques, in which more than two fields are used to deliver 3D-CRT or are defined by modulated beam intensities (e.g., IMRT), or which make use of a rotating gantry irradiating from a continuum of directions (e.g., VMAT), the concept of field borders becomes less clear and meaningful. To overcome this situation, the

American Association of Physicists in Medicine recommends the use of the 50% isodose line (i.e., tissues receiving 50% of the prescription dose to the tumour) as the field border for IMRT and VMAT [K30]. Alternatively, the treated volume has recently been increasingly used when considering peripheral organ doses. According to the ICRU Report 83 [H37], the treated volume is the volume of tissue enclosed within a specific isodose envelope. The 95% and 98% isodose lines have been proposed as the threshold to determine the treated volume in both photon and proton radiotherapy [H37, W3] for 3D-CRT and IMRT/VMAT, respectively.

155. There is no general consensus on the definitions of high, medium, or low dose regions in radiotherapy, and all terms used in this section may be found in the relevant literature on second primary cancer risk. Furthermore, the concepts of high, medium and low dose in radiotherapy typically do not conform to standard definitions applied in the fields of radiation protection or radiobiology.

# C. Absorbed dose calculations

156. The TPS is a computerized tool which predicts and optimizes the distribution of the absorbed dose in the PTV as well as in surrounding tissue, given a determined configuration of the treatment beam employed, and the CT scan of the patient. In recent years, MR images have been used, following their conversion to a pseudo-CT scan as required by the TPS. Typically, CT scans cover only the part of the patient body required for treatment planning and the determination of doses to organ at risk (OAR), according with therefore complying with the as low as reasonably achievable (ALARA) principle [G17]. Concurrently, the algorithms for the calculation of absorbed doses employed in the TPS are not, in general, designed to provide accurate results far from the treated volume [H48, W15]. Because of the lack of a whole-body geometrical description of the patient and the insufficient accuracy in the dose calculation algorithms when computing the distribution of the absorbed dose in regions where the absorbed dose is below 5% of the prescription dose, the optimization of the treatment plan considers only the treated volume and the surrounding tissue. It follows that the validity range of algorithms for the calculation of doses are limited to the region enclosed by the 5% isodose envelope. In the following two subsections, a brief summary of the uncertainty and accuracy associated to each type of algorithms used in the calculation of absorbed doses inside and outside the validity range is given. This information is useful to assess the uncertainty and validity associated with the organ dosimetry which supports epidemiological studies of cancer risks that rely on computations performed with TPS.

## 1. Calculation of the absorbed dose in the validity range of the treatment planning system

157. Algorithms employed in the calculation of absorbed doses can be classified into three categories: *(a)* correction-based algorithms; *(b)* model-based algorithms; and *(c)* Monte Carlo-based algorithms.

### (a) *Correction-based algorithms*

158. Correction-based algorithms are semi-experimental in design and rely on dosimetry measurements in water. The introduction of the tissue-air ratio [A10], tissue-phantom ratio [K5], and tissue-maximum ratio [H41] allowed to enhance the accuracy for these algorithms when computing the absorbed dose in homogeneous media with densities similar to that of water. For a heterogeneous system such as the human

body, the presence of tissues with a density different from water, such as bone and lung, reduces the accuracy of these algorithms as they do not consider lateral scatter when the radiation beams are transported within the media. The Clarkson [C13], Batho [B8], irregular field [C20, K15] and equivalent tissue-air ratio [W20] algorithms are examples of correction-based approaches still used for manual dose calculations, and to replicate dosimetry checks in some commercial software. Carrasco et al. [C5] showed that the Batho, irregular field, and equivalent tissue-air ratio algorithms correctly predicted the absorbed dose in a lung-water phantom, within 2%, everywhere, with the exception of regions very close to the interfaces between the two materials where differences of up to 24% were found for 18 MV photon beams. Treatment planning systems that apply correction-based algorithms are, in general, not in current use in clinical practice.

### (b) *Model-based algorithms*

159. Model-based algorithms compute the dose distribution from physics models. When radiation interacts with a tissue, it releases energy which is either locally deposited or carried away by particle scatter. In this process, secondary radiation may be created (e.g., the scattered photon in a Compton photon interaction). These processes can be simplified by convoluting the photon energy fluence of the primary beam with a kernel describing the contribution of secondary photons and electrons. This method is called pencil beam convolution (PBC) [A17]. In PBC, the lateral scattering is considered to be homogeneous, and the correction for tissue inhomogeneity is only applied in the longitudinal direction. An evolution of PBC is the convolution-superposition algorithms in which heterogeneous electronic density in the material media is considered. Examples of commercial implementations of convolution-superposition algorithms are the analytical anisotropic algorithm (AAA) (Eclipse, Varian Medical System), and the collapsed-cone convolution-superposition algorithm (CCCS) (Pinnacle, Philips) [A17]. Computational solvers of the linear Boltzmann transport equation (LBTE) use two physics models and discretize variables to solve the LBTE explicitly. The Attila algorithm [W9] is an example of this type of algorithm which has been commercially implemented as Acuros XB [V10] (Eclipse, Varian Medical System).

### (c) *Monte Carlo-based algorithms*

160. Monte Carlo algorithms for radiation transport simulate individual particle histories by following the successive interactions that those particles experience in matter. This process involves the random sampling of the interaction type and the outcome of that interaction by use of the underlying differential probability cross sections. Interactions might produce loss of energy, changes in the flight direction, nuclear reactions—for some particle species and energies—and the creation of secondary particles. Monte Carlo codes for radiation transport can be divided into three types:

*(a)* General-purpose codes, which are aimed at the simulation of a wide range of energies and materials, (e.g., EGS [K9], PENELOPE [S29], GEANT4 [A6], MCNP [A24], FLUKA [A8] and PHITS [S8]);

*(b)* Fast codes, in which the range of energy and materials available are those relevant to radiotherapy (e.g., VMC [K8], VMC++ [K10], XVMC [F8], DPM [S30]), therefore achieving faster simulation speed at the expense of reducing accuracy with respect to general-purpose codes; and

*(c)* Pre-computed codes, which simulate radiation transport relying on pre-calculated look-up tables obtained with general-purpose codes (e.g., MMC), therefore achieving an even faster simulation speed at the expense of a further reduction in dose accuracy [B51].

161. In radiotherapy, general-purpose Monte Carlo codes are mainly used for research due to the required simulation time that exceeds what is acceptable for routine clinical practice. On the other hand, fast codes have found application during the last two decades in TPS. Examples include Monaco (Elekta) and iPlan (BrainLab) in photon therapy, Eclipse (Varian) in electron and photon therapies, and RayStation (RaySearch) in photon and proton therapies. To speed up the computation of the absorbed dose in the patient's CT scan, which serves as the virtual anatomic model of the patient, Monte Carlo TPS rely on virtual source models that produce the particle beams according to a mathematical model that represents the irradiating device. As a result, the Monte Carlo simulation of radiation transport is only performed inside the virtual geometry of the patient defined by the planning CT scan. In contrast, the Monte Carlo dose verification system PRIMO simulates the radiation transport in the actual medical linear accelerator geometry using a general-purpose code, and subsequently computes the distribution of the absorbed dose in the patient's CT scan with either a general-purpose or a fast Monte Carlo code [R9]. Monte Carlo radiation transport codes are widely regarded as the gold standard in terms of dosimetric accuracy [H9]. This particularly holds true for the results obtained from general purpose codes.

### (d) *Algorithm accuracy*

162. Many investigations have been performed to evaluate the above-mentioned dose calculation algorithms and the hierarchy of accuracy is considered as follows: Monte-Carlo algorithm → LBTE solvers → CCCS → AAA → PBC → correction-based methods [H18, H50, K26, L24, M2]. Although convolution and convolution-superposition algorithms yield more accurate results than correction-based algorithms when calculating distributions of absorbed doses in the presence of large density gradients or small fields, there are still cases in which discrepancies with experimental data can be larger than 3%. When comparing Acuros XB with AAA, Fogliata et al. [F10] found that the gamma pass rates [L28] for a rectangular medium-sized 6 MV field irradiating normal lung, with respect to a reference dose distribution, using a distance to agreement criterion of 3 mm/3%, were 100% for the former and 86% for the latter. When comparing the clinical target volume D100% for 20 LungPLANs calculated by three algorithms (AAA, Acuros dose to medium, Dm,m and Acuros dose to water Dw,m) the maximum difference for AAA to Acuros Dm,m, was 14.12% and the average was 5.39%, and for Acuros Dw,m, it was 3.68% with an average of 1.45% [C1]. Other examples showing similar results were published by Shaw et al. [S35] who looked at dose in bone for spine stereotactic radiotherapy, and Kry et al. [K31] who looked at a particular assembly of lung and bone tissue. Treatment planning system based on AAA, CCCS, and LBTE dose calculation algorithms are presently among the most widely utilized dose-engine algorithms for planning EBRT with photon beams, either for conventional treatments such as 3D-CRT or for modern delivery techniques such as IMRT and VMAT.

## 2. Absorbed doses outside the validity range of the treatment planning system

163. Treatment planning systems are neither conceived nor developed for computing distributions of absorbed doses outside their validity range. This apples to both photon and proton therapy. This section gives an account of the reported discrepancies between TPS results and reference dosimetry data according to the type of dose calculation algorithms employed.

164. Model-based algorithms used in photon treatments show large discrepancies between TPS results and reference data, even in locations close to the boundary of the validity range. Howell et al. [H48] showed that AAA used in 3D-CRT calculation can underestimate the absorbed dose at the local level by 40% in locations between 3 cm and 11 cm from the field edge. Wang and Ding [W5] also reported large

dose underestimates (40–80% relative to local dose) when using AAA to compute the absorbed dose in organs located far from the PTV for IMRT plans. These underestimations occur mainly at locations where local doses are less than 1% of the prescription dose. Conversely, the Tomotherapy TPS (Accuray) provides calculated doses in agreement with measurements up to 30 cm from the field edge [C16, S18]. The better accuracy of this TPS far from the treated volume stems from the fact that it includes experimental data measured at a distance of up to 30 cm from the field edge for obtaining the dose point kernels. The more accurate out-of-field doses can be estimated by multiplying doses calculated by Eclipse by a factor of 2 at the local dose level of 1% or less of prescribed dose [W6]. For proton therapy, and at positions farther from the field edge, neutron doses are not considered in commercial TPS, limiting their ability to predict accurate out-of-field dose.

165. Monte Carlo-based algorithms that are used in commercial TPS consider the simulation of the treatment device via virtual source models. These models are not conceived to consider the scattered radiation directed away from the treatment field, which contributes to the dose distribution in the patient far from the treated volume. At the same time, the fast Monte Carlo codes that are implemented in the commercial TPS do not simulate neutron transport. Neutrons produced within the patient are responsible for most of the absorbed dose far from the treated volume in proton therapy with scanning beams [S20]. In photon therapy with 10 MV or higher energy beams, most neutrons are produced in the treatment head of the linear accelerator. In the case of photon treatments, Sánchez-Nieto et al. [S6] demonstrated that Monaco underestimates out-of-field absorbed doses and does not accurately describe the dose distribution beyond the border of the field. Large underestimates in absorbed doses in tissue have also been reported by Schneider et al. [S18] and by Colnot et al. [C15] for the Monte Carlo algorithm implemented in Multiplan TPS (Accuray). Verbeek et al. [V14] irradiated a water phantom with a single pencil proton beam and found discrepancies relative to local dose of around 50% between the dose computed by RayStation and experimental data at 7.5 cm from the beam axis.

## D. Biological weighting of absorbed dose with the endpoint of cancer induction

166. In the field of radiological protection, considerations of differences in radiation quality—via the relative biological effectiveness (RBE) for stochastic effects—are addressed by weighting the absorbed dose to an organ or tissue by radiation weighting factors. For radiation quality weighting, the ICRP recommends using a maximum value which is usually obtained at low values of absorbed dose and particle energy (<5 MeV) which is not applicable to clinical situations. In the context of dosimetry for second primary cancers following radiotherapy, it is generally not recommended to apply the radiation weighting factor concept given by the ICRP [I5]. Indeed, the values of RBE generally increase with decreasing absorbed dose, absorbed dose per fraction, absorbed dose rate, and particle energy and with increasing number of fractions. Nevertheless, for comparison between clinical radiation qualities, RBE is used to calculate the iso-effective dose also called biological effective dose expressed in gray (Gy) or gray-equivalent (GyE) [C9]. However, for photons and electrons with energies used for clinical procedures, a weighting factor of 1.0 is generally acceptable.

167. For therapeutic applications, protons usually have energies between 70 MeV and 250 MeV. The ICRP weighting factor of 2.0 for protons [I5] would perhaps only be applicable at very low energies in the tail of the Bragg peak. The main contribution to the normal tissue integral dose, however, comes from the plateau region of the spread-out Bragg peak due to the protons passing through normal tissue to reach the target volume. This portion of the Bragg curve consists predominantly of absorbed dose deposited by

protons with and energy much higher than 5 MeV and therefore has a weighting factor of 1.0. In contrast, in the normal tissue distal to the target volume, although the weighting factor may be higher, the irradiated volume by protons is very much smaller, and the deposited absorbed dose is lower owing to the finite maximum range of protons in tissue. Therefore, it is safe to assume that the weighting factor of protons irradiating the vast majority of normal tissues is close to 1.0 [B1, N5].

168. For carbon ions, the situation is similar to that of tissues lying in the plateau region of the spread-out Bragg peak. In this case, a weighting factor of 1.0, similar to that for protons, can be recommended [B21, H17]. However, carbon ions with a range of around 2 cm (100 MeV/u) correspond to a RBE for cellular transformations of about 2.5. At lower energies in the Bragg peak and in the fragmentation tail (range: 0.05 cm), the RBE for cellular transformations can increase up to 12 and can be larger than that for cell kill. Therefore, a careful evaluation of the radiation weighting factor is required, and the use of LET distributions can be helpful in this regard.

169. For neutrons, if a conservative estimate of the risk of second primary cancer is required, the energy dependent radiation weighting factor of ICRP Publication 103 [I5] can be used as a first approximation. This function is a revision of the one originally given in ICRP Publication 60 [I4].

## 1. Tissue specific weighting: simple risk models based on radiological protection concepts

170. Simple models to predict risk of radiation-induced cancer at the levels of absorbed doses used for radiotherapy are based on conventional concepts from radiological protection [I5]. These models are based on the linear non-threshold model, mainly extrapolated from studies on the cohorts of the Japanese atomic bombing survivors, allowing the use of the effective dose for the estimation of the stochastic risk. Basic risk factors can be modified by an absorbed dose and dose rate effectiveness factor for the application to low absorbed dose rates. This linear non-threshold model is only recommended within the framework of radiological protection for absorbed doses up to around 1–2 Gy and the evaluation of the whole-body effective dose and as such, this model is not applicable to the distributions of doses in radiotherapy. Similarly, the use of the effective dose is not appropriate for risk analysis in radiotherapy. Despite these criticisms, radiological protection models have been used by several authors [F11, H6, K28, R14, V15].

171. It should be noted that the dose distribution can be separated into two parts. The primary dose distribution is generated by radiation impinging on the patient through the opening of the beam aperture and includes patient scattering mainly produced by Compton scattering (photons) and multiple Coulomb scattering or inelastic nuclear interactions (protons and ions). The distribution of the scatter dose is generated by radiation scattered from the treatment head, leakage radiation through the collimators, and neutrons produced either in the machine or in the patient. Radiological protection models can be applied exclusively to the dose originating from scattered radiation [F11, H6, K28, R14, V15]. In principle, a linear risk to dose response model can be applied to absorbed doses below around 1 Gy, which is the maximum scatter absorbed dose applied during radiotherapy. The result of using a linear model is that cancer risk is not a function of the integral absorbed dose, but proportional to the amount of scatter absorbed dose.

172. Studies that apply a linear model to scattered radiation result in an estimated increase of cancer risk by a factor of 2 to 10 for both IMRT and passive proton radiotherapy due to greater scatter and neutron absorbed doses compared to photon 3D-CRT. Application of radiological protection models may be appropriate as the absorbed doses investigated by these studies all fall within the lower range;

however, the main drawback is that the primary distribution of absorbed doses (>1 Gy) is excluded. Therefore, risk estimates based on scatter absorbed dose would only include second cancer induction far away from the treated site. Only around 20% of all radiation-induced cancers are located far from the treated volume [T17]. Additionally, an epidemiological study [X2] involving more than 450,000 former radiotherapy patients, has shown a comparable second primary cancer risk between 3D-CRT and IMRT, suggesting that the application of radiological protection models to scattered induced radiation doses is not appropriate.

173. Other researchers have applied linear models to the complete distribution of the absorbed dose including the primary portion, but excluding the absorbed dose delivered to the target volume [M23, M34, S13]. One result of these modelling studies is that risk estimates are in general proportional to the integrated absorbed dose and not dependent on the treatment technique—risk after IMRT is approximately the same than after non-IMRT. However, these studies estimated cancer risks substantially different for different types of radiation; and observed a considerable risk reduction for protons when compared to photons [M23, M34, S13]. While the advantage of these studies is the consideration of the complete dose distribution, the main disadvantage is the application of the linear model to very large absorbed doses.

174. In summary, the application of radiological protection models to radiotherapy patients yields contradictory results since the linear model is applied in two completely different ways to the distribution of the absorbed dose. Risk estimates based on such models have to be applied with extreme care, since above 1 Gy, the application of risk factors from the Japanese atomic bombing survivors data is questionable. In addition, those models are estimating cancer mortality, instead of cancer incidence, and they do not consider fractionation effects. Finally, imaging doses are not considered either.

## 2. Tissue specific weighting: models for cancer risk estimates including fractionation

175. Dose fractionation is common in radiotherapy. Fractionation effects can be important for carcinogenesis and can be included into the risk models by considering cell repopulation between dose fractions. Repopulation tends to counteract cell killing and accounts for the large discrepancies between the standard model for cancer induction that does not include fractionation at high doses and data on second primary cancer [H7].

176. Sachs and Brenner [S1] and Shuryak et al. [S37] developed the Columbia model for calculating the excess relative risk (ERR) of cancer induction over a wide dose range. Their model integrates, into a single formalism, mechanistic analyses of pre-malignant cell dynamics on two different timescales: short-term during radiotherapy and recovery; and long-term during the entire lifespan. The model was applied to nine solid cancer types (stomach, lung, colon, rectum, pancreas, bladder, breast, CNS, and thyroid) using data on radiotherapy-induced second primary cancers, on Japanese atomic bombing survivors, and on the statistics of cancer incidence in the general population of the United States.

177. Another approach is the UZH model [S15], which can be used to calculate EARs. In the UZH model tumour induction is modelled in such a way that each transformation process results in a tumour cell. The microscopic transformation parameter was chosen such that in the limit of low dose and acute exposure, the parameters of the linear-no-threshold model for tumour induction were approached. The long-term time patterns were modelled using a combination of the data for the Japanese atomic bombing survivors and radiotherapy data. The model describes carcinoma and sarcoma induction separately after fractionated radiotherapy as an analytical function and was fitted to the several solid cancers: female breast, lung, rectum, colon, mouth and pharynx, stomach, small intestine, liver, cervix, bladder, skin,

brain and CNS, salivary gland, bone, and soft-tissue sarcoma [S14, S17]. In treatment planning, the normalized dose-response relationship for the organs of interest is used to calculate a weighted average dose called the organ equivalent dose which is proportional to the risk of cancer induction in that organ.

178. It should be noted that the Columbia model and the UZH model are based on dose-response relationships for radiation-induced cancer which can take any shape including non-linear functions of absorbed dose. As a direct consequence, cancer risk is, in general, not proportional to average or integral absorbed dose and therefore either a three-dimensional distribution of the absorbed dose or the absorbed dose-volume histograms of the organs of interest have to be available for the analysis.

179. If the dose-response relationship for radiation-induced cancer is assumed to be linear, then the average organ doses can be used to estimate the incidence of second primary cancers by applying appropriate risk factors (ERR or EAR per unit dose). The meta-analysis performed in chapter V of this annex provides risk factors which are valid for the fractionation schedules which were applied to the radiotherapy cohorts used for epidemiological analysis.

# E. Distributions of the absorbed dose across treatment modalities and imaging procedures

## 1. External beam radiotherapy

### (a) *Photon radiotherapy*

180. Out-of-field organ doses in EBRT with linear accelerators have three origins: (*a*) photon scattering inside the patient, the treatment couch, and the immobilization devices; (*b*) collimator scatter defined as particles exiting the linear accelerator through the irradiated field but with directions pointing outside the field; and (*c*) head leakage defined as particles emitted through the head shielding. At distances of about 20 cm or less from the treated volume, the dominant effect is internal patient scatter, while at larger distances, scatter due to head leakage is the most significant contributor to the organ dose [K6]. The parameters that most influence the out-of-field organ dose are, in decreasing order of importance, the distance to the treated volume and the field size, and, for modulated treatments, the modulation. Conversely, beam energies in the range of 4 to 25 MV and depth in the patient have little influence on the magnitudes of the out-of-field organ dose [K30]. For beam energies larger than about 10 MV, the out-of-field organ dose due to neutrons becomes relevant. In the case of photon therapy, most neutrons are produced in the head of the linear accelerator. Neutron production depends on both the beam energy and on the specificities of the linear accelerator design [F12]. Varian linear accelerators produce more neutrons than those manufactured by Siemens or Elekta [H46]. Most of the neutron dose is deposited by fast neutrons. Since neutrons are rapidly thermalized due to the interactions with the patient's body, the depth in the body is relevant for this component of the out-of-field organ dose [K29].

181. The magnitude of the out-of-field organ dose depends on the modality employed. A 3D-CRT tends to produce lower out-of-field organ doses when compared with IMRT. Conversely, non-targeted in-field organ doses are larger. Tomotherapy yields out-of-field organ doses similar to those of IMRT [M16] up to a distance of 30 cm from the field edge, while for larger distances, out-of-field organ doses from tomotherapy are lower than those from IMRT [H3]. Out-of-field organ dose in VMAT are usually lower

than those in IMRT [D4, H3], and for some organs, absorbed doses are even lower than in 3D-CRT [M4]. De Saint-Hubert et al. reported that the out-of-field organ dose per unit of prescribed dose to the tumour ranged from 11 mGy/Gy at 6 cm to the isocentre to 1 mGy/Gy at 50 cm for an IMRT irradiation of a brain glioma. The same organ dose per unit of prescribed dose to the tumour was reported with a VMAT irradiation at 6 cm, but the out-of-field organ dose decreased more rapidly, reaching 0.1 mGy/Gy at 50 cm [D4]. Differences in the out-of-field organ dose according to the technique become more evident at distances larger than 20 cm from the treated volume. The out-of-field organ dose rates when the flattening filter, aimed at homogenizing the energy of the photon beam, is removed increase and doses are of the same order of magnitude as for 3D-CRT [H3].

182. Only small differences in the neutron dose equivalent were observed when comparing 3D-CRT, IMRT, and VMAT on Varian linear accelerators using 15 MV photons. Conversely, substantial differences in neutron dose equivalent were reported for different linear accelerator models using IMRT with energies above 15 MV [H5]. The overall dose, additional to the therapeutic dose induced by secondary neutrons, was found to be of the same order of magnitude as the dose from a wide range of contemporary imaging modalities used for image guided radiotherapy [H5] and an order of magnitude lower than the stray dose of a corresponding treatment with a 6 MV photon beam [H5].

183. It must be stressed, however, that the relationship between techniques and out-of-field organ dose is affected by many factors such as the complexity of the target volume and consequently, the degree of modulation and monitor units applied. However, there are not only differences in the out-of-field dose between the various irradiation techniques, but also in the out-of-target dose. Essentially, the out-of-target dose is distributed over a larger volume when intensity-modulated techniques are used with a similar integral dose. This effect is further enhanced by using non-coplanar irradiation techniques, such as CyberKnife. This effect has no influence on the induction of second primary cancer as long as the dose-response relationships for radiation-induced cancer are linear. In the case of a non-linear dose-response relationship for organs at risk, this could lead to an increased cancer incidence after intensity-modulated radiotherapy. Conversely, the dose distributions in intensity-modulated treatments are much more conformal compared to 3D-CRT and 2D treatments. Therefore, for organs that are very close to the target volume, a reduction in second primary cancer risks with IMRT and VMAT would be expected.

## (b) *Electron radiotherapy*

184. Most of the far out-of-field organ doses in electron radiotherapy are produced by the scatter form the collimator that occurs at the head of the linear accelerator and at the scrappers of the electron applicator. For this reason, the magnitude of the organ dose is highly dependent on the model of the linear accelerator employed. It has been observed that Elekta and Siemens linear accelerators produce more out-of-field organ dose than those from Varian. Owing to the low penetration of electrons in the human body, these out-of-field organ doses are relatively superficial (i.e. down to a depth of about 2 cm to 6 cm, depending on the beam energy) [K30]. However, it must be pointed out that for high electron energies (E>10 MeV), bremsstrahlung causes an X-ray tail and, as a result, even tissues that are deeper than the range of the respective electron fields are irradiated by this X-ray tail.

## (c) *Proton radiotherapy*

185. Most out-of-field organ doses in proton therapy originate from neutrons, and to a lesser extent from photons and secondary protons. For passive scattering proton therapy (e.g., single scattering, double scattering), externally produced neutrons are prevalent over those produced within the tissues of the patient. Conversely, for active scanning techniques (e.g., pencil beam scanning), neutrons produced in

the patient contribute most significantly to the out-of-field organ doses. In both cases, organ doses due to neutrons decrease in magnitude with the distance due to their thermalization inside the patient.

186. The neutron dose equivalent is higher in general for proton treatments, compared to that for photon treatments and apparently higher for scattered protons compared to spot-scanned ones. Inside and close to the primary radiation field, the measured neutron dose is similar to that for active scanned proton beams rather than for passive scattering proton beams. For distances far away from the treatment field, the neutron dose equivalent from active scanning is of the same order as, or lower than that from a Varian 15 MV IMRT irradiation [H5]. The neutron doses from passive scattering proton therapy are smaller than those from photon therapy when the neutron and the photon stray dose are combined [H5].

187. De Saint-Hubert et al. [D5] performed the irradiation for a brain glioma of an anthropomorphic paediatric material phantom (CIRS, ATOM) using proton pencil beam scanning (PBS). The phantom was filled with $^{6}$Li and $^{7}$Li thermoluminescent and bubble detectors for measuring the dose contribution of photons and neutrons. The experimental data on organ doses compared well with Monte Carlo simulations performed with the GEANT4 [A6] wrap-up software TOPAS [P3]. When they compared the out-of-field equivalent organ doses with those obtained from photon treatments (3D-CRT, IMRT, VMAT) with the same PTV and OAR dose objectives, they found that out-of-field organ doses in photon techniques were higher by factors of 5 to 15 than those from PBS proton therapy [D6]. The difference between the organ equivalent doses increased with the distance to the isocentre, depending on the beam quality. Larger discrepancies between the out-of-field equivalent organ dose were found by Knežević et al. when comparing 3D-CRT, IMRT, and Gamma Knife against PBS, as well as for brain irradiation treatment [K23].

## (d) *Carbon radiotherapy*

188. For both carbon ion and proton radiotherapy, non-target doses are, in general, lower than those delivered by photon radiotherapy. When comparing carbon ion with proton radiotherapy, the former has a lower entrance dose but performs more poorly at the distal regions of the Bragg peak owing to the production of light ion fragments. For this reason, secondary charged particles should be considered more carefully in carbon ion radiotherapy than in proton radiotherapy. Carbon ion fragments are mainly composed of protons, deuterons, tritons, and helium ions. As in the case of proton radiotherapy, passive scattering and active scanning modalities exist, with the latter producing lower magnitude out-of-field organ dose than the former. Mohamad et al. [M25] shows that carbon ion beam therapy has a secondary cancer incidence rate comparable to surgery and is significantly less frequent than X-ray therapy.

189. Matsumoto et al. reported that the out-of-field neutron organ dose equivalent per unit prescribed dose to the tumour is lower in carbon ion radiotherapy than in proton radiotherapy [M9]. For a carbon ion radiotherapy used in the treatment of paediatric brain cancer (cerebellar ependymoma), the organ dose equivalents per unit prescribed dose to the tumour was between 0.01 mSv/Gy (RBE) for the bladder and 1.45 mSv/Gy (RBE) for the thyroid using either passive scattering or active scanning techniques. The patient, the malignancy, and the treatment objectives considered by Matsumoto et al. were similar to those included in the study by De Saint-Hubert et al. [D6] with proton PBS, who reported organ dose equivalents per unit prescribed dose ranging from 0.02 mSv/Gy (RBE) for the bladder to 1.2 mSv/Gy (RBE) for the thyroid. The organ dose equivalents per prescribed tumour dose obtained with active scanning proton therapy were similar to or smaller than those obtained with passive scattering carbon ion radiotherapy, but they were larger than those obtained with active scattering carbon ion radiotherapy.

## 2. Brachytherapy

190. Brachytherapy involves the delivery of sealed radionuclide sources within the tumour manually or via automated tissue insertion. Radionuclides selected for brachytherapy applicators include lower-energy beta and X-ray emitters, such as $^{103}$Pd, $^{125}$I, and $^{131}$Cs, and higher-energy photon-emitting radionuclides, such as $^{60}$Co, $^{137}$Cs, $^{192}$Ir and $^{198}$Au, encased in seeds, pellets, thin wires, or tubes. Dose delivery is primarily by beta-particles with small tissue ranges such that the gradients of the absorbed dose at the tumour edges are steep, therefore minimizing non-target doses to surrounding tissues and body structures [Y7]. For ocular tumours, radiation delivery is performed via coated plaques. Intracavity brachytherapy involves insertion into a body cavity such as the trachea for lung cancer or the vagina for cervical or endometrial cancer [H15]. Interstitial brachytherapy involves insertion of a catheter into or near the tumours or within an organ such as the prostate [Z2]. High dose-rate brachytherapy treatment takes several minutes per fraction. In contrast, low dose-rate brachytherapy insertions may extent from several hours to days to even permanent placement with in situ radionuclide decay. Percent depth dose distances for a point source of $^{192}$Ir range from about 1.3 cm at 50% of maximum dose to approximately 3.1 cm at 10% of maximum dose.

191. An alternative form of localized dose delivery can be achieved via electronic brachytherapy sources which involve a miniature X-ray tube operated at tube potentials from 40 to 50 kV to include a gold anode target [G1]. The resulting photon spectrum is composed of bremsstrahlung X-rays and characteristic X-rays between 8 and 15 keV. At 50 kV, the effective energy is 20.4 keV with a half-value layer of 0.64 mm of aluminium at a depth of 1 cm in water. These devices may deliver dose rates at a depth of 1 cm up to 2–3 Gy min$^{-1}$. Contact radiotherapy for skin and rectum cancer uses an X-ray generator with the same tube potential as electronic brachytherapy. the X-ray generator is positioned close to the area to be treated, with or without applicators, so that X-ray penetration limits the dose to the surrounding healthy tissue [C19].

## 3. Radiopharmaceutical therapy

192. For advanced stages of cancer, in which the presence of tens to hundreds of individual tumours in the cancer patient precludes the use of EBRT, systemically delivered therapies are instead employed. They include chemotherapy, immunotherapy, or radiopharmaceutical therapy [S32]. The advantage of radiopharmaceutical therapy is that it can target the tumour with high specificity and can invoke direct kill of the tumour cells, therefore avoiding induced therapy resistance stemming from alterations in cell signalling pathways [G20]. Radiopharmaceuticals are formed by the chemical binding of a radionuclide and a pharmaceutical agent with the latter designed for specific tumour localization. Mechanisms of localization may include cell sequestration, phagocytosis, passive diffusion, active transport, capillary blockage, perfusion, chemotaxis, antibody-antigen complexation, receptor binding, and physiochemical adsorption [B55]. Radionuclides used in cancer therapy have historically been beta particle emitters, yet currently there is an increased interest in the use of alpha particle emitters. The high LET of alpha particles over beta particles can allow for reduced levels of administered activity provided that targeting of the tumour is maintained [S33].

193. For widely disseminated disease, radiopharmaceuticals are typically administered intravenously to the patient and are initially present in the general blood circulation prior to uptake by the tumour or urinary excretion [D21]. Consequently, radioactive decay occurring while the pharmaceutical is in blood circulation results in irradiation of tissues throughout the body, with an absorbed dose that depends primarily on the photon component of the radionuclide decay scheme. Higher levels of radiation dose are also observed in the non-tumour bearing tissues of the organ in which the targeted tumours are located, and in organs and tissues which additionally express cell-surface markers that provide non-targeted

localization of radiopharmaceutical—defined as source organs under the schema of the Medical Internal Radiation Dose (MIRD) Committee of the Society of Nuclear Medicine. An example of the former is the haematopoietically active bone marrow (ABM) in regions of trabecular spongiosa with bone metastases; while an example of the latter is the salivary and lacrimal glands of patients treated with radiopharmaceuticals targeting prostate specific membrane antigen. In all cases, the distribution of the absorbed dose within these source tissues is highly dependent on both the non-uniformity of the localization of the radiopharmaceutical at the tissue level; and the ranges of the emitted radiation particles. For a given radiopharmaceutical spatial distribution, the resulting distribution of the radiation dose is increasingly more uniform as the radiation particle energy, and therefore tissue range, increases.

194. At present, the estimation of absorbed doses due to internal irradiation from radiopharmaceutical therapy is not performed routinely, as a rule; instead, activities of administered pharmaceuticals are recorded. This is different from the practice of EBRT and brachytherapy, in which the planning and performance of radiotherapy is based on estimates of the absorbed doses to the targeted tumours, in line with the basic principles of clinical radiobiology. To ensure that radiopharmaceuticals are developed and implemented in a manner that maximizes tumour response while minimizing the adverse toxicity to normal tissues, it is essential to adopt strategies analogous to external beam radiation therapy and brachytherapy where dosimetry-based treatment planning play a critical role. Optimization of radiopharmaceutical therapy is hindered by the absence of standardized practices for prescribing, reporting, and recording dosimetric quantities related to this type of radiotherapy. The ICRU Report 96 [I7] provides basic information necessary to standardize corresponding techniques and procedures related to absorbed doses from internal irradiation, which is important to minimize the possible risk of second primary cancers.

## 4. Imaging procedures in radiotherapy

195. Imaging procedures with ionizing radiation are extensively used in modern radiotherapy for diagnostic purposes and during radiotherapy. CT scans are employed for the purpose of treatment planning, while kV X-ray images and CBCT are used for patient positioning. Electronic portal imaging devices are also employed for positioning with portal images of the exit fluence. These procedures deliver a dose to the imaged region that, although small when compared with the dose delivered to the tumour during treatment, should be considered in full when accounting for both in-field and out-of-field organ doses of the radiotherapy patient. The American Association of Physicists in Medicine produced an extensive report on the management of doses from imaging in radiotherapy in which these aspects are carefully considered [M36]. This report was recently complemented by the report from the American Association of Physicists in Medicine (AAPM) Task group 180 [Ding et al., 2018]. In particular, it contains updated dose data resulting from current image acquisition procedures and modalities. Recommendations for reducing the imaging dose are also provided. It should also be mentioned that the work of the ICRP TG 116 "Radiological Protection Aspects of Imaging in Radiotherapy" will lead to an ICRP publication on imaging in radiotherapy in the near future.

196. In the study of out-of-field organ doses in passive-scatter proton radiotherapy in paediatric patients undergoing brain radiotherapy, Zacharatou Jarlskog et al. [Z1] found that the ratio of therapy to organ equivalent doses caused by a CT scan varied from 15 to 32 for the thyroid, from 3 to 8 for the lungs, and from 0.7 to 7 for the stomach wall, for patients ranging in age from 9 months to 14 years. Halg et al. [H4] have experimentally shown that daily kilovoltage planar imaging and Tomotherapy megavoltage fan beam CT for patient setup result in very low additional dose to the patient compared to the effective dose resulting from therapeutic stray radiation outside the treatment volume. The dose from cone beam computed tomography (kilovoltage or megavoltage beams) were significant when compared to the stray dose delivered during treatment outside the treatment volume. The dose burden from typical planning CT

scans is smaller; however, when gating methods are employed to obtain the CT scan, the dose can be 10 times larger compared to the therapeutic stray dose outside the treatment volume [H4]. De Saint-Hubert et al. evaluated the complete patient exposure for the case of paediatric brain glioma considering 3D-CRT, IMRT, VMAT and proton PBS irradiations together with the whole-body dose distribution produced by the required imaging procedures during the treatment, namely one CT scan and one positioning image (kV X-ray or CBCT) per fraction [D6]. They reported that in the case of photon irradiation, the total imaging dose equivalent per organ was between one to three orders of magnitude smaller than the total therapeutic dose equivalent per organ for all organs located at more than 7 cm from the treated volume. However, for the proton PBS treatment, the imaging dose was comparable to the therapeutic dose in this out-of-field region. Post primary cancer treatment diagnostic scans should also be considered with regards to out-of-field organ exposure since the primary cancer is considered cured only five years after the end of radiotherapy exposure.

## F. Reconstruction of dose for epidemiological studies on second primary cancer induction

### 1. Review of software for the estimation of whole-body doses in radiotherapy and imaging procedures

197. As discussed previously, treatment planning systems yield inaccurate results when computing the organ dose beyond their validity range. Computational programs exist, however, that can cope with the computation of organ doses in tissue regions far from the treatment field with varying degrees of accuracy. At the present time though, the majority of these codes have not been released for general use and are restricted for research purposes. They can be classified into those which use mathematical models to compute the organ dose, and those which are based on Monte Carlo radiation transport simulations.

#### (a) *Mathematical models*

198. For conventional photon radiotherapy, François et al. [F14] developed a model based on measurements in which the organ dose in the range between 10 and 50 cm from the field border is parametrized as a function of depth, distance from the border, and field size and shape. This model was improved by Diallo et al. [D16] and was used in several epidemiological studies [D17, S25, S34]. A semi-empirical model was also developed by McParland and Fair [M13] for calculating the peripheral dose received at points outside the collimated field edge for incident photon beams with energies below 10 MV. The dependence of the peripheral dose on depth, distance, field shape and size, and tissue heterogeneities are accounted for by this model, which, however, is not accurate for complex radiation treatment fields.

199. Van der Giessen [V1] developed a software called PERIDOSE based on published data from many authors. The code distinguishes between orthogonal and tangential beams and accounts for patient thickness and has an accuracy in patients of ± 60% (2 standard deviations). The code cannot be used for either IMRT or VMAT treatments.

200. Jagetic and Newhauser [J1] developed a physics-based analytical model to calculate the total absorbed dose deposited both inside and outside the treatment field for 6 MV radiotherapy. The model

includes separate terms for absorbed dose from primary, leakage, and scattered photons. Although it provides in principle more accurate results than previously mentioned codes, it is not capable of computing organ doses during IMRT or VMAT treatments. Similarly, Schneider et al. [S12] developed a physics-based analytical model of total absorbed dose from primary, scattered, and leakage radiation for square fields from a 6 MV beam at any arbitrary point in a phantom, including in-plane, cross-plane, and out-of-plane locations. The model agreed well with measurements and Monte Carlo simulations of absorbed dose in a water-box phantom and was validated for field-sizes ranging from 2×2 cm to 20×20 cm. This model, however, is not able to compute doses from IMRT and VMAT treatments.

201. The model developed by Hauri et al. [D4, H22, H23] predicts the out-of-field dose from 20 cm to 90 cm from the isocentre for 3D-CRT, IMRT, and VMAT. The discrepancy with experimental data is less than 50%.

202. Regarding proton radiotherapy, Zhang et al. [Z4] developed an analytical model of neutron ambient dose equivalent and equivalent dose for 250 MeV passive scattering proton beams which were within 17% of the results of a Monte Carlo simulation on average. In 2013, Pérez-Andújar et al. [P2] further generalized this model to predict equivalent doses from external neutrons for passive scattering proton beams at energies ranging from 100 to 250 MeV. Schneider et al. [S20] extended this analytical model for the assessment of neutron equivalent doses per unit absorbed dose delivered during radiotherapy. Predicted values of this quantity from the analytical model agreed well with results from a Monte Carlo simulation for energies in the range from 100 to 250 MeV (10% difference on average) and with measurements at 75 MeV (15% difference on average).

### (b) *Monte Carlo codes*

203. General purpose and fast Monte Carlo codes are, in general, capable of accurately computing distributions of whole-body doses providing that an accurate description of the radiation source is coded, a whole-body CT scan of the patient exists, and sufficient simulation time is allowed to reduce the statistical uncertainty inherent to the results of Monte Carlo simulations. Usually, not all of this information is available, and therefore Monte Carlo models are mainly used to compare radiotherapy techniques and to predict and compare the risks of second primary cancer for a specified tumour treatment and patient.

204. Lee et al. [L7] established a streamlined procedure to conduct Monte Carlo-based dose calculations by importing Digital Imaging and Communications in Medicine radiation treatment files into a TPS and then computing whole-body organ doses with the fast Monte Carlo code XVMC. This code does not account for the leakage and scattered radiations from the treatment head of the linear accelerator.

205. De Saint-Hubert et al. [D4] computed the out-of-field dose to all organs in an anthropomorphic paediatric phantom irradiated with IMRT and VMAT modalities for a brain glioma. The doses measured by 150 thermoluminescent detectors inserted within the CIRS ATOM phantom compared well with the results obtained from the Monte Carlo dose verification system PRIMO [R9], yielding discrepancies relative to the local dose with the experimental data smaller than 20%.

## G. Summary and discussion

206. Accurate knowledge of the distribution of whole-body doses is necessary to perform epidemiological studies in which the tumour site of the second cancer can be correlated to the radiation dose delivered during the treatment of the first primary cancer. However, distributions of whole-body doses are not routinely computed during treatment because of lack of whole-body CT scans of patients and of adequate software for performing this type of computation in clinical practice. Moreover, the possibility of relating the whole-body dose distribution with risk of second primary cancer induction is conditional on the accuracy of biological weighting factors and the risk projection models themselves.

207. Research has shown that, in general, the accuracy of modern TPS when computing dose distributions outside the treated volume and the surrounding OARs decreases rapidly with increasing distance from the primary field. Studies using dedicated research software for out-of-field dose computation indicate that the out-of-field dose distribution in case of EBRT with photon beams is larger than that obtained with proton or heavy ion beams. Monte Carlo radiation transport simulations have never been used in epidemiological studies on a patient specific basis for the purpose of retrospectively estimating organ doses for second primary cancer risk assessment.

208. In some cases, the measurements of neutrons are of ambient dose equivalent determined in air, at or near the patient, this quantity can overestimate the organ dose equivalent, or of personal dose equivalent inside phantoms.

# V. EPIDEMIOLOGY

## A. Introduction

209. In this chapter, information on the risks of second primary cancer after radiotherapy for several cancer types was reviewed. Relevant data were combined in a meta-analysis to provide overall estimates of the risk per unit organ dose at the site of the second primary cancer.

210. The Committee's evaluation of estimates of the risks per unit organ absorbed dose was based on data taken from an extensive literature review carried out to identify risks for second primary cancer per unit organ dose for several cancer sites. It also built on previous review work that provided overall meta-analytic estimates of the risk per unit organ dose at the site of the second primary cancer.

211. An overview of the studies that were considered for inclusion in the meta-analysis to derive a pooled estimate of the risks of various types of second primary cancer can be found in section V.B. The section also contains published results on cumulative incidence, latency and risks of second primary cancers following radiotherapy by site of the second primary cancer. Models developed based on the meta-analytic results presented in section V.B for the particular purpose of predicting the risk of second primary cancer after radiotherapy are described in section V.C, while section V.D contains and assessment of the population-level impact of the risk of second primary cancer after radiotherapy. Section V.E of this chapter compares risks associated with second primary cancer after radiotherapy with those from other sources of radiation exposure, while section V.F examines the limitations, major uncertainties of existing evidence, and gaps in knowledge.

# 1. Past reviews of information on second primary cancer after radiotherapy

212. Several publications that provided reviews of epidemiological studies of risks of second primary cancers were reviewed and relevant citations given in these past reviews were included in the literature search undertaken for this evaluation. This section provides a summary of these publications

213. Coleman [C14] provided an overview of the knowledge on the effect of radiotherapy and chemotherapy on second primary cancers from the early 1980s. Lin and Teitell [L17] evaluated the risk of second primary cancer after treatment for paediatric Hodgkin lymphoma by examining retrospective analyses. They presented a summary of these findings in a meta-analysis of six papers on second primary cancers after radiotherapy. The authors provided relative risks (RRs) for several second primary cancers (e.g. breast, thyroid, gastrointestinal tract and lung cancer, sarcoma, melanoma, cancer of the urogenital tract, brain and CNS cancer). Doi et al. [D23] published a meta-analysis of the risk of second primary cancer after childhood cancer, based on nine studies. These studies investigated different sites for second primary cancer and the ERRs showed a degree of heterogeneity ($p<0.001$). An overall ERR per unit dose of 0.53 (95% CI: 0.22, 1.31) $Gy^{-1}$ was calculated using a random effects model.

214. Berrington de Gonzalez et al. [B19] provided a systematic review of epidemiological studies of second primary solid cancers in organs that received a dose of more than 5 Gy from radiotherapy, and for which dose-response relationships were estimated from specific doses to individual organs. Overall, there was little evidence that the dose-response relationship was non-linear with a downturn in risk, even at organ doses exceeding 60 Gy. Thyroid cancer was the only exception with evidence of a downturn in risk at doses higher than 20 Gy.

215. Giannoula et al. and Lorenz et al. [G4, L26] published a systematic review of risks of secondary thyroid cancer following radiotherapy for childhood cancer. Overall, 21 studies with 100,818 children with childhood cancer were included in this review. All but one study reported increased risks of second primary thyroid cancer associated with radiation exposure for childhood cancer. Turcotte et al. reviewed and summarized recent literature on epidemiological studies on second primary cancer, including data on risk factors, surveillance, and intervention relevant to survivors of childhood cancer [T28]. Such survivors were reported to have a long-term increased risk of second primary cancer compared with the general population, with a greater than twofold increase in the risk for solid tumours extending beyond age 40 years, but no information was provided on the share of this cancer burden causally linked to radiotherapy. A dose dependent increased risk was reported for solid tumours after radiotherapy, with the highest risks for tumours occurring in or near the treatment field (e.g., greater than fivefold increased risk for breast, brain, thyroid, skin, bone, and soft-tissue cancers).

216. In 2011 the National Council on Radiation Protection and Measurements (NCRP) published NCRP Report No.170, a comprehensive review of information on second primary cancers and cardiovascular disease after radiotherapy [N6]. The studies identified in the NCRP Report No.170 were also captured by the literature review used in this evaluation.

217. Journy et al. conducted a systematic review of clinical and epidemiological studies investigating the effect of the irradiated volume or dose-volume distribution to the remaining volume at risk on the incidence of second primary cancer [J4]. Eighteen studies comparing risks of second primary cancer according to the irradiated volume were identified. An increased risk of second primary cancers—mainly breast and lung cancers—with extended radiotherapy was observed among patients treated for Hodgkin lymphoma or childhood cancers.

## 2. Types of epidemiological study design

218. In the literature on the risks of second primary cancers following radiotherapy included in this evaluation, several types of risk measures were derived from mainly two types of studies:

(a) *Case-control study*: an epidemiological study in which cases with the health outcome of interest are selected from the population and compared in terms of exposure with controls free of the outcome, at the time of diagnosis of the corresponding case that represent the exposure distribution in the population. The effect of exposure on the outcome in a case-control study is expressed as an odds ratio (OR) [U7];

(b) *Cohort study*: an epidemiological study in which people free of the health outcome of interest are selected and subdivided into groups based on exposure of interest. The occurrence of the health outcome (typically incidence or mortality) is recorded for the entire study population and the effect of exposure on the risk of the outcome is expressed as a rate ratio, risk ratio, or hazard ratio (HR). Follow-up for development of disease cases since the start of study period can be prospective or retrospective [U7].

219. Many studies of second primary cancers after radiation treatment are of a hybrid design, such as case-cohort studies, in which cases are extracted from a cohort and a selection of controls is made from the cohort before the cases develop, nested case-control studies, in which cases are recruited from patients who had radiotherapy, and not typical case-control studies. For reference, a nested case-control study is defined as "a type of case-control study in which cases and controls are drawn from the population in a fully enumerated cohort. Typically, some data on some variables are already available about both cases and controls and thus concerns regarding differential misclassification of these variables can be reduced" [P10].

## 3. Types of second primary cancer risk measures

220. A risk measure is a statistical quantity that summarizes the data and characterizes the probability of occurrence of the outcome of interest, such as risk associated with second primary cancer. Several types of risk measures and study designs were identified in the course of the evaluation and are described in this chapter. Further explanation on the different risk measures is given in the glossary. Definitions in the glossary were taken mainly from NCRP Report No.185 [N7], UNSCEAR 2019 Report [U7] and WHO report on basic epidemiology (2nd edition) [B37].

221. In the literature on the risks of second primary cancers following radiotherapy, several types of absolute and relative risk measures are applied. Risk types reported from the literature in this chapter are as quoted in the original publications and no attempt was made to convert different risk types to a single risk measure (i.e., a hazard ratio reported in the original study was not converted to an excess relative risk) for quoting the original papers results. Numerical estimates of risk measures are also reported with the same number of decimal places that were reported in the original publication. Organ specific doses reported from the literature in this chapter—mainly absorbed organ doses at the site of the second primary cancer—are as quoted in the original publications and no attempt was made to apply a dosimetric conversion factors. Details on the organ dosimetry methods in each study are summarized in tabular form in electronic attachment B.

## 4. Meta-analysis methodology

222. A meta-analysis is a formal quantitative methodology based on analysing data taken from independent studies on subjects with similar characteristics. The methodology involves calculating a weighted mean value of results of individual studies, by applying an appropriate weight to the results of each study, to derive an overall estimate of the risk for a specific cancer site. Conducting a meta-analysis usually requires substantially fewer resources than pooling the data from independent studies and analysing the pooled data to obtain overall results. The important features of a meta-analysis methodology, relevant to this annex, are the ability to systematically assess and investigate the degree of consistency of the effects of the treatment (i.e. radiotherapy) on chosen detrimental health outcomes (second primary cancer after radiotherapy) between studies and to improve the uncertainty of the estimate of the effect of the treatment by obtaining a pooled result which is less uncertain than any of the contributing results from the individual studies investigated. Pooled ERRs are reported in this chapter with confidence intervals (CI) and prediction intervals. The prediction interval is based on past risks and provides an estimate of the range of values within which future risks may fall, with a certain probability.

### (a) *Method for excluding risks from publications based on overlapping cohort data*

223. Although studies with overlapping data are not frequent, the inclusion of two or more sets of risks in the meta-analysis based on partially the same cohort data would incorrectly lead to an underestimation of the overall variability in the risk per unit dose and therefore the needed to be excluded from the meta-analysis. There are several ways to avoid using overlapping data such as those for a particular cohort referred to in more than one publication:

*(a)* Only the most recent publication from a specific cohort study could be considered;

*(b)* Only the publication with the highest statistical power from a specific cohort study could be considered;

*(c)* The primary investigators of a specific study could be contacted to advise on which paper may be most appropriate to include; or

*(d)* Each study with overlapping data could be assessed to determine which one to include on the basis of expert opinion.

224. The most recent publication from a specific cohort study may include data on only a sub-cohort, whereas earlier papers may contain information on the full cohort and, therefore, they may have more statistical power to determine an effect. It was agreed that option *(d)* should be applied for this evaluation.

### (b) *Meta-analytic method for combining reported risk per unit organ dose at the second primary cancer site*

225. One of the aims of the meta-analysis is to evaluate the contribution to the second primary cancer risk from radiotherapy independently from the contribution of chemotherapy and their combined effects. This was accomplished by basing the subgroup analysis on risks that have either been calculated for groups without chemotherapy or by including in the meta-analysis risks that were adjusted for chemotherapy. The calculations were performed under the posit that the mean estimate of the risk provided by pooling individual studies represents a more comprehensive value than those provided by

any individual study. The dose estimates at the second primary cancer site were taken as given by the source papers without further investigation. The meta-analysis included studies for which:

*(a)* A value of the linear ERR per unit dose and a confidence interval (CI) could be extracted; or

*(b)* Values of RR, OR, or HR along with CI per dose category could be determined; or

*(c)* Values of RR or incidence rate ratios (IRR) and corresponding CI derived from comparison with an unexposed external control group could be extracted together with a mean organ dose to the exposed group.

226. For studies that did not report an estimate of the linear ERR per unit dose but reported dose category RRs, a linear ERR per unit dose estimate with associated 95% CI was derived using a method described below. For studies only reporting RR or IRR values based on external comparison, the linear ERR per unit dose ($ERR_{pd}$) with associated 95% CI was derived from the RR and the mean organ dose ($\bar{D}_O$) using the equation:

$$ERR_{pd} = \frac{RR - 1}{\bar{D}_O}$$

227. The primary analysis only used studies without potentially overlapping data. When studies reported risk estimates from models that adjusted for other effects, these were preferred over results from unadjusted models. For studies only reporting dose category RRs, the primary analysis for second thyroid cancers only considered dose category RRs from categories with reference organ doses below 15 Gy. Only linear dose-response models were considered for pooling, while quadratic, linear-quadratic, as well as linear-exponential models were not included for pooling, due to a lack of suitable methodology in the literature. The meta-analysis was carried out only when at least three studies could be pooled.

228. Estimates of individual ERR per unit dose estimates from all studies were combined using a random effects meta-analysis [V17] and the restricted maximum likelihood method. Heterogeneity of risks was assumed due to different ages at exposure and different attained age at diagnosis of the second primary cancer. Following various studies [D2, D24, L29], the meta-analysis estimate for the ERR per unit dose was calculated by first transforming study specific estimates of ERR per unit dose ($ERR_{pd}$) to RR per unit dose ($RR_{pd}$) using the equation:

$$\ln(RR_{pd}) = \ln(ERR_{pd} + 1)$$

The quantities $ln(RR_{pd})$ were then pooled in a random-effects meta-analysis, yielding a value of $ln(RR_{pd})_{pool}$, which was transformed back to a pooled estimate of the ERR per unit dose using the equation

$$ERR_{pd}^{pool} = e^{ln(RR_{pd})_{pool}} - 1$$

When calculating a pooled value, estimates of $RR_{pd}$ were weighted in inverse relation to their standard deviations. These standard deviations were derived from the widths of the CIs of the ERR per unit dose from the individual studies, taking into account the CI level (90% or 95%), and assuming a log-normal distribution of the $RR_{pd}$ values. For some studies, in which the lower boundary of the CI values was missing or was indicated as being less than zero, symmetry of the CI on the logarithmic scale around the estimate of the ERR per unit dose point was assumed. When the lower boundary of the CI for a study specific ERR per unit dose was less than −1, its standard deviation was estimated on the original scale instead.

229. The meta-analysis was characterized using Cochran's Q test [S56], the number of degrees of freedom, the $p$ value for test of heterogeneity, and the $I^2$ measure for inconsistency [H32].

Leave-one-out-diagnostics were used to assess the change in the pooled estimate of the ERR per unit dose when each study was excluded in turn, and pooling was done on the remaining studies. Funnel plots were used to examine asymmetry and infer publication bias as a possible explanation for asymmetry. For quality assurance, the meta-analysis was independently implemented using different types of software based on the study specific estimate of the ERR per unit dose with associated 95% CI. An alternative method to carrying out the meta-analysis, suggested by Richardson et al. [R7], was not implemented, since the required information on maximum organ dose was missing from many studies. Furthermore, the method was developed and validated using studies on populations with much lower radiation exposure than those received by patients treated with radiotherapy for first primary cancer.

230. Two sensitivity analyses were carried out. In the first sensitivity analysis, ordinary linear regression instead of generalized least squares regression was used to derive study specific estimates of the ERR per unit dose from studies that only reported dose category RRs. In the second sensitivity analysis, study specific estimates of the ERR per unit dose were pooled directly on the original scale using random-effects meta-analysis as in Hauptmann et al. [H21]. Standard deviations of study specific estimates of the ERR per unit dose were derived from the widths of their respective CIs, assuming a normal distribution.

231. The general methodology of the meta-analysis described for the primary analysis was also applied to several subgroups of studies:

*(a)* All identified studies, even those with potentially overlapping data;

*(b)* Only studies that reported linear ERR per unit dose estimates with corresponding CIs;

*(c)* Only studies with exposures exclusively during childhood or adolescence;

*(d)* Only studies with exposures exclusively during adulthood;

*(e)* Only studies with data from patients without chemotherapy;

*(f)* Only studies with data from patients with chemotherapy where results were adjusted for chemotherapy; and

*(g)* Only studies with exclusively external exposure (no brachytherapy).

For second primary thyroid cancers, one subgroup was defined by including studies with dose category RRs with reference organ doses only up to 5 Gy. A further subgroup for second thyroid cancer was defined by setting this limit to 20 Gy. It was not possible to define subgroups based on distinct periods of time since exposure. Similarly, it was not possible to define subgroups based on different fractionation patterns, in particular hypofractionation.

232. In pooling individual estimates of ERR per unit dose for neoplasms of the haematopoietic and lymphoid tissues, gastrointestinal cancer, and all solid cancers combined, some publications contributed more than one value of ERR per unit dose. These studies presented results separately for cancer entities that were assigned to the same cancer site in this annex. The chosen meta-analysis methodology did not include a random effect for the contributing study as the individual patients were assumed to be distinct between different study endpoints, such that there were no overlapping data. This approach entails the assumption that the analysis does not need to distinguish between the heterogeneity in estimates of the effect due to different studies, and the heterogeneity due to different cancer outcomes within the same cancer site.

233. Point doses at the site of the second primary cancer are not required for radiation epidemiology studies, although they are preferred [G19], and many studies only report mean organ doses. Reasonable risk estimates can still be obtained using mean organ doses; however, this can lead to a distortion of the dose-response relationship [S21], especially when there is a large dose gradient across the organ in which the second primary cancer developed.

### (c) *Deriving a linear excess relative risk per unit dose estimate from risk for specific dose categories*

234. For each study that did not report an estimate of the linear ERR per unit dose, but instead provided RRs for separate dose categories, an estimate of the linear ERR per unit dose was obtained together with its 95% CI following the method from Doi et al. [D24]. Firstly, a reference dose representing each dose category had to be determined. Takahashi and Tango [T2] studied how this step may affect subsequent trend analyses. The reference dose was set to the mean or median dose of the category, or to the midpoint of the interval when mean or median values for the category were not reported. When the highest dose category had no upper boundary, its reference dose was set to the values of its lower boundary plus the length of the second highest interval. The reference dose of the lowest category was subtracted from the reference doses of the remaining categories.

235. For each study that only reported dose category RRs, an estimate of the linear ERR per unit dose was derived as the slope from a linear regression without an intercept and an offset of 1.0 where the category RRs were the modelled outcome and the category reference doses served as covariates [L21]. To account for different uncertainties in the RRs for each category, as well as for the positive correlation between RRs from the same study due to having a shared reference category, the method from Berrington and Cox [B18] was followed. The linear regression was implemented with a generalized least squares model using the Linear and Nonlinear Mixed Effects Models (NLME) package [P7] for the statistical environment R [R1]. The variance function was defined as proportional to the standard errors of the logarithm of the dose category RRs. Assuming a log-normal distribution of the category RRs, these standard errors were derived from the widths of the respective CIs. Based on a simulation study, the correlation between RR estimates from different dose categories was assumed to be equal to 0.5. The CI for the ERR per unit dose estimate was calculated using parametric bootstrapping with 2,000 replicates assuming a multivariate log-normal distribution for the category RR with a mean value equal to the observed $ln(RR)$, and the covariance matrix constructed using the standard errors for $ln(RR)$ and correlations as defined above. When the generalized least squares model did not converge, an estimate of the ERR per unit dose with associated CI was derived from ordinary linear regression with parametric bootstrapping instead, as in Doi et al. [D24].

236. The approach to estimate a linear ERR per unit dose based on dose category RRs was validated using studies that reported an estimate of the linear ERR per unit dose with CI as well as RRs per category dose. Estimates of the linear ERRs per unit dose and their CIs as reported in the study were compared to estimates of the ERRs per unit dose derived from category RRs using the method detailed above.

## B. Results of literature review and meta-analysis

237. The literature review undertaken for this evaluation involved a detailed search of three databases (PubMed, EMBASE and Web of Science) and identified 995 non-duplicate articles up to September 2020 with no earliest publication date cut-off. Additional searches were made through the references cited in the past reviews summarized in section A.1 of this chapter. After an initial screening of the abstracts,

335 articles were retained and reviewed in detail for information on radiation risk of second primary cancer in patients who were treated with radiotherapy for a first primary cancer. Information on cumulative radiation risk and latency periods for second primary cancer was extracted. Particular attention was given to extracting information on risk of second primary cancer in patients who were treated for a first primary cancer with radiotherapy from studies in which estimates of organ doses at the sites of the second primary cancers were available. The data extracted showed that seven second cancer sites had sufficient information on which to base a meta-analysis: haematopoietic and lymphoid tissues (considered together), sarcoma, breast, lung, gastrointestinal organs, thyroid, and brain.

238. Insufficient information to justify a meta-analysis could be extracted on second primary skin cancer, and some other second primary cancer sites including cancers of the kidney and male and female genitalia. Additional information was extracted from the literature and reported below on risks of all types of second primary cancer when grouped together (e.g., all solid cancer). However, a meta-analysis of risks per unit reference dose of all solid cancers was not performed for this overall group, due to the heterogeneity in the reference dose applied to the risks. Nonetheless, the calculated risks for the six individual solid cancer sites, for which site-specific meta-analyses were performed in this study, were combined to obtain an overall risk per unit dose for solid cancers.

239. This section provides details of data and information in the studies included in the literature review on cumulative incidence, latency information, and the risk of second primary cancer following radiotherapy by cancer site. Each subsection describes the major studies that provided risks per unit organ dose at the site of one of the seven second primary cancers singled out in this evaluation. Information on the shape of the dose response relationship, particularly on reported non-linearities in the response at very high doses, are also provided. The role of sex and age at time of cancer diagnosis and other factors as risk modifying factors are also considered, where relevant information is available. Electronic attachment C to this annex includes detailed information extracted from the studies considered for this evaluation.

## 1. Neoplasms of the haematopoietic and lymphoid tissues

240. Comparisons of studies of second primary neoplasms of haematopoietic and lymphoid tissues are complex because these neoplasms represent a very heterogeneous group of malignancies originating in cells with different radiosensitivity and the classification of haematological neoplasms have changed over time. In addition, studies of second primary neoplasms of haematopoietic and lymphoid tissues adopt very different criteria for selection and classification of outcomes, mostly grouping them into leukaemia—often excluding chronic lymphocytic leukaemia (CLL)—and non-Hodgkin lymphoma, sometimes separating acute leukaemia. Many studies, especially in the past, do not separate leukaemia subtypes that were linked to chemotherapy. A further problem is that the target tissue for these malignancies may differ and may include the active bone marrow for haematopoietic malignancies and lymphatic tissues for lymphoid malignancies.

### (a) *Studies with information on cumulative risks*

241. Studies that reported information on cumulative incidence of second primary neoplasms of haematopoietic and lymphoid tissues are summarized in table 3. Only three of them contained information on radiation dosimetry of individual patients [T25]. Data from studies with at least 20 cases of second primary malignancies show that the cumulative incidence for second primary leukaemia at times since the first primary cancer between 15 and 20 years is between 0.3% to 5.4%. The cumulative incidence of second primary leukaemia increased when combined treatment modalities (radiotherapy and chemotherapy) were applied for primary cancer, compared to radiotherapy alone.

Table 3. Studies with information on cumulative incidence of second primary haematopoietic or lymphoid cancer after radiotherapy

RT: Radiotherapy; NHL: Non-Hodgkin lymphoma

| *Study reference* | *Country of study* | *Type of primary cancer* | *Number of survivors* | *Age at treatment (years)* | *Diagnosis time frame* | *Second primary cancer incidence* | *Cumulative incidence (%, 95% CI or ±SD)* | *Time frame[a] (years)* |
|---|---|---|---|---|---|---|---|---|
| Fisher et al. [F9] | United States | Breast carcinoma | 8 483 | All ages | 1971–1985 | Leukaemia: 25 (5 among those with regional RT) | 1.39 (±0.49) after regional RT | 10 |
| Tucker et al. [T25] | Europe and North America | Childhood cancer | 9 170 | <15 | 1936–1979 | Leukaemia: 22 | 0.8 (±0.2); 4.2 (±1.9) after Hodgkin lymphoma | 20 |
| Lavey et al. [L4] | United States | Breast carcinoma | 1 139 | All ages | 1970–1981 | Leukaemia: 4 (1 after RT+CT) | 3.0 (0.0, 8.9) after RT+CT | 10 |
| Hawkins et al. [H25] | United Kingdom | Childhood cancer | 16 422 | <15 | 1962–1983 | Leukaemia: 22 | 0.2 (±0.0) | 5 |
| Sont et al. [S46] | Netherlands (Kingdom of the) | Hodgkin lymphoma | 482 | All ages | 1969–1988 | Leukaemia: 6<br>NHL: 5 | Leukaemia: 1.5<br>NHL: 15 | 5 |
| Bhatia et al. [B22] | Europe and North America | Childhood cancer | 1 380 | <15 | 1955–1986 | Leukaemia: 26<br>NHL: 6 | After RT alone: Leukaemia: 0 NHL: 0.4 (0, 1.2)<br>After RT+CT: Leukaemia: 3.4 (1.8, 4.9) NHL: 0.9 (0, 1.9) | 15 |
| Diamandidou et al. [D18] | United States | Breast carcinoma | 1 474 | All ages | 1974–1989 | Leukaemia: 14 | All: 1.5 (0.7, 2.9)<br>After RT+CT: 2.5 (1.0, 5.1) | 10 |
| Brusamolino et al. [B52] | Italy | Hodgkin lymphoma | 1 659 | 12–83 | 1975–1992 | Acute leukaemia: 36 | RT alone: 0.3; RT+CT: 5.4 | 15 |
| Dunst et al. [D29] | Germany | Ewing sarcoma | 674 | All ages | 1981–1991 | AML/MDS: 5 | After CT+RT 36–45 Gy: 1.8 (0, 4.6)<br>After CT+RT 45–60 Gy: 2.2 (0, 5.4) | 15 |
| Slanina et al. [S43] | Germany | Hodgkin lymphoma | 1 588 | All ages | 1940–1991 | AML: 3<br>NHL: 9 | NHL: Leukaemia:<br>0.1 0.1<br>0.9 0.3<br>1.4 0.6<br>1.9 0.6 | <br>5<br>10<br>15<br>20 |
| Chaplain et al. [C7] | France | Breast carcinoma | 3 093 | <85 | 1982–1996 | Acute leukaemia: 10 | After RT alone: 0.14 (±0.1)<br>After RT+CT: 1.12 (±0.4) | 4 |

| *Study reference* | *Country of study* | *Type of primary cancer* | *Number of survivors* | *Age at treatment (years)* | *Diagnosis time frame* | *Second primary cancer incidence* | *Cumulative incidence (%, 95% CI or ±SD)* | *Time frame*[a] *(years)* |
|---|---|---|---|---|---|---|---|---|
| Green et al. [G21] | United States | Hodgkin lymphoma | 182 | <20 | 1960–1989 | Leukaemia: 2<br>NHL: 3 | Leukaemia: 1.98 (±1.48)<br>NHL: 4.62 (±3.21) | 30 |
| Van Leeuwen et al. [V5] | Netherlands (Kingdom of the) | Hodgkin lymphoma | 1 253 | <40 | 1966–1986 | Leukaemia: 18<br>NHL: 16 | Leukaemia: 3.3 (2.1, 5.3)<br>NHL: 3.5 (2.0, 6.0) | 25 |
| Bhatia et al. [B23] | North America | Childhood cancer | 8 831 | <15 | 1983–1995 | AML/MDS: 14<br>NHL: 6 | 0.21 (0.1, 0.3)<br>0.27 (0.1, 0.4)<br>0.08 (0.01, 0.2)<br>0.19 (0.01, 0.4) | 10<br>15<br>10<br>15 |
| Lenz et al. [L9] | Germany | Low grade lymphoma | 431 | <65 | 1996–2002 | t-MDS (ASCT arm): 5<br>IFN-α arm: 0 | 3.8 (0.4, 7.3)<br>0 | 5 |
| Haddy et al. [H1] | France and United Kingdom | Childhood cancer | 4 204 | <16 | 1947–1986 | Leukaemia: 11 | 0.21 (0.10, 0.42)<br>0.76 (0.3, 2.0) | 15<br>40 |
| Richiardi et al. [R8] | Australia, Canada, Europe, Singapore | Germ cell testicular cancer | 29 511 | All ages | 1943–2000 | Myeloid leukaemia: 15 | 2.4 (1.4, 4.1)/1 000 (diagnosed since 1980) | 10 |
| Howard et al. [H45] | Europe and North America | Testicular cancer | 42 722 | Mean: 35 | 1943–2001 | Leukaemia: 89 (excluding CLL) | Excess cumulative risk:<br>0.14<br>0.23 | <br>15<br>30 |
| Navid et al. [N4] | United States | Ewing sarcoma | 237 | ≤20 | 1979–2004 | Leukaemia: 8 | 3.0 (±1.1)<br>3.0 (±1.1) | 5<br>10 |
| Seo et al. [S31] | South Korea | Thyroid cancer | 103 741 | All ages | 2008–2013 | Leukaemia: 46 | 16.5 (12.1, 22.0) | 6 |
| Kushner et al. [K32] | United States | Breast carcinoma | 755 | 25–89 | 1992–2001 | Leukaemia: 6<br>Lymphoma: 5 | Leukaemia:<br>0.4 (0.1, 1.4)<br>0.6 (0.2, 1.6)<br>0.9 (0.4, 2.1)<br>1.4 (0.6, 3.70)<br>Lymphoma:<br>0.2 (0.0, 1.1)<br>0.5 (0.2, 1.6)<br>0.7 (0.3, 1.9)<br>1.4 (0.5, 3.9) | <br>5<br>10<br>15<br>20<br><br>5<br>10<br>15<br>20 |

[a] Time since first primary cancer (unless otherwise specified).

### (b) *Studies with information on latency periods between treatment for first primary cancer and occurrence of second primary neoplasm of the haematopoietic and lymphoid tissues*

242. Studies that reported latency periods between the radiation treatment for first primary cancer and the occurrence of second primary neoplasm of the haematopoietic and lymphoid tissues are summarized in table 4. The shortest latency period for second primary neoplasm of the haematopoietic and lymphoid tissues reported by most studies was 1–2 years while the median or mean latency periods were about five years after treatment.

Table 4. Studies with information on latency periods between radiation treatment of first primary cancer and occurrence of second primary haematopoietic or lymphoid cancer

EBV: Epstein-Barr virus; NHL: Non-Hodgkin lymphoma; ANLL: Acute nonlymphocytic leukaemia

| *Study reference* | *First primary cancer* | *Second primary cancer* | *Latency period (years)* | | |
|---|---|---|---|---|---|
| | | | *Mean* | *Median* | *Range* |
| Fisher et al. [F9] | Breast carcinoma | | | 6.6 | |
| Murohashi [M35] | Cervical and ovarian cancer | | 4.9 | | 1.8–11.1 |
| Curtis et al. [C21] | Breast carcinoma | | 12.0 | | 1.6–27.0 |
| Travis et al. [T12] | NHL | ANLL | | 5.1 | 0.2–13.3 |
| Diamandidou et al. [D18] | Breast carcinoma | | | 5.5 | 1.8–9.4 |
| Brusamolino et al. [B52] | Hodgkin lymphoma | | | 5.0 | 1.2–12.0 |
| Metayer et al. [M19] | Hodgkin lymphoma | Acute leukaemia | | 6.6 | >1 |
| Bhatia et al. [B23] | Childhood cancer | AML/MDS | | 3.1 | 0.9–10.7 |
| | | NHL | | 3.1 | 1.5–2.7 |
| Ng et al. [N13] | Breast carcinoma | Leukaemia | | 5.3 | |
| | | NHL | | 7.1 | |
| Casagranda et. al. [C6] | Childhood cancer | Leukaemia | | 3.2 | 1.6–9.1 |
| | | Lymphoma[a] | | 2.2 | 0.8–3.8 |

[a] All induced by EBV.

### (c) *Studies with risk estimates with respect to active bone marrow or lymphatic tissue doses*

243. Studies identified in the literature review that reported risks for second malignant neoplasm of the haematopoietic and lymphoid tissues with respect to the relevant bone marrow dose are briefly described in this section. A summary of the main features of these studies is given in table 5, while full details are available in electronic attachment C. Some of these studies were found to have overlapping data since they analysed data for the same cohort. To prevent a biased meta-estimator of the combined risks in the meta-analysis, references with overlapping data are identified in the paragraphs below.

244. Storm and Boice [S48] conducted a cohort study of 25,718 women diagnosed with invasive cervical cancer from the Danish Cancer Registry; more refined analyses were performed using nested case-control approach with 27 leukaemia cases and 106 matched controls. The mean ABM dose was 7.83 Gy, while the dose outside the pelvis was 1.76 Gy. When the average ABM dose was used, the absolute risk of leukaemia was estimated at 0.24 per $10^6$ PY-rad; however, it was 1.09 per $10^6$ PY-rad if only the doses received outside the pelvis were considered. Neither of these estimates was statistically significantly increased. The risk of acute leukaemia and myelogenous leukaemias rose and remained elevated over all dose categories. The relative risk increased up to about 8 Gy and then decreased among those who received more than 9 Gy. There is most probably some degree of data overlap between the studies of Boice et al. and Blettner et al. [B29, B32, B33] since Danish cases were included in the international case-control study.

245. A cohort study of 60,949 uterine corpus cancer patients in the SEER program registries who had survived for more than one year [L23] reported mean ABM doses of 20 Gy from combination therapy, 19 Gy from external beam, and 1.3 Gy from brachytherapy. The IRR for leukaemia following external beam therapy with a latency period of one year was 2.03 (range: 1.34–3.10). The IRR for non-Hodgkin lymphoma after combination radiotherapy with a latency period of five years was 2.00 (range: 1.36–2.95).

246. Boice et al. [B33] conducted an international case-control study of cervical cancer patients from Canada, Europe, and the United States of all types of second primary cancer, including leukaemia. The weighted mean dose to total bone marrow was estimated to be 7.1 Gy. This study was based on the same data analysed in Blettner and Boice [B29, B32], where the ERR per unit dose for leukaemia was estimated at 0.88 $Gy^{-1}$; the RR for acute leukaemia and chronic myelogenous leukaemia at 1 Gy was 1.14 (90% CI: 1.00, 1.45); the EAR was 0.10 per $10^4$ PY-Gy (90% CI: 0.00, 0.31); RR for non-Hodgkin lymphoma was 2.5 (90% CI: 0.8, 7.6) following an ABM dose of 7.1 Gy versus a dose for the reference category of <2 Gy; unmatched RRs following the same ABM dose of 7.1 Gy were 0.63 (90% CI: 0.2, 2.6) for Hodgkin lymphoma and 0.26 (90% CI: 0.0, 2.6) for multiple myeloma.

247. A case-control study nested within the French CCSS cohort of about 15,000 patients [A21] reported a significant trend in the increase of leukaemia risk (35 leukaemia cases, 140 matched controls) with radiation dose ($p$=0.046). However, this trend became non-significant after adjustment for chemotherapy and other factors was included in a multi-variate analysis ($p$=0.388).

Table 5. Studies identified in the literature review with risk estimates for second primary haematopoietic or lymphoid cancer

| *Study details* | *Treatment modality, fractionation details, dose type reported, RBE details* | *Number of subjects* | *Second primary cancer cases (controls, matched, unmatched)* | *Type of first primary cancer* | *Period of first cancer diagnosis* |
|---|---|---|---|---|---|
| A prospective study of 1 572 women treated with RT for cervical (1 478) and ovarian cancer (94) at the National Institute of Radiological Sciences of Japan [M35] | Cervical patients – external RT, intracavitary irradiation, or both<br>Ovarian – external RT alone | 1 572 | 5 Nonlymphocytic leukaemia cases | Cervical and ovarian cancer | 1961–1975 |
| A cohort of 25 718 women diagnosed with invasive cervical cancer from the Danish Cancer Registry; nested case control (CACO) [S48] | External beam, intracavitary application of radium or a combination of both in fractions of several hundred rad; ABM doses calculated | 25 718 women | 27 Leukaemia cases, 20 acute leukaemia + myelogenous leukaemia,<br>106 matched controls | Cervical cancer | 1943–1977 |
| Nested CACO within the International Radiation Study Group on Cervical Cancer of women from Europe and North America (17 registries, 16 clinics) [B32]<br>Total overlap with study by Blettner and Boice [B29] | External beam, (orthovoltage (58%), followed by $^{60}$Co (22%), betatrons (7%), and other megavoltage units (6%)<br>Brachytherapy or a combination of both<br>Unable to evaluate the influence of fractionation effect | Over 150 000 women | 195 cases of leukaemia identified (143 acute leukaemia + CML), 745 controls selected<br>133 acute leukaemia+ CML cases and 489 matched controls treated with RT | Invasive cancer of the uterine cervix | The earliest reported treatment was in 1920, but most women were treated between 1940 and 1970 |
| Late Effects Study Group from 13 major paediatric oncology centres from North America and Europe<br>Two or more years' survivors of childhood cancer [T25] | Mostly orthovoltage, some megavoltage units in more recent years<br>Individual dose was averaged over the total AMB | 9 170 | 22 Leukaemia cases in the cohort; 25 cases and 90 matched controls in CACO | All childhood cancers | 1936–1979 |
| International study of cervical cancer patients from Canada, Europe, and the United States<br>Nested CACO of all types of SPC, including leukaemia [B33]<br>Total overlap for leukaemia risks with study by Boice and Blettner [B32] | Brachy- and teletherapy, external irradiation mostly by orthovoltage, $^{60}$Co, supervoltage | Approximately 150 000 | 193 All leukaemia cases and 744 matched controls<br>AL + CML=141 and 545 matched controls<br>HL=14 and 27 matched controls<br>NHL=94 and 187 matched controls<br>MM=49 and 104 matched controls | Invasive cervical cancer | The earliest reported treatment was in 1920, but most women were treated between 1940 and 1970 |

| *Study details* | *Treatment modality, fractionation details, dose type reported, RBE details* | *Number of subjects* | *Second primary cancer cases (controls, matched, unmatched)* | *Type of first primary cancer* | *Period of first cancer diagnosis* |
|---|---|---|---|---|---|
| Nested CACO within the cohort of women with invasive breast cancer from Connecticut (United States) [C21] | External beam therapy (orthovoltage machines, $^{60}Co$, megavoltage machines) | 22 753 women | 48 Leukaemia cases included, 38 non-CLL (ANLL=27, ALL=3, CML=8) and 10 CLL, 97 matched controls | Invasive breast cancer | 1935–1972 |
| Nested CACO within the International Radiation Study Group on Cervical Cancer of women from Europe and North America (17 registries, 16 clinics) [B29, B32, S50]<br>Overlap with Boice et al. [B32], supplementary statistical considerations were presented here | External beam, brachytherapy; radiation absorbed dose to 14 different parts of the skeleton estimated and mean dose to the ABM calculated using weights | Over 150 000 women | 195 cases of leukaemia identified (143 AL+CML), 750 controls selected<br>133 cases and 500 matched controls with enough information for dose calculation | Invasive cervical cancer | The earliest reported treatment was in 1920, but most women were treated between 1940 and 1970 |
| Nested CACO within the cohort of women with invasive breast cancer from five United States areas [C22, S50] | Megavoltage units, $^{60}Co$, individual dose to 16 sections of ABM estimated (methods cited in Stovall et al. [S50]). | 82 700 women | 90 Cases: 84 Leukaemia cases excluding CLL (ANLL=74, ALL=3, CML=7) and 6 MDS, 264 matched controls | Invasive breast cancer | Between 1973 and 1985 |
| A cohort of 16 422 one-year survivors of childhood cancer diagnosed in the United Kingdom, nested CACO conducted [H25] | Absorbed radiation dose to the ABM | 16 422 | 22 Leukaemia cases in the cohort study<br>26 Leukaemia cases and 96 matched controls in CACO | All childhood cancers | Cohort study between 1962 and 1983; CACO – between 1940 and 1983 |
| Nested CACO within the cohort of 110 000 women with invasive cancer of the uterine corpus from nine cancer registries in Canada, Denmark, Finland, Norway, and the United States [C23] | External-beam therapy, brachytherapy or a combination of both; brachytherapy usually consisted of radium implants (73%), caesium 137 (19%), cobalt (3.5%), and radon seeds (2.2%)<br>Doses were computed for 17 sections of ABM | 110 000 women | Included 153 leukaemia patients with RT, 61 with no RT<br>459 Matched controls with RT, 290 with no RT | Invasive cancer of the uterine corpus | 1935–1985; most patients were diagnosed between 1960s and 1970s |
| Cohort of two-year survivors of NHL diagnosed between 1965 and 1989 from selected registries in Canada, Netherlands (Kingdom of the), Sweden and the United States, nested CACO [T14] | Primarily megavoltage | 11 386 | 35 ANLL case patients and 140 matched controls | NHL | 1965–1989 |

| *Study details* | *Treatment modality, fractionation details, dose type reported, RBE details* | *Number of subjects* | *Second primary cancer cases (controls, matched, unmatched)* | *Type of first primary cancer* | *Period of first cancer diagnosis* |
|---|---|---|---|---|---|
| Population based CACO nested among all patients with endometrial cancer diagnosed in Ontario, Canada [C23, H39]<br>Overlap with Curtis et al. [C23] | External beam, brachytherapy and combination of both; absorbed radiation dose to the ABM | 13 843 | 20 ANLL, 2 ALL, 10 CML and 15 CLL cases, 148 matched controls | Endometrial cancer | 1964–1987 |
| A cohort study of 4 400 childhood cancer three-year survivors from eight centres in France and the United Kingdom [H1] | $^{60}$Co, X-rays produced by low energy X-ray machines, photons or electron beam energies produced by medical linear accelerators; Dos-EG used for dose estimations | 4 204 patients with RT and chemotherapy included | 11 Leukaemia patients: AML=6, ALL=3, AL=2 unspecified, no CML | All childhood cancers | 1947–1986 |
| CACO of French paediatric patients treated for solid cancer [A19] | Electrons, cobalt, one patient with $^{192}$Ir wires; Dos-EG software; weighted mean of the RBM doses received at 128 skeleton points; accounted for fractionation | | 61 Leukaemia or myelodysplasia cases with MLL rearrangement but not t(9;11) or t(11;19),<br>196 matched controls, only 34 cases and 85 controls received RT | Paediatric solid cancers | 1980–1999 |
| A cohort of ≥one-year survivors of uterine corpus cancer in SEER program registries [L23] | Weighted doses to the ABM, calculated for three types of therapy, not individual but based on typical protocols | 60 949 women | 335 Lymphatic and haematopoietic cancers surgery only<br>303 Surgery+RT<br>172 NHL, surgery only<br>155 Surgery+RT<br>7 HL, surgery only<br>9 Surgery+RT<br>59 MM, surgery only<br>44 Surgery+RT<br>60 Non-CLL leukaemia, surgery only<br>71 Surgery+RT | Uterine corpus cancer | 1973–2003 |
| French Childhood Cancer Survivors Study (CCSS) cohort; extension of the French part of the Euro2K cohort; nested CACO conducted [A21, V16] | ABM doses were derived according to Veres [V16] (patient's skeleton was segmented into 19 compartments) | About 15 000 | 35 Leukaemia cases<br>28 AML<br>1 Acute undifferentiated leukemia<br>3 ALL<br>3 CML<br>140 matched controls | Solid childhood tumours | 1964–2000 |
| Pooled analysis, four CACO [A23] | | | 147 Leukaemia cases,<br>522 matched controls | Childhood cancers | 1930–2000 |

248. A prospective study of 1,572 women treated with radiotherapy for cervical (1,478 patients) and ovarian cancer (94 patients) at the National Institute of Radiological Sciences of Japan [M35] reported a leukaemia risk of 0.43 excess cases per $10^6$ PY–rad. The mean dose to the bone marrow in leukaemia patients (five cases) was estimated at 11.4 Gy (range: 8.5–13.7 Gy), compared to an average dose of 11.8 Gy for all patients.

249. A case-control cohort study nested in the cohort of two-year survivors of non-Hodgkin lymphoma diagnosed between 1965 and 1989 from selected registries in Canada, the Kingdom of the Netherlands, Sweden and the United States included 35 case patients with acute nonlymphocytic leukaemia and 140 matched controls [T14]. The median doses to the ABM of patients who received radiotherapy only were 9.3 Gy for the cases and 5.1 Gy for the controls; doses to patients who received radiotherapy combined with chemotherapy were 15.2 and 8.6 Gy, respectively. The RR of acute nonlymphocytic leukaemia among patients receiving a dose of 6.35 Gy or more was 3.1 ($p$=0.13) compared with patients who were not treated with radiotherapy or received a dose of less than 6.35 Gy.

250. A case-control study carried out by the Late Effects Study Group from 13 major paediatric oncology centres from Europe and North America identified 25 cases and 90 matched controls selected to be included in the case-control study [T25]. The mean dose to the ABM was 10 Gy (range: 038 Gy). No increase in risk was observed among the subjects treated only with radiotherapy.

251. Hawkins et al. conducted a case-control study nested within the cohort of 16,422 one-year survivors of childhood cancer diagnosed in the United Kingdom [H25]. Twenty-six leukaemia cases and 96 matched controls with individual estimated doses from radiotherapy were included in the analysis. The study found that the RR for all leukaemia types for patients treated with radiotherapy only was 8.4 (95% CI: 0.9, 81), while the RR for patients treated with radiotherapy in combination with chemotherapy was 62.9 (95% CI: 3, 1, 368). There was evidence of a trend in the RR of second primary leukaemia with increasing average dose to the patients' ABM. However, the 95% CIs of the estimates of relative risk associated with specific levels of exposure were wide.

252. A cohort study of 4,400 childhood cancer patients who had survived for 3 years from eight centres in France and the United Kingdom identified 11 leukaemia cases [H1] The estimated ABM doses ranged between 3 and 6.6 Gy. The linear ERR per unit dose, adjusted for chemotherapy, was 0.31 (95% CI: –0.32, 0.94) $Gy^{-1}$ and increased to 0.96 (95% CI: –1.6, 3.5) $Gy^{-1}$ if a negative exponential term was taken into account.

253. Curtis et al. reported results from the international case-control study nested within the cohort of 110,000 women with invasive cancer of the uterine corpus from nine cancer registries in Canada, Denmark, Finland, Norway and the United States [C23]. The mean ABM dose was 5.2 Gy for all modalities combined, 1.7 Gy for brachytherapy, and 9.7 Gy for external-beam therapy. The study included 153 leukaemia cases and 459 matched controls among patients who had been treated with radiotherapy and 61 cases and 290 matched controls among patients who had not received radiation treatment. The ERR per unit dose for all leukaemia patients except CLL following any external beam with linear model, was 0.13 (95% CI: 0.04, 0.27) $Gy^{-1}$, while the ERR per unit dose following brachytherapy with a linear-exponential model was 4.69 (95% CI: 1.10, 13.4) $Gy^{-1}$. It was estimated that about 14 extra leukaemia cases per 10,000 women over a 10-year follow-up period occurred.

254. A case-control study nested within the cohort of 22,753 female patients with invasive breast cancer from Connecticut (United States) [C21] identified 48 leukaemia cases and selected 97 controls. The mean dose to the ABM from primary radiotherapy was 5.3 Gy; the range of doses was 1.3–10 Gy for the cases and 1.2–8.5 Gy for the controls. The RR for leukaemia from primary radiotherapy was 1.16 (90% CI: 0.6, 2.1). There was no indication that risk varied over categories of radiation dose.

255. A larger case-control study nested within the cohort of 82,700 women with invasive breast cancer from five areas in the United States identified 90 cases of leukaemia and selected 264 controls [C22]. The mean and median dose to the whole ABM were 7.5 Gy and 7.2 Gy, respectively. The RRs of radiotherapy adjusted for alkylating agent were 1.8 (95% CI: 1.0, 3.2) for all leukaemias, except CLL, and MDS; 2.1 (95% CI: 1.1, 3.5) for acute nonlymphocytic leukaemia and MDS and 2.1 (95% CI: 0.3; 14.1) for ALL and chronic myelogenous leukaemia. There was a significant trend in the risk of leukaemia with increasing radiation dose after adjustment for the effects of alkylating agents. Patients who received doses above 9 Gy showed a 7-fold elevated risk.

256. Boice et al. [B32] reported results from a case-control study nested within the International Radiation Study Group on Cervical Cancer of female patients from Europe and North America (17 registries, 16 clinics). The cohort included over 150,000 female patients with invasive cancer of the uterine cervix. Overall, 195 cases of leukaemia were identified, 143 of which were acute leukaemia and chronic myelogenous leukaemia, and 750 controls were selected. The weighted mean dose to the total bone marrow was 7.1 Gy, with (range: 1–25 Gy). For acute leukaemia and chronic myelogenous leukaemia, the ERR per unit dose was 0.009 $cGy^{-1}$, in the low dose range where cell killing is negligible. A reduction of risk was observed when dose increases: the RR at 1 Gy for the linear exponential function was 1.7 (the CI was not given). The risk increased with increasing dose up to a mean dose of about 4 Gy and then decreased at higher doses.

257. Blettner and Boice [B29] presented supplementary statistical considerations from the same case-control study nested within the International Radiation Study Group on Cervical Cancer, overlapping with Boice et al. [B32]. The observed RR was obtained using matched analysis for data, grouped by mean dose to the ABM. The maximum fitted value was given at a mean dose of 5 Gy and yielded an estimated RR of 2.5. The confidence intervals for parameters of the linear-exponential model, which provides the best fit, are also given in the study.

258. Allard et al. [A19] reported results of a case-control study of French paediatric patients treated for solid cancer during the period between 1980 and 1999: 61 leukaemia or MDS cases were reported, with mixed-lineage leukaemia rearrangement but not t(9;11) or t(11;19), and 196 matched controls were selected. Only 34 cases and 85 controls received radiotherapy. The mean dose to the ABM was 2.4 Gy with a standard deviation of 4.1 Gy, a median of 0 Gy and a range of 0 to 28 Gy. No dose-response relationship was found irrespective of the model or adjustments used.

259. In a population-based case-control study nested among 13,843 patients with endometrial cancer diagnosed from 1964 to 1987 in Ontario, Canada, 47 confirmed cases of leukaemia were identified and four control subjects were matched to each patient [H39]. For subjects who received only brachytherapy, the median dose to the ABM was 1.8 Gy (range: 0.5–5.8 Gy). For external beam therapy alone or external beam therapy combined with brachytherapy, the median ABM dose was 9.9 Gy (range: 6.4–19.6 Gy). An elevated risk of all leukaemias other than CLL was observed, but only within the first 10 years after endometrial cancer treatment (OR=12.0; (90% CI: 2.8, 52.1)), whereas there was no evidence of an effect of radiotherapy after 10 or more years after treatment.

### (d) *Results of the meta-analysis of risks of second primary neoplasms of the haematopoietic and lymphoid tissues*

260. The review of papers on neoplasms of the haematopoietic and lymphoid tissues found that there was some heterogeneity between the studies in terms of the outcomes. The majority of studies included all non-CLL, while Hawkins et al. [H25] and Haddy et al. [H1], were for acute leukaemia. The study by Allodji et al. [A23] included all non-CLL and MDS. The primary random effects meta-analysis model

pooled the log RR from the results of 11 non-overlapping individual studies on risks of second primary neoplasms of the haematopoietic and lymphoid tissues. The studies by Murohashi [M35, S48, T14, T25], Storm and Boice [M35, S48, T14, T25], Travis et al. [M35, S48, T14, T25] and Tucker et al. [M35, S48, T14, T25] were excluded from the meta-analysis because they either did not report ERRs per unit dose or specific risks for dose categories, or only reported SIRs or mean doses. The meta-analysis was only based on data for ERRs since information on EAR per unit dose in the studies reviewed was insufficient. The pooled ERR per unit dose estimated in the primary analysis was 0.01 (95% CI: –0.00, 0.03) $Gy^{-1}$. The primary analysis also estimated a value for $I^2$ of 0.0% (95% CI: 0.0, 60.2) indicating that none of the variability in the estimates of the effect size of radiotherapy is due to the differences between studies. The Cochran's Q test (*p*=0.68), indicating no statistically significant heterogeneity between the individual studies. ERRs of second primary neoplasms of the haematopoietic and lymphoid tissues per unit dose estimated in the primary analysis are shown in figure XIV.

Figure XIV. Risks of second primary neoplasms of the haematopoietic and lymphoid tissues per unit dose to the bone marrow estimated in the primary analysis

The CIs were derived from the standard error of the point estimate based on the assumption of a log-normal distribution and may not coincide with the CI as originally published

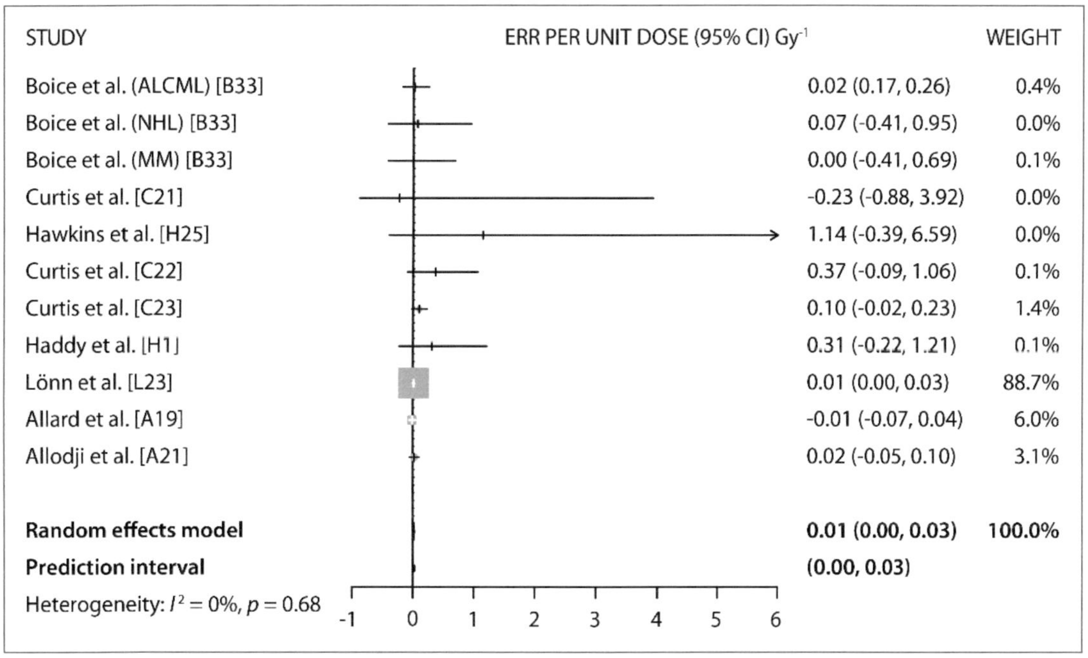


261. A sensitivity analysis pooling on the same ERRs for the 11 studies on the original scale resulted in the same (to two decimal places) pooled estimate of the ERR per unit dose. All subsequent results for the pooled estimate of the ERR per unit doses in further sensitivity and subgroup analyses are based on pooling the log RR. In leave-one-out diagnostics, the pooled ERRs per unit dose for the 10 remaining studies varied between 0.01 and 0.03, with the highest calculated risk of 0.03 (95% CI: –0.02, 0.08) $Gy^{-1}$ being obtained by leaving out the Lönn et al. [L23] study. This study was unusual in having an unexposed surgery-only comparison group on which the IRR risk estimate was based. With 88.7%, this study by far had the highest weight in the meta-analysis. In the sensitivity analysis that applied the linear uncorrelated model relevant to studies only publishing specific risk for dose categories, the pooled estimate of the ERR per unit dose was 0.03 (95% CI: 0.00, 0.07) $Gy^{-1}$, $I^2$ was 29.5% (95% CI: 0.0, 65.3) and the *p*-value for the Cochran's Q test was 0.16. When the results of all the 16 studies were included in the meta-

analysis without eliminating studies with potentially overlapping data, the pooled estimate of the ERR per unit dose was 0.01 (95% CI: 0.00, 0.03) $Gy^{-1}$.

262. Three studies published ERRs per unit dose which were estimated directly; the pooled estimate of the ERR per unit dose from these studies was 0.04 (95% CI: –0.06, 0.15) $Gy^{-1}$. The risks reported in four studies, without potentially overlapping data, pertained only to exposures in childhood or early adolescence, and the pooled estimate of the ERR per unit dose based on these studies was 0.00 (95% CI: –0.04, 0.05) $Gy^{-1}$. Seven papers contained risks related only to exposures in adulthood, and the pooled estimate of the ERR per unit dose from these papers was 0.04 (95% CI: –0.02, 0.10) $Gy^{-1}$. Three papers reported risks pertained to exposed patients without chemotherapy, and the pooled estimate of the ERR per unit dose was 0.02 (95% CI: –0.15, 0.23) $Gy^{-1}$; and seven studies gave risks related to exposures adjusted for chemotherapy, and the pooled estimate of the ERR per unit dose for these studies was 0.03 (95% CI: –0.03, 0.09) $Gy^{-1}$. Six studies reported risks pertained to only external exposures, and the pooled estimate of the ERR per unit dose was 0.01 (95% CI: –0.00, 0.03) $Gy^{-1}$. Ten papers published risks derived without external comparison groups, and the pooled estimate of the ERR per unit dose was 0.03 (95% CI: –0.02, 0.08) $Gy^{-1}$. The results are summarized in figure XV. Results were not reported if fewer than three studies contributed to the meta-analysis.

Figure XV. Comparison of risks per unit dose estimated in the primary analysis and in the sensitivity analyses for second primary neoplasms of the haematological and lymphoid tissues

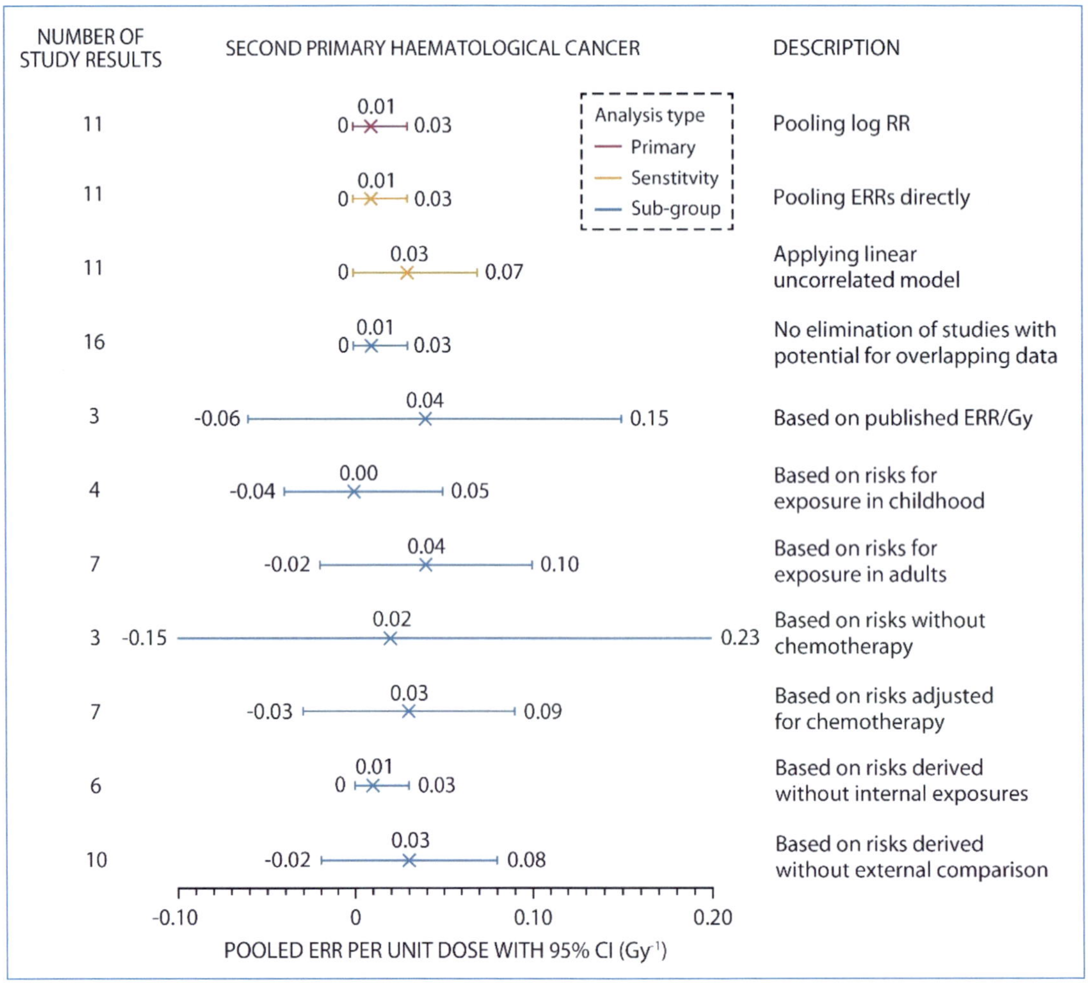

### (e) *Summary*

263. Sixteen studies selected in the literature review provided information on risks of second neoplasms of haematopoietic and lymphoid tissues after radiotherapy, although there were some inconsistencies in the outcomes included in the studies. The studies had either cohort or nested case-control designs and all but one of them were conducted in countries in North America or Europe. Four studies dealt with patients who were children or adolescents at the time of the radiation treatment. It was not always possible to clearly distinguish the follow-up period from the cancer diagnosis period. Most studies reported results based on patients treated with radiation and chemotherapy together, but most of the studies reporting risks per unit ABM dose provided risks adjusted for chemotherapy.

264. The cumulative incidence rates of second neoplasms of haematopoietic and lymphoid tissues after first cancer treatment that mostly included a combination of therapies, rather than radiotherapy alone, were usually reported in the literature at 5 to 15 years after the first primary cancer, with a few studies reporting it at 20 to 30 years after. The cumulative incidence rates for leukaemia reported in the studies varied between 0.13 and 5.4%, with no major difference according to the first primary cancer or time from diagnosis. For non-Hodgkin lymphoma, the range of cumulative incidence rates was wider, and varied from 0.1% to 15%, with the highest value observed in a study on Hodgkin lymphoma as the first primary cancer.

265. In several studies, most of the acute leukaemia cases were diagnosed in the first 10 years following the treatment of the first primary cancer. In some case the relative risk remained elevated 20 years after the treatment.

266. The reported results from this review were based on an increasing linear dose-response relationship between radiation doses and the risk of second neoplasms of haematopoietic and lymphoid tissues. The evidence is very mixed with respect to whether a linear curve is an adequate approximation of the actual dose-response relationship for second primary cancers in tissues of the haematopoietic and lymphoid system, and therefore results from a linear meta-analysis should be treated with caution. Most of the studies reviewed for second primary cancers in tissues of the haematopoietic and lymphoid system included therapy-related myeloid neoplasms arising in patients with cytotoxic chemotherapy for prior neoplastic disease, that are only recently recognized as distinct diagnostic entities. These myeloid neoplasms may differ from the types of neoplasms included in other epidemiological studies on radiation exposures other than radiotherapy because of the potential role of cancer predisposition in their development and the role of chemotherapy, both of which can mask the effect of radiation.

## 2. Sarcoma

267. Sarcomas are malignant tumours that arise in connective tissues and have multiple histologically heterogeneous subtypes. They are classified into two major groups: bone sarcomas and soft-tissue sarcomas, which are generally analysed together in most studies. Malignant bone tumours accounted for 4–5% of cancers among children and adolescents in the United States from 2014 to 2018 [S39]. The aetiology of sarcomas is poorly understood; however, there is sufficient evidence of an association between radiotherapy and bone cancer in humans [I2].

### (a) *Studies with information on cumulative risks*

268. Studies that reported information on cumulative incidence of second primary sarcoma incidence after radiotherapy are summarized in table 6. The table shows that cumulative incidence of second primary sarcoma after the first cancer was treated with radiotherapy is generally low but varies widely (range: 0.01–6.5%) depending on the first primary cancer, age at treatment, interval from diagnosis, and hereditary factors. In particular, the risks are high among retinoblastoma patients (up to 50%), which indicates that genetic predisposition may affect risks associated with high doses from therapeutic radiation exposure.

Table 6. Studies with information on cumulative incidence of second primary sarcoma after radiotherapy

| *Study reference* | *Country of study* | *Type of primary cancer* | *Number of survivors* | *Age at treatment (years)* | *Diagnosis time frame* | *Second primary incidence* | *Cumulative incidence (%, 95% CI or ±SD)* | *Time frame[a] (years)* |
|---|---|---|---|---|---|---|---|---|
| Tucker et al. [T24] | Canada and the United States | Childhood cancer | 9 170 | | 1936–1979 | 48 | 2.8% | 20 |
| Hawkins et al. [H26] | United Kingdom | Childhood cancer | 13 175 | | 1940–1983 | 55 | 0.9% | 20 |
| Kuttesch et al. [K33] | United States | Ewing sarcoma | | 14.6 | 1963–1990 | 266 | 3%<br>6.5% | 10<br>20 |
| Wong et al. [W19] | United States | Retinoblastoma | 1 604 | | 1914–1984 | 84 | 51% hereditary<br>5% non-hereditary | 50 |
| Huang and Mackillop [H51] | United States | Breast carcinoma | 19 4798 | <45 (14.5%)<br>45–<65 (31.6%)<br>≥65 (53.9%) | 1973–1995 | 135 | 78.6/100 000 | 10 |
| Yap et al. [Y5] | United States | Breast carcinoma | 274 572 | <20 (0.0%)<br>20–39 (7.2%)<br>40–59 (38.0%)<br>60–79 (43.7%)<br>≥80 (11.1%) | 1973–1997 | 263 | 3.2/1 000 | 15 |
| Menu-Branthomme et al. [M18] | France and the United Kingdom | Childhood cancer | 4 400 | 6.4 | 1942–1985 | 23 | 0.6% (95% CI: 0.3, 1.0) | 20 |
| Rubino et al. [R19] | France | Breast | 6 597 | <5 (16.7%)<br>5–9 (66.6%)<br>≥10 (16.7%) | 1954–1983 | 12 | 0.28% (95% CI: 0.10, 0.45) | 15 |
| Virtanen et al. [V19] | Finland | Breast, cervix uteri, uterine corpus, lung, ovary, prostate, rectum, lymphoma | 295 712 | <55 (48.3%)<br>≥55 (51.7%) | 1953–2000 | 147 | 0.03% (95% CI: 0.017, 0.037) | 10 |
| Henderson et al. [H28]) | United States and Canada | Childhood cancer | 14 372 | 8 (1–20) | 1970–1986 | 104 | 0.37% (95% CI: 0.24, 0.50)<br>0.89% (95% CI: 0.68, 1.10)<br>1.37% (95% CI: 0.97, 1.78) | 10<br>20<br>30 |

| *Study reference* | *Country of study* | *Type of primary cancer* | *Number of survivors* | *Age at treatment (years)* | *Diagnosis time frame* | *Second primary incidence* | *Cumulative incidence (%, 95% CI or ±SD)* | *Time frame[a] (years)* |
|---|---|---|---|---|---|---|---|---|
| Jenkinson et al. [J2] | United Kingdom | Childhood cancer | 16 541 | | 1926–1987 | 39 | 0.23% | 20 |
| Kleinerman et al. [K20] | United States | Retinoblastoma | 963 | <1 (56.6%)<br>1 (27.7%)<br>2–7 (15.7%) | 1914–1984 | 69 | 13.1% (95% CI: 9.7, 17.0) | 50 |
| Schwartz et al. [S25] | France and the United Kingdom | Childhood cancer | 4 171 | France: 4<br>UK: 5 | 1942–1986 | 39 | 0.3% (95% CI: 0.1, 0.4)<br>0.9% (95% CI: 0.6, 1.2)<br>1.1% (95% CI: 0.8, 1.5). | 10<br>20<br>35 |
| Bazire et al. [B9] | France | Breast | 17 745 | | 1981–2000 | 49 | 4.4% (95% CI: 3.0, 5.8) | 15 |
| Bright et al. [B47] | Europe | Childhood cancer | 69 460 | 0–4 (47.1%)<br>5–9 (16.7%)<br>10–14 (24.2%)<br>15–19 (12.0%) | 1940–2008 | 301 | 1.4% (95% CI: 1.1, 1.6) | 45 |
| Hird et al. [H36] | Canada | Abdomen or pelvis | 173 580 | 18–49 (6.9%)<br>50–59 (18.0%)<br>60–69 (33.6%)<br>70–79 (28.7%)<br>≥80 (12.8%) | 2002–2017 | 332 | 0.3% | 13 |
| Smith et al. [S44] | United States | Head and neck squamous cell cancer | 70 476 | <59 (48.2%)<br>60+ (51.8%) | 1975–2018 | 64 | 0.08%<br>0.13% | 10<br>20 |

[a] Time since primary cancer (unless otherwise specified).

## (b) *Studies with information on latency periods between treatment for first primary cancer to occurrence of second primary sarcoma*

269. Studies that reported latency periods between the radiation treatment for first primary cancer and the occurrence of second primary sarcoma are summarized in table 7. A systematic review for radiation-induced sarcomas of the central nervous system was performed by Yamanaka and Hayano [Y2]. The latency period of second primary soft tissue cancers was generally longer than that of the second primary bone malignancies; the median latency period for second primary cancer of the soft tissue ranged from 5 to 17 years. However, there was no significant difference in the latency period for the second primary sarcoma between groups who received radiation treatment and those who did not.

Table 7. Studies with information on latency periods between radiation treatment of first primary cancer and occurrence of second primary sarcoma

| *Study reference* | *First primary cancer* | *Second primary cancer* | *Latency period (years)* | | *Shortest latency period (years)* |
|---|---|---|---|---|---|
| | | | *Mean* | *Median* | |
| Bechler et al. [B11] | Osteosarcoma | | 10.0 | | 5 |
| Kuttesch et al. [K33] | Ewing sarcoma | | | 7.6 | 3.5 |
| Wong et al. [W19] | Retinoblastoma | | | 15.0 | 1.6 |
| Karlsson et al. [K4] | Breast cancer | | | 10.1 | 1.1 |
| Rubino et al. [R16] | Thyroid cancer | | 15.0 | | 2 |
| Menu-Branthomme et al. [M18] | Childhood cancer | | 15.0 | | 6 |
| Rubino et al. [R19] | Breast cancer | | | 7.0 | 3.3 |
| Svahn-Tapper et al. [S57] | Childhood cancer | | 13.0 | | 8.1 |
| Jenkinson et al. [J2] | Childhood cancer | | 16.0 | | |
| Kalra et al. [K3] | All cancer | | 17.0 | | 4 |
| Armstrong et al. [A27] | CNS tumour | Bone cancer | | 8.2 | |
| | | Soft-tissue sarcoma | | 12.6 | |
| Friedman et al. [F16] | Childhood cancer | Bone cancer | | 9.8 | 5.3 |
| | | Soft-tissue sarcoma | | 15.2 | 5.3 |
| Henderson et al. [H31] | Childhood cancer | | | 11.8 | 5.3 |
| Wu et al. [W23] | All cancer | | 10.7 | | 5 |
| Yamanaka and Hayano [Y2] | CNS tumour | | 12.4 | | |
| Hird et al. [H36] | Abdomen or pelvis | | 5.7 | | 2.2 |
| Smith et al. [S44] | Head and neck squamous cell cancer | | 10.4 | | 3.2 |

### (c) *Studies with risk estimates with respect to ABM doses*

270. Studies identified in the literature review that reported risks for second primary sarcoma with respect to the relevant ABM dose are briefly described in this section. A summary of the main features of these studies is given in table 8, while full details are available in electronic attachment C. A few studies provided estimates of second primary bone and soft-tissue sarcoma risk per unit dose, whereas several studies reported relative risks by dose category.

271. Tucker et al. [T24] conducted a case-control study, nested within a cohort of 9,120 two-year childhood cancer survivors in the Late Effects Study in the United States and Canada. This study included 64 secondary osteosarcomas and 209 matched controls. The radiation dose to the site of the bone tumour was estimated for each patient. For each control, the radiation dose to the equivalent site of the bone cancer in the matched case was calculated. The risk of bone cancer rose sharply with increasing radiation doses, reaching a 40-fold increase when the doses were above 600 rad. The relative risks adjusted for alkylating agent treatment were 1, 0.6, 6.0, 16.9, 21.2, 38.3 for the dose categories of 0 , <1,000, 1,000–2,999, 3,000–3,999, 4,000–5,999 and >6,000 rad, respectively. The risk appeared to decrease when the doses exceeded 8,000 rad. An increased risk of bone cancer did not appear to be associated with other drugs used for the treatment of the first cancer.

272. Henderson et al. [H31] conducted a nested case-control study of secondary sarcoma—105 cases and 422 controls—within the Childhood Cancer Survivor Study of 14,372 childhood cancer survivors. Radiation doses at the site of the second primary cancer were estimated from medical records. The odds ratios (OR) were 1.3, 15.6, 160.0, and 114.1 at doses of <10, 10–29.9, 30–49.9 and >50 Gy, respectively. A significant linear dose-response relationship was reported, and the excess odds ratio (EOR) per unit dose was 1.32 (95% CI: 0.44, 4.22) $Gy^{-1}$ after adjusting for chemotherapy. There was no evidence of a downturn in the risk for the highest dose category. No effect modification by sex, age at diagnosis, or attained age was observed.

Table 8. Studies identified in the literature review with risk estimates for second primary sarcoma

| *Study details* | *Treatment modality, fractionation details, dose type reported, RBE details* | *Number of subjects* | *Second primary sarcoma cases (controls, matched, unmatched)* | *Type of first primary cancer* | *Time period of first cancer diagnosis* |
|---|---|---|---|---|---|
| Late Effects Study Group – two-year survivors of childhood cancer; CACO analyses also performed [T24] | Orthovoltage, $^{60}Co$, megavoltage, betatron, electrons, brachytherapy; dose is estimated to the site of bone tumour | 9 170 | 64 cases of bone sarcoma and 209 matched controls included in CACO | All childhood cancers | 1936–1979 |
| International study of cervical cancer patients from Canada, Europe, and the United States; nested CACO of all types of SPC including bone sarcomas and others [B33] | Orthovoltage, any (200–400 kVp), megavoltage, external beam, brachytherapy<br>Doses to bone were estimated for a 1 $cm^3$ volume at the site of the bone<br>Adjustment was made for the difference in bone absorption associated with different energy beams | Over 150 000 women | Overall, 4 188 second cancers and 6 880 controls<br>15 bone carcinomas and 155 matched controls | Cervical cancer | Earliest RT date 1920, extended to late 1970s; 25% treated prior to 1950 |
| British cohort of three-year survivors of childhood cancer diagnosed<br>Study outcome: bone cancer, also addressed in CACO (almost nested) [H26] | Dose to the bone cancer site calculated on the basis of measurements in a water and an anthropomorphic phantom and included collimator scatter, head leakage, and scatter within the patient<br>Dose estimates were corrected for differential absorption in bone based on radiation energy | 13 175 | 55 cases in the cohort study, 59 case subjects and 220 matched controls included in CACO | All malignant neoplasms in childhood | 1940–1983 |
| U.S. study of three-year survivors after Ewing's sarcoma from three centres (NCI, St Jude Children's Research Hospital and University of Florida Medical School) [K33] | Megavoltage radiation therapy | 266 | 10 cases of sarcomas | Ewing's sarcoma | 1963–1990 |
| A cohort study of one-year survivors of retinoblastoma from Boston and New York; CACO conducted<br>Note: Dose-response analyses performed separately for STS only and for STS bone sarcoma combined (leg and trunk sarcomas were excluded) [W19] | Absorbed dose to bone or soft tissue at each sarcoma site was estimated by measurements and computer simulations; random assignment of "site matched doses" in controls<br>Orthovoltage radiation mostly before 1960, $^{60}Co$, betatron or other megavoltage, electron beam thereafter | 1 729 | 52 cases of bone sarcoma (36 included in the final analysis), 31 cases of STS, 89 unmatched controls | Hereditary and non-hereditary retinoblastoma | Boston, MA, 1937–1984; and New York, NY, 1914–1984 |

| *Study details* | *Treatment modality, fractionation details, dose type reported, RBE details* | *Number of subjects* | *Second primary sarcoma cases (controls, matched, unmatched)* | *Type of first primary cancer* | *Time period of first cancer diagnosis* |
|---|---|---|---|---|---|
| Three-year survivors diagnosed with solid cancer during childhood in eight centres in France and the United Kingdom, also a nested CACO [L5] | $^{60}$Co gamma rays, orthovoltage X-rays, megavoltage X-rays and electrons; mean fractions for cases 25(0-51), for controls 21(0-131).<br>The absorbed doses to 151 anatomical sites, including 91 points of the skeleton was estimated in the cohort study, in CACO-cumulative absorbed dose at the site of the osteosarcoma | 4 257 | 32 cases and 160 matched controls | All solid childhood cancers | 1942–1986 |
| Cohort study of three-year survivors of a first solid cancer diagnosed during childhood in France or the United Kingdom, also a nested CACO [M18]<br>Note: The study outcome is STS | $^{60}$Co gamma rays, orthovoltage X-rays, megavoltage X-rays and electrons; the dose of radiation to the site of STS estimated<br>Fractionation effect: Less or more than 0.5 Gy per fraction | 4 400 | 16 STS in the cohort study, 23 STS and 111 matched controls in CACO study | All solid childhood cancers | 1942–1985 |
| Cohort of breast cancer patients treated at the Institut Gustave Roussy (IGR), also CACO<br>Note: STS and bone sarcomas were combined [R19] | External beam radiotherapy, cumulative dose absorbed at the site of sarcoma<br>Fractionation effect: less or more than 0.75 Gy per fraction | 6 597 | 12 cases of sarcoma in the cohort study<br>14 cases of STS and bone sarcoma and 98 matched controls in CACO | Breast cancer | 1954–1983 |
| A population-based case-control study of patients with cancer in childhood or adolescence in five Nordic countries [S57] | Absorbed dose received at the site of second primary cancer | | Overall, 196 cases and 567 controls<br>14 Bone carcinomas, 42 controls | All malignancies diagnosed before 20 years | 1960–1987 |
| Three-year survivors of childhood cancer in Britain, diagnosed with primary cancer before age of 15 years [J2]<br>Note: The study outcome is STS, diagnosed before 31 Dec 1990, CACO includes 53 cases of STS, of which 39 were diagnosed before 31 Dec 1990 and were nested within the cohort study. The remaining 14 cases were diagnosed during 1991–1996 | Dose to STS site calculated: within areas treated for the first tumour, standard RT data were used to estimate dose.<br>For sarcomas outside such areas, doses were based on radiation measurements in water phantoms | 16 541 | 53 Cases of STS and 179 matched controls | All malignant neoplasms in childhood | 1926–1987 |

| *Study details* | *Treatment modality, fractionation details, dose type reported, RBE details* | *Number of subjects* | *Second primary sarcoma cases (controls, matched, unmatched)* | *Type of first primary cancer* | *Time period of first cancer diagnosis* |
|---|---|---|---|---|---|
| It is a CACO nested within the CCSS, study period 01 Jan 1970–01 Feb 2005 [H31, S51]<br>Note: Compared with the 2007 cohort study by Henderson et al., [H28] nine additional sarcomas were identified, 12 sarcomas in the previous report were reclassified and therefore not included in this analysis | Dose to sarcoma site calculated, more details could be found in the study by Stovall et al. [S51] | 14 372 | 105 secondary sarcomas, 422 matched controls | Leukaemia, CNS malignancy, HL, NHL, neuroblastoma, STS, kidney cancer, or bone cancer | 1970–1986, before age of 21 |
| A cohort study of children with solid cancer in France and the UK followed until the end of 2010 in France and until the end of 2006 in the UK [S25] | External RT only, for cohort analysis, radiation dose was estimated in 59 bones<br>For CACO, only the local dose to the bone sarcoma site was estimated | 4 171 | 39 patients with BS but only 34 case-patients included in CACO and 170 matched controls | All solid childhood cancers | 1942–1985 |

273. Boice et al. [B33] conducted an international case-control study on second primary cancers in women treated for invasive cancer of the uterine cervix in Canada, Europe, and the United States between 1920 and 1965. The radiation doses for the selected organs were reconstructed from the original radiotherapy records. The mean dose to the bone and to the connective tissue were 22 and 7.0 Gy, respectively. For the study on bone cancer, 15 patients and 155 controls were matched according to race, age, and calendar year. An overall RR of 1.34 (90% CI: 0.3, 5.6) was reported, reaching a threefold increase at bone doses greater than 10 Gy (dose to the reference group was <1 Gy). For the study on soft-tissue sarcoma, 46 cases and 598 controls were matched by race, age, and calendar year. The overall RR associated with radiotherapy was 0.68 (90% CI: 0.2, 1.9); however, the risk increased among 10-year survivors receiving doses greater than 10 Gy (RR=2.8) and among those exposed after age 45 (RR=2.9).

274. Wong et al. [W19] conducted a case-control study for sarcoma nested in a cohort study of one-year survivors of retinoblastoma in Boston and New York. Out of 1,604 retinoblastoma patients who survived for at least one year after diagnosis, the study included 52 bone sarcoma cases with 89 controls, and 31 soft-tissue sarcoma cases with 89 controls. The mean doses to the site of the bone tumour were 32.8 Gy for the cases and 20.0 Gy for the controls. For soft-tissue sarcomas, the mean doses to the site were 20.4 Gy for the cases and 10.6 Gy for controls. The ERR per unit dose was 0.166 (95% CI: 0.02, 16.3) $Gy^{-1}$ for soft-tissue sarcoma and 0.191 (95% CI: 0.136, 0.315) $Gy^{-1}$ for all sarcomas combined. The ERR for all sarcomas combined did not differ according to the age at the time of the diagnosis of retinoblastoma.

275. Rubino et al. [R19] conducted a cohort study of 6,597 patients treated for breast cancer in France between 1954 and 1983, as well as a case-control study of 14 cases of bone or soft-tissue sarcoma and 98 matched controls. The mean dose at the site of the sarcoma was 38.8 Gy for the cases and 18.9 Gy for the controls. The odds ratios (ORs) were 1.6 (95% CI: 0.2, 11.0) and 30.6 (95% CI: 4.9, 611) for women who received 14–44 Gy and at least 45 Gy at the site of sarcoma, respectively. The EOR per unit dose for bone and soft-tissue sarcomas combined was 0.05 (95% CI: indeterminate to 1.18) $Gy^{-1}$. The best model fit was obtained using a quadratic dose-response function, without a negative exponential term for cell killing at high doses. The fit of the model did not improve significantly when the dose rate was considered in the analysis.

276. Hawkins et al. [H26] conducted a nested case-control study of 59 cases of second bone cancer and 220 controls with reconstruction of the individual doses to the organs of interest from a cohort of 13,175 three-year survivors of childhood cancer diagnosed in the United Kingdom between 1940 and 1983. The RRs adjusted for alkylating agent exposure were 0.7 (95% CI: 0.2, 2.2), 12.4 (95% CI: 0.9, 163.3), 93.4 (95% CI: 6.8,1 285.4), 64.7 (95% CI: 3.8, 1 103.4) at dose categories of 0, 1–9.9, 10–29.9, 30–49.9 and >50 Gy, respectively. There was a significant linear trend in RRs across different levels of radiation exposure, although a decrease in risk was observed at doses equal to or greater than 50 Gy.

277. Menu-Branthomme et al. [M18] conducted a nested case-control study of soft-tissue sarcoma in a British-French cohort of 4,400 three-year childhood cancer survivors. A total of 25 cases and 121 controls were included. Individual doses to the site of the soft-tissue sarcoma were also calculated. The mean doses to the site of the soft-tissue sarcoma were 26 Gy for the cases and 12 Gy for the controls. Adjusted for chemotherapy, the ORs were 0.9 (0.1–5.8), 1.5 (0.2–9.9), 3.0 (0.6–14.6), 10.7 (1.9–61.4) at dose categories of 0–1, 1–10, 10–30 Gy and >30 Gy compared with the group who did not receive radiotherapy.

278. Svahn-Tapper et al. [S57] conducted a second primary cancer case-control study nested within a Nordic cohort of 25,120 cancers diagnosed before the age of 20 between 1960 and 1987, followed up until 1991. Individual site-specific doses were reconstructed using phantom simulation and treatment data. For the second primary bone tumour, 14 cases were matched with 42 controls according to sex, age, and calendar year of diagnosis. The estimate for the ERR per unit dose was 3.1 (95% CI: 0.1, 53.0) $Gy^{-1}$

and the RRs were 1.7 (95% CI: 0.2, 12.9), 24.5 (95% CI: 1.2, 497.6), 38.4 (95% CI: 1.4, 1 066.2), 34.7 (95% CI: 1.3, 957.3) at dose categories of 0–1, 1–5, 3–30 and >30 Gy, respectively (dose to reference group was 0 Gy). For the second connective tissue cancers, 17 cases matched with 47 controls, and the estimate for the ERR per unit dose was 0.8 (95% CI: 0.0, 21.4). The RRs were 1.0 (95% CI: 0.1, 13.2), 9.9 (95% CI: 0.6, 169.7), 5.8 (95% CI: 0.1, 471.5) and 29.6 (95% CI: 1.8, 490.8) for 0–1, 1–5, 3–30 and >30 Gy, respectively (dose to the reference group was 0 Gy).

279. Jenkinson et al. [J2] conducted a case-control study nested in a cohort of 16,541 three-year survivors of childhood cancer in the United Kingdom, diagnosed with primary cancer before the age of 15 between 1926 and 1987. The dose was estimated from radiotherapy data and radiation measurements in phantoms. The RRs adjusted for exposure to alkylating agent were 1.28 (95% CI: 0.35, 4.63), 3.73 (95% CI: 0.78, 17.7), 37.08 (95% CI: 4.45, 309.3), 51.35 (95% CI: 5.97, 441.5) at dose levels of 0.1–49, 50–999, 1,000–2,999 and >3,000 cGy.

280. Schwartz et al. [S25] conducted a cohort study of 4,171 British and French survivors of solid childhood cancer treated between 1985 and 1995. The mean absorbed dose to the bone surfaces estimated using patient data from medical records was 6.8 Gy. The ERR per unit dose for bone sarcoma following radiotherapy was 1.78 (95% CI: 0.62, 5.94) $Gy^{-1}$, adjusting for sex, age at first cancer diagnosis, type of first cancer, and chemotherapy administration. The risk of bone sarcoma increased slowly up to a cumulative organ absorbed dose of 15 Gy and then more significantly at higher radiation doses compared with patients not treated with radiotherapy.

## (d) *Results of the meta-analysis of risks of second primary sarcoma*

281. The primary random effects meta-analysis model pooled the log RR from the results of 11 non overlapping individual studies of risks of second primary sarcoma. Studies by Bechler et al. [B11]. Rubino et al. [R16], Strong et al. [S53] and Kuttesch [K33] were excluded from the meta-analysis because they either did not report ERRs or other risks, or because the ERRs were estimated based on activity rather than organ doses. The meta-analysis was only based on data for ERRs since information on EAR per unit dose in the studies reviewed was insufficient. The pooled ERR per unit dose estimated in the primary analysis was 0.01 was 0.33% (95% CI: 0.09, 0.63) $Gy^{-1}$. The primary analysis also estimated a value for $I^2$ of 47.7% (95% CI: 0.0, 73.9), indicating that this percentage of the variability in the estimates of the effect size of radiotherapy is due to the differences between studies. The Cochran's Q test ($p$=0.04) indicated a statistically significant heterogeneity between the individual studies. ERRs of second primary sarcoma per unit dose estimated in the primary analysis are shown in figure XVI.

282. A sensitivity analysis pooling the ERRs from the same 11 studies on the original scale resulted in a different, but statistically compatible, pooled estimate of the ERR per unit dose of 0.14 (95% CI: 0.04, 0.25) $Gy^{-1}$. All subsequent results for the pooled estimates of ERR per unit dose in further sensitivity and subgroup analyses were based on pooling the log RR. In leave-one-out diagnostics, the pooled ERRs per unit dose for the 10 remaining studies varied between 0.17 and 0.48 $Gy^{-1}$. In the sensitivity analysis that applied the linear uncorrelated model relevant to studies only publishing risk for specific dose categories, the pooled estimate of the ERR per unit dose was 0.35 (95% CI: 0.10, 0.66) $Gy^{-1}$, $I^2$ was 48.6% (95% CI: 0.0, 74.3) and the p-value for the Cochran's Q test was 0.04. None of the 11 studies were recognized to have potentially overlapping data.

Figure XVI. Risks of second primary sarcoma per unit dose to the bone marrow estimated in the primary analysis

The CIs were derived from the standard error of the point estimate based on the assumption of a log-normal distribution and may not coincide with the CI as originally published

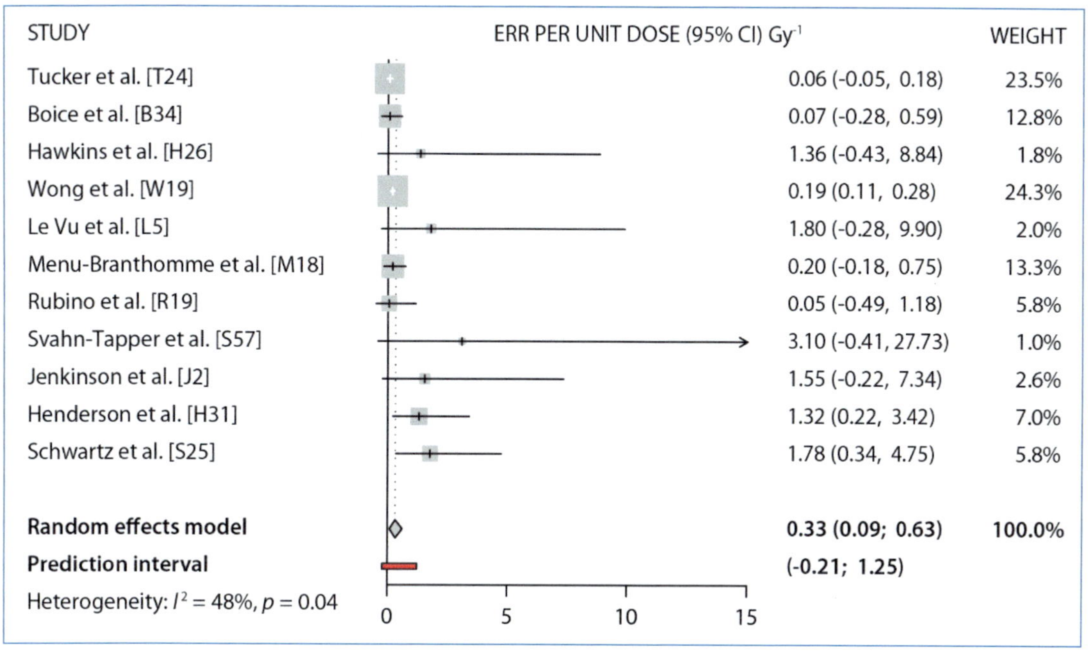


Figure XVII. Comparison of risks per unit dose estimated in the primary analysis and in sensitivity analyses for second primary sarcoma

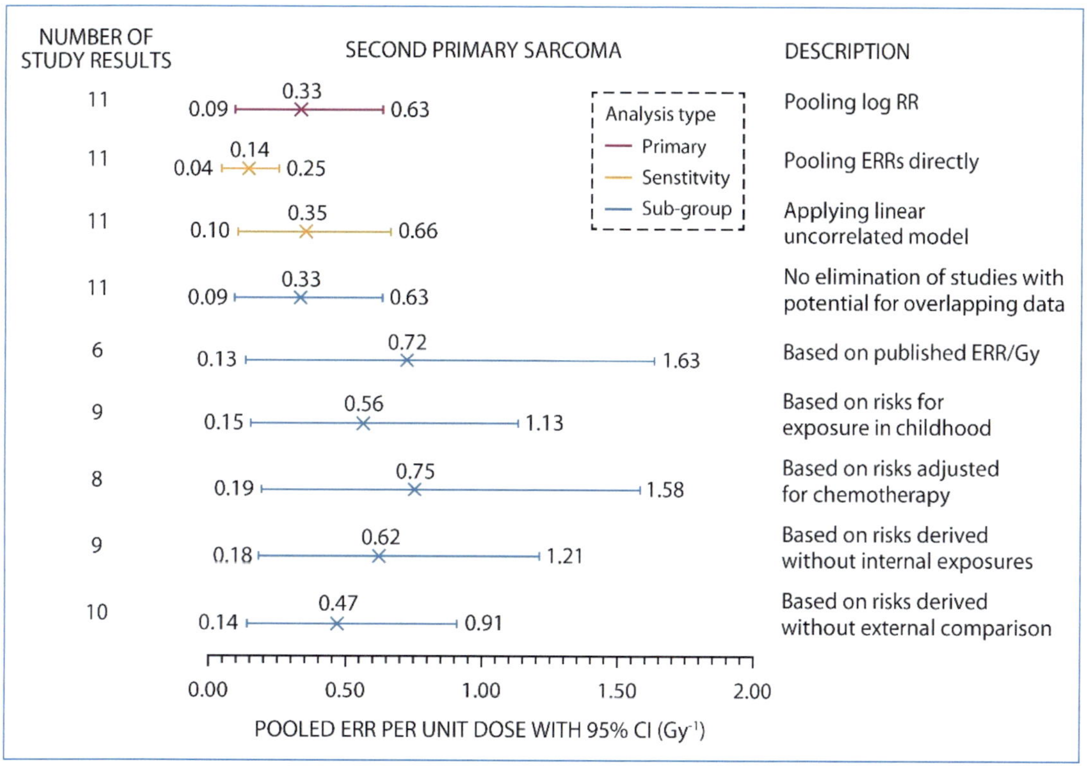

283. Six studies published ERR per unit dose estimates directly. Based on these risks alone, the pooled estimate of the ERR per unit dose was 0.72 (95% CI: 0.13, 1.63) $Gy^{-1}$. Nine studies, without potentially overlapping data, published risks only related to exposures in childhood or early adolescence. The pooled estimate of the ERR per unit dose based on data from these studies was 0.56 (95% CI: 0.15, 1.13) $Gy^{-1}$. Just two studies published risks related only to exposures in adulthood. Eight studies published risks related to exposures adjusted for chemotherapy, and the pooled estimate of the ERR per unit dose was 0.75 (95% CI: 0.19, 1.58) $Gy^{-1}$. Only two published risks related to exposed patients without chemotherapy. Risks in nine study were related to external exposures only and pooled estimate of the ERR per unit dose was 0.62 (95% CI: 0.18, 1.21) $Gy^{-1}$. If only the 10 studies without external comparisons of baseline rates were included in the subgroup analysis, the pooled estimate of the ERR per unit dose was 0.47 (95% CI: 0.14, 0.91) $Gy^{-1}$. These results are summarized in figure XVII. Results were not reported if fewer than three studies contributed to the meta-analysis.

### (e) *Summary*

284. Eleven studies selected for the literature review provided information on second primary sarcomas after radiation treatment. However, these studies generally included limited data due to the rarity of these cancers and heterogeneity of tumours. All studies described had cohort or nested case-control designs and were conducted in countries in North America or Europe. Several studies had not clearly distinguished the follow-up period from cancer diagnosis or radiation treatment. Some patients were treated with radiation and chemotherapy in combination. However, only a few studies classified patients according to the specific type of treatment.

285. The findings from this review confirmed that the dose-response relationship of second primary bone and soft-tissue sarcomas increases with radiation doses, as described in the UNSCEAR 2000 and 2006 reports [U3, U4]. Nevertheless, the dose-response relationships reported in the literature suggest a decrease in the relative risk at very high doses.

## 3. Female breast cancer (ipsilateral and contralateral)

### (a) *Studies with information on cumulative risks*

286. Studies that reported information on cumulative incidence of second primary breast cancer after radiotherapy are summarized in table 9. They include both studies in which radiation dosimetry of individual patients was not undertaken [B6, D3, E1, G21, M33, R4, S11, T4], and studies in which radiation dosimetry of individual patients was performed [G25, H42]. A systematic review of relevant publications was performed by Henderson et al. [H29]. Data from studies with more than 20 second primary breast cancers show that at time since first primary cancer of 30 years, the cumulative incidence is about 20%.

Table 9. Studies with information on cumulative incidence of second primary breast cancer after radiotherapy

| *Study reference* | *Country of study* | *Type of primary cancer* | *Number of survivors* | *Age at treatment (years)* | *Diagnosis time frame* | *Second primary cancer incidence* | *Cumulative incidence (%, 95% CI or ±SD)* | *Time frame[a] (years)* |
|---|---|---|---|---|---|---|---|---|
| Green et al. [G21] | United States | Hodgkin lymphoma | 182 | <20 | 1960–1989 | 4 | 11.89 (± 6.05) | 30 |
| Guibout et al. [G25] | France and the United Kingdom | Childhood cancer | 1 814 | <17 | 1946–1986 | 16 | 2.8 (1.0, 4.5)<br>5.1 (2.1, 8.2) | 30<br>by age 40 |
| Taylor et al. [T4] | United Kingdom | Hodgkin lymphoma | 383 | <15 | 1940–1991 | 16 | 12.2(4.3, 20.1) | 25 |
| Basu et al. [B6] | United States | Hodgkin lymphoma | 398 | <19 | 1960–1990 | 29 | 24 | 30 |
| Reulen et al. [R4] | United Kingdom | Childhood cancer | 8 093 | <15 | 1940–1991 | 81 | 0.2<br>1.2<br>2.7 | by age 30<br>by age 40<br>by age 50 |
| Hooning et al. [H42] | Netherlands (Kingdom of the) | Breast cancer | 7 425 | <71 | 1970–1986 | 503 contralateral | 6.5<br>12.3<br>17.1 | 10<br>20<br>30 |
| De Bruin et al. [D3] | Netherlands (Kingdom of the) | Hodgkin lymphoma | 1 122 | <51 | 1965–1995 | 120 | 19 (16, 23) overall<br>26 (19, 33) treated at age <21 | 30 |
| Henderson et al. [H29] | International | Review on childhood, adolescent or young adult cancer | over 14 000 | ≤30 | 1966–2008<br>1960–2000 | 422 | 13<br>20 | by age 40<br>by age 45 |
| Schellong et al. [S11] | Germany | Hodgkin lymphoma | 590 | <18 | 1978–1995 | 26 | 19 (12, 29) | 30 |
| Moskowitz et al. [M33] | United States | Childhood cancer | 1 230 | <21 | 1970–1986 | 203<br>(62 deaths) | 12 (8, 18) (mortality)<br>19 (13, 25) (mortality) | 5<br>10 |
| Ehrhardt et al. [E1] | United States | Childhood cancer | 1 467 | <23 | to 2010 | 68 | 1 (unexposed)<br>8 (≥10 Gy) | by age 35 |

[a] Time since primary cancer (unless otherwise specified).

## (b) *Studies with information on latency periods between treatment for first primary cancer and second primary breast cancer occurrence*

287. Studies that reported latency periods between the radiation treatment for first primary cancer and the occurrence of second primary breast cancer are summarized in table 10. The table shows that the shortest latency period for second primary breast cancers reported was generally five years and that the latency period extended to 37 years after treatment.

Table 10. Studies with information on latency periods between radiation treatment of first primary cancer and occurrence of second primary breast cancer

| *Study reference* | *First primary cancer* | *Latency period (years)* | |
|---|---|---|---|
| | | *Median* | *Range* |
| Travis et al. [T18] | Hodgkin lymphoma | | >5 |
| Rubino et al. [R17] | Neuroblastoma | 14, 27 and 37[a] | 14 |
| van Leeuwen et al. [V6] | Hodgkin lymphoma | 18.7 | |
| Guibout et al. [G25] | Hodgkin lymphoma | | >16 |
| | Neuroblastoma | | 13–37 |
| | Wilms' tumours | | 18–36 |
| Basu et al. [B6] | Paediatric Hodgkin lymphoma | | 9.4–36.1 |
| Hooning et al. [H42] | Breast cancer | 7.7 | 1.0–27.3 |
| Inskip et al. [I10] | Childhood cancer | 19.4 | 6.7–29.6 |
| Schellong et al. [S11] | Hodgkin lymphoma | 20.7 | >14.3 |
| Krul et al. [K27] | Hodgkin lymphoma | 21.9 | 16.9–26.8[b] |
| Veiga et al. [V13] | Childhood cancer | 15, 39[c] | |

[a] Original values for 3 cases reported in the paper.

[b] Inter-quartile.

[c] For first and second diagnosis, respectively.

## (c) *Studies with risk estimates with respect to breast doses*

288. Studies identified in the literature review that reported risks for second primary breast cancer (ipsilateral, contralateral) with respect to the relevant breast dose are described in this section. A summary of the main features of these studies is given in table 11, while full details are available in electronic attachment C. Some of these studies were found to have overlapping data since they analysed data for the same cohorts. To prevent a biased meta-estimator of the combined risks in the meta-analysis, references with overlapping data are identified in the paragraphs below.

Table 11. Studies identified in the literature review with risk estimates for second primary breast cancer

| *Study details* | *Treatment modality, fractionation details, dose type reported, RBE details* | *Number of subjects* | *Second primary breast cancer cases (controls, matched, unmatched)* | *Type of first primary cancer* | *Period of first cancer diagnosis* |
|---|---|---|---|---|---|
| International study of women with invasive cancer of the uterine cervix [B34] | Brachytherapy, EBRT, combined | 150 000 | 953 (1 806 matched) | Cervix | 1920–1965 |
| Connecticut breast cancer survivors, nested CACO [B35] | Several types | 41 109 | 655 (1 189 matched) | Breast | 1935–1982 |
| Danish breast cancer survivors cohort, nested CACO [S49] | Postoperative X-ray irradiation, McWhirter method | | 529 (529 matched) | Breast | 1943–1978 |
| French cohort [A4] | Brachytherapy, $^{131}$I, EBRT | 2 365<br>1 149<br>175<br>122 | 48<br>19<br>5<br>1 | Thyroid | 1934–1995 |
| French and British childhood cancer survivors study. [R17] | External beam radiotherapy, orthovoltage X-rays (200–250 kV), $^{60}$Co gamma rays, megavoltage X-rays (4–25 MV) and electrons (6–22 MV) | 544 | 3 | Neuroblastoma | 1948-1986 |
| Population-based cancer registry study in Iowa (US), Denmark, Finland, Sweden, Ontario (CA), and the Netherlands (Kingdom of the), CACO [T18] | Mantle only, mantle and subdiaphragmatic, mantle and additional supradiaphragmatic fields | 3 817 | 105 (266 matched) | Hodgkin disease | 1965–1994 |
| Netherlands Hodgkin survivors cohort, nested CACO [V6] | High-energy photons, usually 8 MeV in 78% of patients, remainder treated with orthovoltage X-rays, $^{60}$Co, or electrons | 770 | 48 (175 matched) | Hodgkins disease | 1965–1988 |
| France and United Kingdom Childhood Cancer Survivors Study [G25] | EBRT with mean of 20 fractions | 1 814 | 16 | Childhood cancer | 1946–1986 |
| Five Nordic countries cohort, CACO [S57] | Some orthovoltage beams, some high voltage X-ray and gamma rays, software package, Dos-EG | 196 + 567 | 24 (71) | Childhood cancer | 1960–1987 |
| Netherlands Late Effects Breast Cancer cohort [H42] | EBRT with electrons or photons (15-25 fractions) | 7 221 | 503 | Breast cancer | 1970–1986 |
| WeCare CACO [S52] | External-beam therapy (CB doses from thermoluminescent dosimeters in tissue-equivalent phantoms) | 1 806 | 708 (1 399 matched) | Breast cancer | 1985–1999 |
| Netherlands Hodgkin lymphoma survivors cohort, nested CACO [K27] | Standard Hodgkin protocols | 3 905 | 174/466 | Hodgkin lymphoma | 1965–2000 |
| North America Childhood Cancer Survivor Study, nested CACO. [V13] | | 14 358 | 271 (1 044 matched) | Childhood cancer | 1970–1986 |

289. Adjadj et al. reported the risk of developing breast carcinoma after treatment for thyroid carcinoma in a cohort of 2,365 women treated in three French cancer centres between 1934 and 1995. Forty-eight women in this cohort developed a subsequent second primary breast carcinoma by the end of 1997 [A4]. The mean absorbed dose delivered to the breasts by $^{131}$I and external radiotherapy was 0.7 Gy. A statistically insignificant second primary breast carcinoma risk per unit breast dose was reported with ERR per unit dose of –0.02 (95% CI: NA, 1.53) $Gy^{-1}$ ($p$=0.8).

290. Boice et al. conducted an international case-control study on second primary breast cancer risk and breast dose in women treated for invasive cancer of the uterine cervix between 1920 and 1965 [B34]. The study identified 953 cases and 1,806 controls drawn from 150,000 patients and the mean breast dose was 0.31 Gy. The RR for those receiving radiation treatment was 0.88 (95% CI: 0.7, 1.2) and three non-zero dose category specific risk estimates were given. Radiotherapy was linked to a significant 35% reduction in breast cancer risk in women with intact ovaries (561 cases, 1,037 controls), which the authors attributed with high likelihood to the cessation of ovarian function. There was a suggestion of a dose-response relationship among women without eanovaries, [B34].

291. In an ongoing series of papers from the British and French CCSS on patients treated between 1946 and 1986, Guibout et al. [G25] investigated the role of treatment and type of first cancer in the risk of second primary breast cancer in 1,814 three-year female survivors. The mean follow-up period was 16 years, and 16 patients developed a second primary breast cancer, 13 of them after radiotherapy. An ERR per unit dose to the breast of 0.13 (95% CI: <0, 0.75) $Gy^{-1}$ was reported and four non-zero dose category specific risk estimates were given. After stratification on castration of the ovaries and attained age, and adjusting for radiation dose, first cancer type, and chemotherapy, a higher risk of a second primary breast cancer was associated with a first cancer type of Hodgkin lymphoma (RR=7.0 (95% CI: 1.4, 30.9)). According to the authors this higher risk appears to be due not only to a higher radiation dose to the breasts, but also to a specific susceptibility.

292. Data from eight British and French treatment centres on 544 patients diagnosed with neuroblastoma between 1948 and 1986 before age 16 years and surviving at least five years, was collected to investigate the development of second primary cancer [R17]. The follow-up period finished at the end of 1990 and 1991 in the French and in the British centres, respectively. The mean radiation dose to the breast received by the three patients who developed a second primary breast cancer was 8.0 Gy compared to 5.0 Gy for the other patients. The corresponding mean ERR per unit dose to the breast dose was reported to be 0.13 (95% CI: <0, 0.7) $Gy^{-1}$. There is a potential data overlap between this study and that of Guibout et al. [G25].

293. Hooning et al. provided information on the roles of radiotherapy, chemotherapy, and family history in the development of second primary contralateral breast cancer in the Kingdom of the Netherlands Late Effects Breast Cancer Cohort, which included 7,221 patients, treated between 1970 and 1986 and followed up at least to the end of 1999 [H42]. A total of 503 cases of second primary contralateral breast cancer were identified with a median follow-up period of 13.8 years. For the subset of patients treated before the age of 45 years, the dose-response relationship for risk of any second primary contralateral breast cancer showed a linear ERR per unit dose of 0.21 (95% CI: 0.01, 0.61) $Gy^{-1}$, whereas the relationship was stronger for the risk of medially located contralateral breast cancers by mean dose to the medial part of the contralateral breast. The linear ERR per unit dose to the breast reported was 0.37 (95% CI: 0.04, 1.78) $Gy^{-1}$.

294. Krul et al. investigated risks of second primary breast cancer after radiotherapy for Hodgkin lymphoma in a nested case-control study among female five-year Hodgkin lymphoma survivors treated before age 41 years [K27]. The study included 174 second primary breast cancer cases and 466 controls. A linear dose-response relationship, based on breast dose with an adjusted EOR per unit dose of 0.061 (95% CI: 0.021, 0.154) $Gy^{-1}$,was reported.

295. Storm et al. reported data for a cohort of 56,540 women diagnosed with invasive breast cancer in Denmark between 1943 and 1978 that was used for a nested case-control study on the risk of second primary contralateral breast cancer occurring eight or more years after first diagnosis [S49]. Radiotherapy was not found to increase the overall risk of contralateral breast cancer. The study reported an RR of 1.04 (95% CI: 0.74, 1.46) and a mean dose to the contralateral breast of 2.51 Gy.

296. Stovall et al. reported results from the Women's Environmental, Cancer, and Radiation Epidemiology study, spanning the period between 1985 and 1999, of a case-control analysis undertaken to investigate risks of second primary breast cancer and doses to the contralateral breast from radiotherapy [S52]. A total of 708 cases (women with asynchronous bilateral breast cancer) and 1,399 controls (women with unilateral breast cancer) were counter-matched on radiation treatment. Participants were less than 55 years of age at first breast cancer. Women less than 40 years of age who received an absorbed dose of less than 1.0 Gy to the specific quadrant of the contralateral breast had a contralateral breast cancer RR of 2.5 (95% CI: 1.4, 4.5) compared to unexposed women. No excess risk was observed in women of more than 40 years of age. A mean dose to the contralateral breast of 1.1 Gy was reported.

297. Svahn-Tapper et al. conducted a case-control study nested in the cohort of patients from five Nordic countries who were less than 20 years old at first cancer diagnosis between 1960 and 1987, with second primary cancers diagnosed up to 1991 [S57]. For the 24 second primary breast cancers and the 71 controls, an additive ERR per unit dose to the breast of 0.7 (95% CI: 0.2, 2.9) $Gy^{-1}$ was reported.

298. Travis et al. [T18] investigated second primary breast cancer following radiotherapy and chemotherapy among young women with Hodgkin lymphoma in a matched case-control study of a cancer registry-based cohort from Canada (Ontario), Denmark, Finland, the Kingdom of the Netherlands, Sweden, and the United States (Iowa) [T18]. It is likely that data for this study overlapped with those in van Leeuwen et al. [V6]) The cohort had 3,817 female one-year survivors of Hodgkin lymphoma, diagnosed at age 30 years or younger, between 1 January 1965 and 31 December 1994, and the study period ended in 1998. Based on 105 cases and 266 controls, a RR of 3.2 (95% CI: 1.4, 8.2) was reported for patients with breast dose of 4 Gy or more compared to patients who received lower doses and no alkylating agents. The mean breast doses in the cases and the controls were 25.1 Gy (95% CI: 12.0, 61.3) and 21.1 Gy (95% CI: <0.1, 56.0), respectively.

299. Van Leeuwen et al. investigated second primary breast cancer following radiotherapy and chemotherapy among young women with Hodgkin lymphoma in a matched case-control study nested in the Netherlands cohort of Hodgkin lymphoma survivors between 1965 and 1988 [V6]. Based on 48 cases and 175 controls, a RR of 12.7 (95% CI: 1.8, 86) was reported for patients receiving 38.5 Gy or more compared to those receiving below 4 Gy. The ERR per unit dose to the breast was reported to be 0.03 (95% CI: 0.002, 0.06) $Gy^{-1}$ and the mean breast doses were 25.2 Gy for the cases and 22.1 Gy for the controls (range: 0.26–56 Gy).

300. In view of the data overlap between the study by Travis et al. [T18] and Van Leeuwen et al. [V6] it was decided that only the data on risks from Travis et al. [T18] would be used in the meta-analysis because of the higher number of cases and controls and the more comprehensive pooled data from six different registry sources contained in this publication.

301. In an ongoing series of papers from the North America CCSS, Inskip et al. [I10] published a case control analysis using data extracted from records on 6,647 women who were five-year survivors of childhood cancer treated between 1970 and 1986 and followed up to the end of 2001. The cases were 120 patients with histologically confirmed breast cancer and 464 matched controls were drawn. The OR for breast cancer rose linearly with radiation dose and reached an 11-fold increase for local breast doses of approximately 40 Gy relative to no radiation ($p$ for trend <0.0001). the risk associated with breast

irradiation was sharply reduced among women who received 5 Gy or more to the ovaries ($p$=0.002). It was reported that ovarian doses greater than 5 Gy appear to lessen the carcinogenic effects of breast irradiation, most likely by reducing exposure of radiation-damaged breast cells to stimulating effects of ovarian hormones.

302. Schonfeld et al. [S24] analysed a subset of 94 cases and 358 controls from the case-control data from Inskip et al. [I10] to compare dose reconstruction methods in the investigation of second primary breast cancers in the North America CCSS. The EORs per unit dose for breast cancer, using a conditional logistic regression, were 0.21 (95% CI: 0.05, 0.77) $Gy^{-1}$ and 0.10 (95% CI: 0.02, 0.44) $Gy^{-1,}$ using location-specific and body-region doses, respectively.

303. Veiga et al. further investigated the role of treatment and type of first cancer in the risk of second primary breast cancer in the North America CCSS [V13]. The study presented a case control analysis using data extracted from records on 14,358 women who were five-year survivors of childhood cancer, diagnosed between 1970 and 1986 and followed up to the end of 2016. There were 271 cases and 1,044 matched controls identified in this study. The OR per 10 Gy for breast cancer was 3.9 (95% CI: 2.5, 6.5) and was found to be similar for oestrogen receptor positive and oestrogen receptor negative cancers. For women who received ovarian doses less than 1 Gy, the OR per 10 Gy to the breast was higher (OR=6.8; (95% CI: 3.9, 12.5)) than for women who received ovarian doses greater than or equal to 15 Gy (OR=1.4; (95% CI: 1.0, 6.4)), confirming the earlier results from Inskip et al. [I10].

304. In view of the data overlap between the studies by Inskip et al. [I10], Veiga et al. [V13] and Schonfeld et al. [S24], it was decided that only the data on risks from Veiga et al. [V13] would be used in the meta-analysis, because of the highest number of cases and controls and the most recent follow-up contained in this publication.

305. Boice et al. conducted a nested case-control study, with 655 cases and 1,189 matched controls, using data from a cohort of 41,109 breast cancer survivors from Connecticut in the United States, treated between 1935 and 1982 [B35]. For women under age 45 years at treatment, a RR for contralateral second primary breast cancer of 1.59 (95% CI: 1.07, 2.35) was reported. Otherwise, the authors stated that there was little suggestion of risk for women aged 45 years and older at treatment. The mean dose to the contralateral breast was 2.82 Gy (3.02 Gy for the cases and 2.67 Gy for the controls) with a maximum dose of 7.10 Gy reported. Risk for specific dose categories were reported but due to a lack of reported CIs for these risks, could not be included in the meta-analysis based on risk for specific dose categories.

### (d) *Results of the meta-analysis of risks of second primary female breast cancer*

306. The primary random effects meta-analysis model pooled the log RR from the results of 12 non overlapping individual studies on risks of second primary female breast cancer. The meta-analysis was only based on data for ERRs since information on EAR per unit dose in the studies reviewed was insufficient. The pooled ERR per unit dose estimated in the primary analysis was 0.12 (95% CI: 0.04, 0.21) $Gy^{-1}$. The primary analysis also estimated a value for $I^2$ of 34.8% (95% CI: 0.0, 67.1) indicating that this percentage of the variability in the radiotherapy effect size estimates is due to the differences between studies. The Cochran's Q test ($p$=0.11) indicated no statistically significant heterogeneity between the individual studies. ERRs of second primary female breast cancer per unit dose estimated in the primary analysis are shown in figure XVIII.

307. A sensitivity analysis pooling the ERRs from the same 12 studies on the original scale resulted in only a slightly different pooled estimate of the ERR per unit dose of 0.09 (95% CI: 0.01, 0.17) $Gy^{-1}$. All subsequent results for the pooled estimate of the ERR per unit dose in further sensitivity and subgroup

analyses are based on pooling the log RR. In leave-one-out diagnostics, pooled ERRs per unit dose for the 11 remaining studies varied between 0.10 and 0.15. Omitting the Lönn et al. [L23] study, which was unusual in having an unexposed surgery-only comparison group, resulted in a ERR per unit dose of 0.12 (95% CI: 0.04, 0.21) $Gy^{-1}$. In the sensitivity analysis that applied the linear uncorrelated model relevant to studies only publishing risk for specific dose categories, the pooled estimate of the ERR per unit dose was 0.12 (95% CI: 0.04, 0.20) $Gy^{-1}$, $I^2$ was 50.5% (95% CI: 4.1, 74.4) and the p-value for the Cochran's Q test was 0.02. If all the results of the 16 studies were included in the meta-analysis without eliminating studies with potentially overlapping data, the pooled estimate of the ERR per unit dose was 0.11 (95% CI: 0.05, 0.18) $Gy^{-1}$.

Figure XVIII. Risks of second primary female breast cancer per unit dose to the breast estimated in the primary analysis

The CIs were derived from the standard error of the point estimate based on the assumption of a log-normal distribution and may not coincide with the CI as originally published

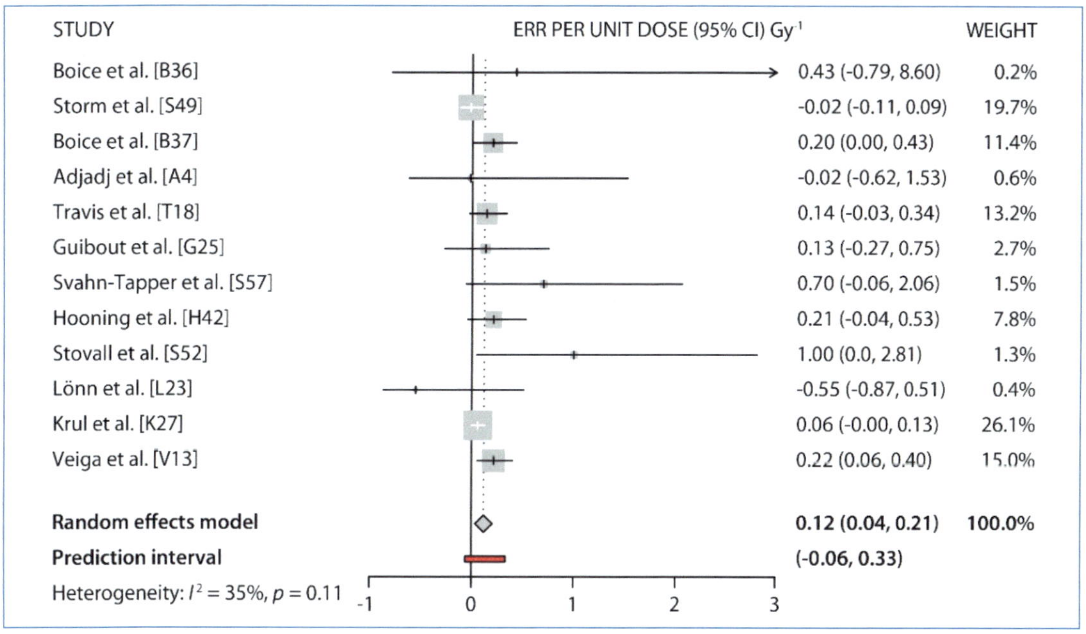


[a] For the Stovall et al. 2008 study, the ERR per unit dose value from a subset of patients with age at exposure <40 years was used but for the whole patient base, the risk is close to zero.

308. Six studies published estimates of ERR per unit dose directly. Based on these risks alone, the pooled estimate of the ERR per unit dose was 0.17 (95% CI: 0.00, 0.36) $Gy^{-1}$. Three studies, without potentially overlapping data, published risks related only to exposures in childhood or early adolescence, and the pooled estimate of the ERR per unit dose from these studies was 0.23 (95% CI: 0.08, 0.40) $Gy^{-1}$. Six studies published risks related only to exposures in adulthood; the pooled estimate of the ERR per unit dose from these studies was 0.13 (95% CI: –0.04, 0.32) $Gy^{-1}$. Three studies published risks related to exposed patients without chemotherapy; the pooled estimate of the ERR per unit dose from these studies was 0.19 (95% CI: 0.00, 0.42) $Gy^{-1}$. Eight studies published risks related to exposures adjusted for chemotherapy; the pooled estimate of the ERR per unit dose from these studies was 0.12 (95% CI: 0.03, 0.21) $Gy^{-1}$. Eleven studies published risks related to external exposures only; the pooled estimate of the ERR per unit dose from these studies was 0.12 (95% CI: 0.04, 0.21) $Gy^{-1}$. If only the 10 studies without external comparisons of baseline rates were included in the subgroup analysis, the pooled estimate of the ERR per unit dose was 0.12 (95% CI: 0.03, 0.21) $Gy^{-1}$. These results are summarized in figure XIX. Results were not reported if fewer than three studies contributed to the meta-analysis.

Figure XIX. Comparison of risks per unit dose estimated in the primary analysis and in the sensitivity analyses for second primary female breast cancer

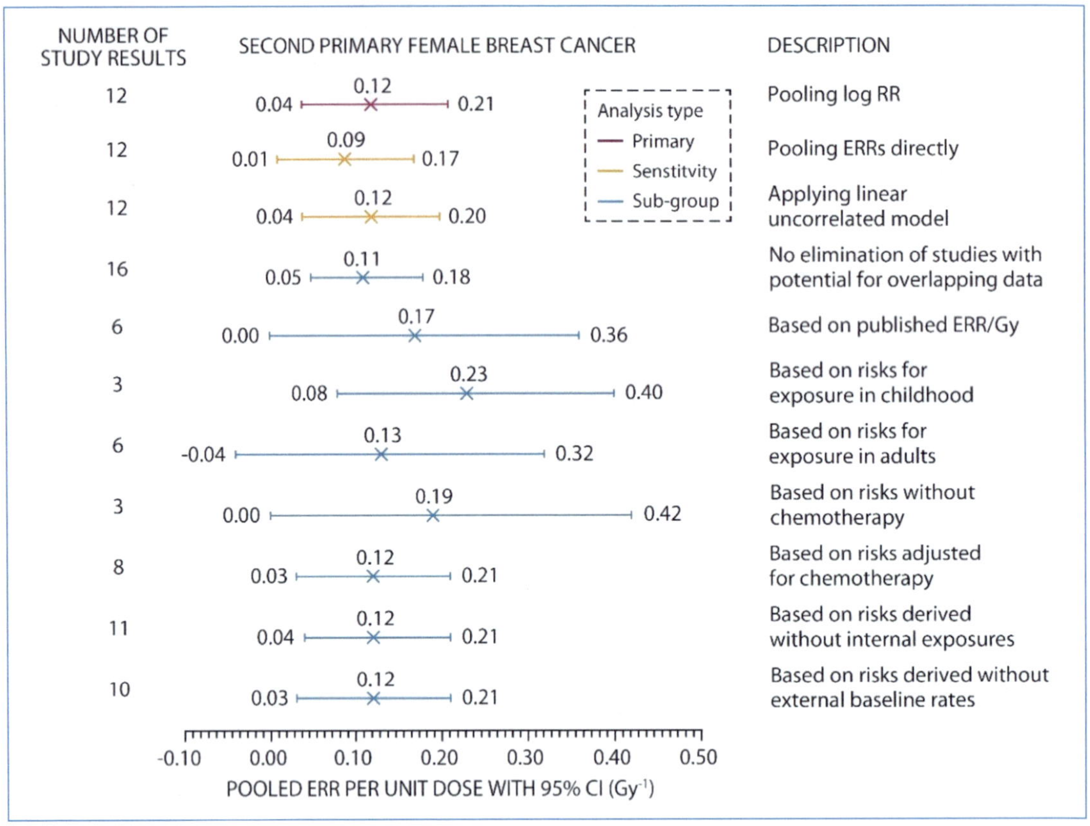


## (e) *Summary*

309. Sixteen studies identified in the literature review provided information on risks of second primary breast cancer to the breast after radiation treatment for first primary cancer. The studies described had either cohort or nested case-control designs and were conducted in countries in North America or Europe. It was not always possible to clearly distinguish the follow-up period from cancer diagnosis period, or period of radiation treatment or age at first radiation treatment—some studies reported that the lowest age at first treatment category was less than 45 years. Therefore, it was not considered feasible to extract a consistent set of information on these factors or to draw conclusions based on risk effect modification by these factors. Most studies reported results based on patients treated with radiation and chemotherapy together; therefore, ERRs per unit dose to the breast reported in these studies were adjusted for chemotherapy.

310. The cumulative incidence rates of second primary breast cancer after radiotherapy, usually reported in the literature at 25 to 30 years after the first cancers, was generally seen to vary widely (ranging from under 3% to 26%) depending on the type of first primary cancer, age at treatment, and interval from diagnosis, with higher values reported for younger age at treatments. In particular, the highest cumulative incidences were reported among Hodgkin lymphoma patients (up to 26%), and the lowest cumulative incidence (generally under 3%) among childhood cancer survivors.

311. The median latency period of second primary breast cancer from the treatment of the first primary cancer ranged from about 8 to 36 years, with most studies reporting latency periods of around 20 years.

The latency period varied depending on the type of first primary cancer type, with the shortest latency being reported for second primary contralateral breast cancer.

312. Although information about the shape of dose-response relationship of second primary breast cancer is limited, available data in the studies included in this evaluation suggest that this relationship is linear and increases with the dose, confirming the findings of the UNSCEAR 2000 and 2006 reports [U3, U4].

## 4. Lung cancer

### (a) *Studies with information on cumulative risks*

313. A study by Brenner et al. [B45] estimated 139 additional solid tumours among the 17,327 prostate cancer patients who underwent radiotherapy and survived more than five years, corresponding to a ratio of 1 per 125 men, or 1 per 465 PY at risk. Lung tumours were among the significant contributors to these additional solid tumours with 54 cases [B45]. The increase in risk of developing second primary lung cancer for prostate carcinoma patients treated with radiotherapy versus surgery only was 11% (95% CI: 1, 21) for all times after diagnosis. More specifically the increased risk was 22% (95% CI: 5, 42) more than five years after diagnosis and 42% (95% CI: 5, 93) more than 10 years after diagnosis.

### (b) *Studies with information on latency periods between treatment for first primary cancer and second primary lung cancer occurrence*

314. Studies that reported latency periods between the radiation treatment for first primary cancer and the occurrence of second primary lung cancer are summarized in table 12. The table shows that latency periods reported in the literature ranged from 1.1 to 30 years after treatment.

Table 12. Studies with information on latency periods between radiation treatment of first primary cancer and occurrence of second primary lung cancer

| *Study reference* | *First primary cancer* | *Latency period (years)* | | |
|---|---|---|---|---|
| | | *Mean* | *Median* | *Range* |
| Kaldor et al. [K2] | Hodgkin lymphoma | | | 3–10[a] |
| Inskip et al. [I8] | Breast cancer | 18 | | |
| Ng et al. [N13] | Hodgkin lymphoma | | | >5 |
| Travis et al. [T17] | Hodgkin lymphoma | 17.0, 10.6, 6.7[b] | | |
| Gilbert et al. [G6] | Hodgkin lymphoma | | | >5 |
| Prochazka et al. [P13] | Breast cancer | 17.6, 13.0[c] | | |
| Kaufman et al. [K7] | Breast cancer | | | 10–30 |
| Grantzau et al. [G19] | Breast cancer | | 12 | 1.1–25.8 |

[a] Inter-quartile.

[b] <40, 40-54, and >54 years old at diagnosis, respectively.

[c] Concordant and discordant lung, respectively.

### (c) *Studies with risk estimates with respect to lung doses*

315. Studies identified in the literature review that reported risks of second primary lung cancer with respect to the relevant lung dose are described in this section. A summary of the main features of these studies is given in table 13, while full details are available in electronic attachment C. Some of these studies were found to have overlapping since they analysed data for the same cohorts. To prevent a biased meta-estimator of the combined risks in the meta-analysis, references with overlapping data are identified in the paragraphs below.

316. Kaldor et al. [K2] conducted a case-control study on cancer registries based on Canadian and European patients from 1960. The study identified98 lung cancer cases, treated with chemotherapy and radiotherapy who had survived at least one year following a diagnosis of Hodgkin lymphoma and 259 matched controls. Among the 44 patients treated with radiotherapy alone, the RR of lung cancer was 1.4 (95% CI: 0.60, 3.55) for mean lung doses of 2.10 Gy (95% CI: 1, 2.5) and 1.6 (95% CI: 0.66, 4.12) for lung doses greater than over 2.5 Gy (mean dose: 3.9 Gy) [K2].

317. The study by Inskip et al. [I8] on lung cancer risk among women treated for breast cancer between 1935 and 1971 who survived for at least 10 years reported 61 cases of lung cancer meeting the criteria for inclusion. The overall RR of lung cancer associated with initial radiotherapy for breast cancer was 1.8 (95% CI: 0.8, 3.8) for 10-year survivors and 2.8 (95% CI: 1.0, 8.2) for 15-year survivors. Mean doses were 15.2 Gy to the ipsilateral lung, 4.6 Gy to the contralateral lung, and 9.8 Gy for both lungs combined. The ERRs per unit dose were 0.20 $Gy^{-1}$ and 0.08 $Gy^{-1}$, based on the mean dose to the affected lung only and to both lungs, respectively.

318. Travis et al. [T17] conducted an international case-control study nested in a population-based cohort of 19,046 Hodgkin lymphoma patients, diagnosed between 1965 and 1994. The study identified 222 patients who developed lung cancer up to 28 years after the initial treatment and 444 matched controls. A lung dose of 5 Gy or more, without alkylating agents, was associated with an increased lung cancer risk (RR=5.9 (95% CI: 2.7, 13.5)). Risk increased with increasing radiation dose ($p<0.001$). Excess risk after radiotherapy began five years after treatment and persisted for more than 20 years. Similar results were reported by Gilbert et al. [G6]. Their study is not, however, independent as it includes 199 cases and 393 controls from the study by Travis et al. [T17], to which 28 cases and 62 matched controls from an earlier study [V4] were added, bringing the total study population to 227 cases and 455 controls. A median radiation lung dose of 32 Gy was associated with an increased lung cancer risk (RR=5.8 (95% CI: 2.8, 13.5)) and an ERR per unit dose of 0.15 (95% CI: 0.06, 0.39) $Gy^{-1}$. Risks were available for a broad range of doses from 0 Gy to >40 Gy as in the study of Travis et al. [T17].

319. The study by Ng et al. [N13] reported the risk of second primary cancers for various organs including the lung in 1,319 patients with Hodgkin lymphoma treated with chemotherapy and/or radiotherapy with a median follow-up of 12 years. The RR was 4.9 (95% CI: 3.0, 7.1) and the EAR was 11 per $10^4$ PY, but the mean dose to the lung was not explicitly stated. The actual ERR per unit dose was not calculated, but as the patients received a standard treatment for Hodgkin lymphoma, the dose could be estimated approximately.

Table 13. Studies identified in the literature review with risk estimates for second primary lung cancer

| *Study details* | *Treatment modality, fractionation details, dose type reported, RBE details* | *Number of subjects* | *Second primary lung cancer cases (controls, matched, unmatched)* | *Type of first primary cancer* | *Period of first cancer diagnosis* |
|---|---|---|---|---|---|
| Canadian and European population-based CACO study [K2] | RT for Hodgkin lymphoma | 25 665 | 44<br>122 (RT only) | Hodgkin lymphoma | After 01 Jan 1960 |
| U.S. registry-based case-referent study [I8] | Adjuvant RT, 30–60 Gy, 5 days/week, 4–6 weeks; individually reconstructed dose distributions | 8 976 | 61<br>120 | Breast cancer | 1935–1971 |
| International population-based CACO study[T17] | RT for Hodgkin lymphoma (mantle, inverted Y), median dose 36 Gy | 19 046 | 222, 444<br>53, 90 (RT only) | Hodgkin lymphoma | 1965–1994 |
| International population-based CACO study[G6] | RT for Hodgkin lymphoma (mantle, inverted Y), median dose 36 Gy | 19 046 | 227<br>455 | Hodgkin lymphoma | 1965–1994? |
| Swedish case-only study [P13] | Different techniques including orthovoltage X-rays, $^{60}Co$ and MV photons<br>Details on total dose, dose/fx and doses to the lungs for all the techniques available | | 191 | Breast cancer | 1958–2000 |
| CACO nested in a United States registry-based cohort [L23] | External beam, brachytherapy and a combination of the two | 60 949 | 73<br>126<br>86 | Uterine corpus cancer | 1973–2003 |
| CACO nested in a Danish population-based cohort study [G19] | 13 Different RT techniques, individually reconstructed dose distributions | 23 627 | 151<br>443 | Breast cancer | 1982–2007 |
| Literature review + meta-analysis [T7] | | | | Breast cancer | |

320. A case-control study using the case only approach by Prochazka et al. [P13] reported a RR of 2.04 (95% CI: 1.24, 3.36) for second lung cancer in women treated for breast cancer (182 cases). The lung on the breast cancer side was considered as exposed, and the contralateral lung was considered as unexposed. The mean lung dose was 17.2 Gy (95% CI: 7.1, 32.0). The ERR per unit dose was 0.11 (95% CI: 0.02, 0.44) $Gy^{-1}$ for all women and 0.23 (95% CI: 0.04, 2.13) $Gy^{-1}$ for smokers.

321. Lönn et al. [L23] undertook a case-control study in 60,949 women treated for uterus cancer by external radiation, brachytherapy, surgery, or combined treatment. Patients were grouped into three dose categories: heavily, moderately, and lightly irradiated sites. Lung and bronchus were included in the lightly irradiated sites. The incidence ratio for patients with a combination of external and brachytherapy to lung and bronchus was 1.58 (95% CI: 1.23, 2.04) with a mean dose to the lung of 0.3 Gy (range: 0.16–0.45 Gy). The IRR was 1.17 (95% CI: 0.95, 1.43) with a mean dose to the lung of 0.3 Gy (range: 0.10–0.30 Gy) for patients treated with external beam therapy to lung and bronchus. For patients with brachytherapy to lung and bronchus, the IRR was 1.08 (95% CI: 0.83, 1.40) with a mean dose to the lung of 0.1 Gy (range: 0.06–0.15 Gy).

322. A more recent study by Grantzau et al., was conducted as a nested case-control study of second lung cancer in a population-based cohort of 23,627 early breast cancer patients treated with post-operative radiotherapy [G19]. The study found that for patients diagnosed with a second primary lung cancer, five or more years after breast cancer treatment, the rate of lung cancer increased linearly with the dose and the ERR per unit dose was 0.085 (95% CI: 0.031, 0.233) $Gy^{-1}$ for a mean lung dose of 8.7 Gy (range: 0.04–52.2 Gy). The cohort included 151 cases diagnosed with second primary lung cancer and 443 controls. The median latency period between breast cancer treatment and second lung cancer diagnosis was 12 years (range: 1–26 years). The OR of lung cancer among patients treated for early breast cancer were reported, including CIs, for 5 dose categories in the range from 0 Gy to >25 Gy and for three lag times between breast cancer diagnostic and the diagnosis of lung cancer.

323. Taylor et al. carried out a review and a meta-analysis of randomized trials of breast cancer radiotherapy including 134 lung cancer cases at least 10 years after radiotherapy reported in 40,781 women enrolled in 75 trials with 1983 as the median year of randomization [T7]. Various radiotherapy regimens were employed, and organ doses were reconstructed assuming a standard anatomy model. The study reported a RR of2.10 (95% CI: 1.48, 2.98) and an ERR per unit dose of 0.11 (95% CI: 0.05, 0.20) $Gy^{-1}$ for a lung mean dose of 5.7 Gy.

### (d) *Results of the meta-analysis of risks of second primary lung cancer*

324. The primary random effects meta-analysis model pooled the log RR from the results of six non overlapping individual studies of risks of second primary lung cancer. The meta-analysis was only based on data for ERRs since information on EAR per unit dose in the studies reviewed was insufficient. The pooled ERR per unit dose estimated in the primary analysis was 0.11 (95% CI: 0.04, 0.19). The primary analysis also estimated a value for $I^2$ of 0.0% (95% CI: 0.0, 74.6) indicating that none of the variability in the estimates of the effect size of radiotherapy is due to the differences between studies. The Cochran's Q test ($p$=0.77) indicated no statistically significant heterogeneity between the individual studies. ERRs of second primary lung cancer per unit dose estimated in the primary analysis are shown in figure XX.

Figure XX. Risks of second primary lung cancer per unit dose to the breast estimated in the primary analysis

The CIs were derived from the standard error of the point estimate based on the assumption of a log-normal distribution and may not coincide with the CI as originally published

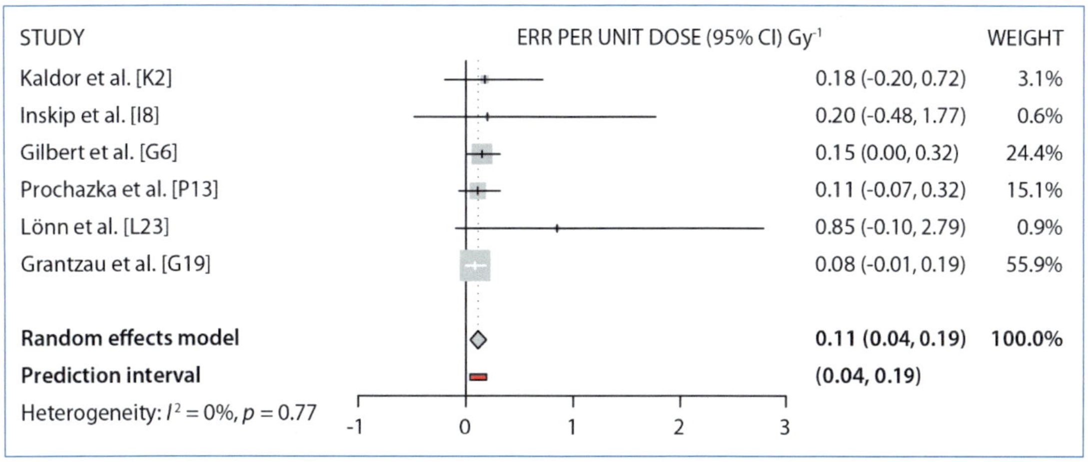


325. A sensitivity analysis pooling the ERRs from the same six studies on the original scale resulted in a very similar pooled estimate of the ERR per unit dose of 0.11 (95% CI: 0.03, 0.19) $Gy^{-1}$. All subsequent results for the pooled estimates of the ERR per unit dose in further sensitivity or subgroup analyses were based on pooling the log RR. In leave-one-out diagnostics, the pooled ERRs per unit dose for the five remaining studies varied between 0.10 and 0.15 $Gy^{-1}$. Omitting the Lönn et al. [L23] study, which was unusual in having an unexposed surgery-only comparison group resulted in an ERR per unit dose of 0.11 (95% CI: 0.04, 0.19) $Gy^{-1}$. In the sensitivity analysis that applied the linear uncorrelated model relevant to studies the reported only specific risks for dose categories, the pooled estimate of the ERR per unit dose was also 0.11 (95% CI: 0.04, 0.19) $Gy^{-1}$; $I^2$ was 0.0 (0.0, 74.6) and the p-value for the Cochran's Q test was 0.75. If all the eight studies were included in the meta-analysis without eliminating studies with potentially overlapping data, the pooled estimate of the ERR per unit dose was 0.11 (95% CI: 0.06, 0.17) $Gy^{-1}$.

326. Four studies published estimates of ERR per unit dose directly. Based on the results of these studies alone, the pooled estimate of the ERR per unit dose was 0.11 (95% CI: 0.03, 0.18) $Gy^{-1}$. There was not sufficient information for subgroup analyses of studies with published risks related to exposures only in childhood, early adolescence, only in adulthood, for patients without chemotherapy or for exposures adjusted for chemotherapy. Six studies published risks related to only external exposures and the pooled estimate of the ERR per unit dose was 0.11 (95% CI: 0.04, 0.19) $Gy^{-1}$. The pooled estimate of the ERR per unit dose of the subgroup analysis of the five studies without external comparisons of baseline rates was 0.11 (95% CI: 0.04, 0.19) $Gy^{-1}$. The results of the sensitivity analyses for second primary lung cancer are summarized in figure XXI. Results were not reported if fewer than three studies contributed to the meta-analysis.

Figure XXI. Comparison of risks per unit dose estimated in the primary analysis and in the sensitivity analyses for second primary lung cancer

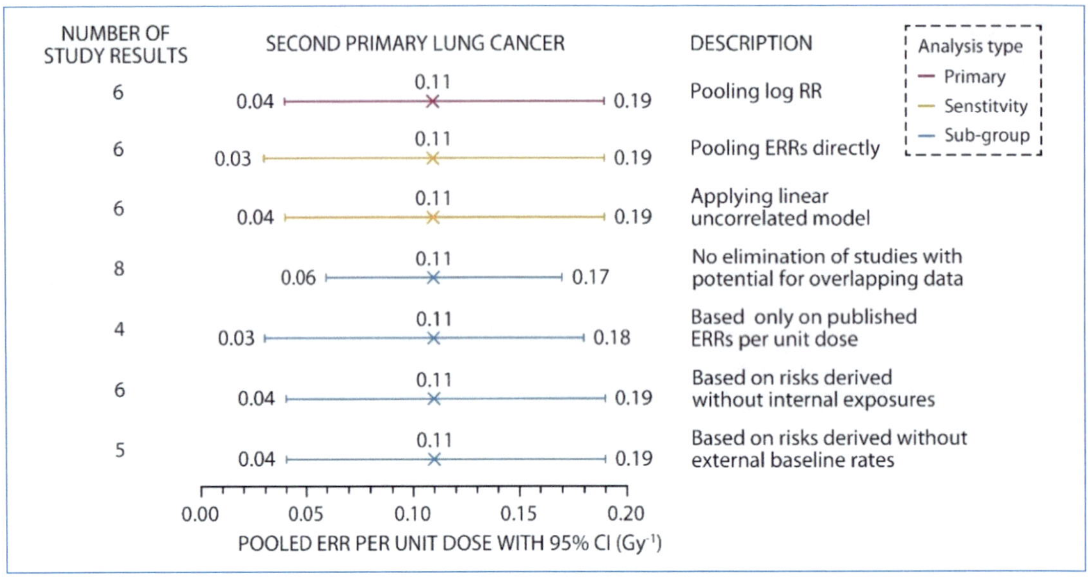


## (e) *Summary*

327. There were no studies on cumulative incidence of second primary lung cancer for which individual patient radiation dosimetry was undertaken. Eight studies provided information on risks of second primary lung cancer per unit lung dose after radiation treatment for first cancers. The studies described had either cohort or nested case-control designs and were conducted in countries in North America or Europe. It was not always possible to clearly distinguish the follow-up period from cancer diagnosis period, period of radiation treatment or age at first radiation treatment and only one study reported the lowest age at first cancer diagnosis as 34 years. Therefore, it was not feasible to extract a consistent set of information on these factors or to draw conclusions based on the modifications of risk effect by these factors or by sex. All studies reported results based on patients treated with radiation and chemotherapy together, but only one out of eight of the studies provided risks per unit lung dose adjusted for chemotherapy. The median latency period of second primary lung cancer from the treatment of the first primary cancer ranged from about 1 to 30 years with no clear differences by first primary cancer type.

328. Although information about the shape of dose-response relationship of second primary lung cancer is limited, available data in the studies included in this evaluation suggest that this relationship is linear and increases with the dose, confirming the findings of the UNSCEAR 2000 and 2006 Reports [U3, U4].

# 5. Gastrointestinal cancer

## (a) *Studies with information on cumulative risks*

329. Studies that reported information on cumulative incidence of second primary gastrointestinal cancer after radiotherapy are summarized in table 14. They include both studies in which radiation dosimetry of individual patients was not undertaken and studies in which radiation dosimetry of individual patients was performed. The table shows that, at time since primary cancer of 30 years, the cumulative incidence for all or a specific site of gastrointestinal cancers is of the order of 1%.

Table 14. Studies with information on cumulative incidence of second primary gastrointestinal cancer after radiotherapy

| *Study reference* | *Country of study* | *Type of primary cancer* | *Number of survivors* | *Age at treatment (years)* | *Diagnosis time frame* | *Second primary cancer incidences* | *Cumulative incidence (%, 95% CI or ±SD)* | *Time frame[a] (years)* |
|---|---|---|---|---|---|---|---|---|
| Van Leeuwen et al. [V3] | Netherlands (Kingdom of the) | Testicular cancer | 1 909 | Mostly (95%) 20 or more | 1971–1985 | 26 including<br>Stomach: 10<br>Colorectal: 8<br>Pancreas: 3 | 3.4 (2.1, 5.6) | 15 |
| Brenner et al. [B45] | United States | Prostate cancer | 51 584 | Mean: 70.3 (age at diagnosis) | 1970–1993 | Colon: 541<br>Rectum: 198 | Colon: 1.0<br>Rectum: 0.4 | Mean (follow-up years): 4.2 |
| Henderson et al. [H30] | United States | Childhood cancer | 14 358 | <21 | 1970–1986 | 45 | 0.64 (0.43, 0.86) | 30 |
| Nottage et al. [N18] | United States | Childhood cancer | 13 048 | <21 | 1960–2009 | Colon and rectum: 19 | 1.4 (± 0.53) | 40 |
| Tukenova et al. [T27] | France, Nordic countries, United Kingdom | Childhood cancer | 4 968 British and French<br>25 120 Nordic | <17 British and French<br><20 Nordic | 1995 British and French<br>1960–1987 Nordic | Colorectum: 19<br>Stomach: 8<br>Pancreas: 4 (British and French) | Colorectum: 0.4<br>Stomach: 0.2<br>Pancreas: 0.1 | Median: 25 |
| Morton et al. [M28] | Canada, Denmark, Finland, Netherlands (Kingdom of the), Norway, Sweden, United States | Hodgkin lymphoma | 17 447 | 11–83<br>Median: 30 | 1953–2003 | Stomach: 89 | 0.39 (0.28, 0.50)<br>0.92 (0.70, 1.13) | 15<br>30 |
| Dores et al. [D25] | Canada, Denmark, Finland, Netherlands (Kingdom of the), Norway, Sweden, United States | Hodgkin lymphoma | 19 882 | 12–76 | 1943–2003 | Pancreas: 43 | 0.16 (0.09, 0.23)<br>0.54 (0.36, 0.72) | 15<br>30 |
| Hauptmann et al. [H19] | Canada, Denmark, Finland, Netherlands (Kingdom of the), Norway, Sweden, United States | Testicular cancer | 22 269 | 18–71<br>Median: 38 | 1959–1987 | Stomach: 92 | 0.30 (0.20, 0.39)<br>1.45 (1.15, 1.74) | 15<br>30 |
| Hauptmann et al. [H20]. | Canada, Denmark, Finland, Netherlands (Kingdom of the), Norway, Sweden, United States | Testicular cancer | 23 982 | 19–73<br>Median: 40 | 1947–1991 | Pancreas: 98 | 0.14 (0.07–0.20)<br>1.08 (0.83, 1.34) | 15<br>30 |
| Kushner et al. [K32] | United States | Ductal or breast cancer | 755 | 25–89<br>Median: 55 | 1992–2001 | 12 | 0.1 (0.0, 1.0)<br>1.0 (0.5, 2.2)<br>1.9 (1.0, 3.6)<br>2.7 (1.5, 5.0) | 5<br>10<br>15<br>20 |

[a] Time since primary cancer (unless otherwise specified).

## (b) *Studies with information on latency periods between treatment for first primary cancer and occurrence of second primary gastrointestinal cancer*

330. Studies that reported latency periods between the radiation treatment for first primary cancer and the occurrence of second primary gastrointestinal cancer are summarized in table 15. The table shows that, in general, the shortest latency period for second primary gastrointestinal cancers reported in the literature was five years or more and that the latency period extended to 42 years after treatment.

Table 15. Studies with information on latency periods between radiation treatment of first primary cancer and occurrence of second primary gastrointestinal cancer

| *Study reference* | *First primary cancer* | *Second primary cancer* | *Latency period (years)* | | |
|---|---|---|---|---|---|
| | | | *Mean* | *Median* | *Range* |
| van Leeuwen et al. [V3] | Testicular cancer | Stomach | | 12.4 | 7–20 |
| De Vathaire et al. [D7] | Thyroid cancer | Digestive tract | 13 | | 2–29 |
| | | Colon | 13 | | 3–31 |
| | | Rectum | 19 | | 8–29 |
| Morton et al. [M27] | Breast cancer | Oesophagus | 13 | | 5–37 |
| Nottage et al. [N18] | Childhood cancer | | | 24.9 | 4.7–38.9 |
| Tukenova et al. [T27] | Childhood cancer | Oesophagus | | 25.0 | 23–28 |
| | | Stomach | | 17.0 | 2–45 |
| | | Colon | | 27.0 | 5–46 |
| | | Pancreas | | 20.5 | >38 |
| Kleinerman et al. [K22] | Cervical cancer | Stomach | | 17 | 5–42 |
| Morton et al. [M28] | Hodgkin lymphoma | | | 15 | |
| Dores et al. [D25] | Hodgkin lymphoma | | | 19 | 6.5–33 |
| Hauptmann et al. [H19] | Testicular cancer | | | 17 | 7–39 |
| Hauptmann et al. [H20] | Testicular cancer | | | 20 | 6–38 |
| Gilbert et al. [G7] | Hodgkin lymphoma, testicular, and cervical cancer | | 17.8 | 16.8 | 5.0–41.8 |

## (c) *Studies with risk estimates with respect to gastrointestinal doses*

331. Studies identified in the literature review that reported risks of second primary gastrointestinal cancer with respect to the relevant organ or tissue dose are described in this section. A summary of the main features of these studies is given table 16, while full details are available in electronic attachment C. Some of these studies were found to have overlapping since they analysed data for the same cohorts. To prevent a biased meta-estimator of the combined risks in the meta-analysis, references with overlapping data are identified in the paragraphs below.

332. In an international study of almost 150,000 patients treated for cancer of the cervix, Boice et al. applied a matched case-control design and reported RRs at 1 Gy of 1.69 (90% CI: 1.01, 3.25) for stomach

(n=348), 1.00 (0.99, 1.02) for colon (n=409), 1.02 (1.00, 1.04) for rectum (n=488) and 1.0 (0.72, 1.62) for pancreas. EARs of 3.16 (90% CI: 0.05, 10.4) per $10^5$ PY-Gy for stomach, 0.01 (-0.09, 0.18) per $10^5$ PY-Gy for colon, 0.06 (0.00, 0.16) per $10^5$ PY-Gy for rectum, and 0.00 (-0.65, 1.43) per $10^5$ PY-Gy for pancreas were also reported [B33].

333. A pooled analysis of the risk of second primary stomach cancer was conducted in an international case-control study of stomach cancer cases among Hodgkin lymphoma, testicular and cervical cancer survivors in Canada, Denmark, Finland, the Kingdom of the Netherlands, Norway, Sweden, and the United States [G7]. Data on 327 stomach cancer cases and 678 matched controls were analysed by conditional logistic regression, and the EOR per unit dose was 0.091 (95% CI: 0.036, 0.20) $Gy^{-1}$ for stomach cancer. EORs per unit dose for different groups of survivors were also reported: 0.049 (95% CI: 0.007, 0.16) $Gy^{-1}$ for survivors of Hodgkin lymphoma, 0.27 (95% CI: 0.054, 1.44) $Gy^{-1}$ for survivors of testicular cancer and 0.096 (95% CI: –0.002, 0.39) $Gy^{-1}$ for those of cervical cancer (*p* for homogeneity=0.25).

334. Risks of second primary stomach cancer were also reported in an international study of testicular cancer survivors in Canada, Denmark, Finland, the Kingdom of the Netherlands, Norway, Sweden, and the United States [H19]. Data on 92 stomach cancer cases and 180 matched controls were analysed by conditional logistic regression, and the EOR per unit dose for stomach cancer was 0.27 (95% CI: 0.054, 1.44) $Gy^{-1}$.

335. The risks of second primary cancer in various sites of the body were quantified in a cohort of 60,949 year or more survivors of uterine corpus cancer diagnosed between 1973 and 2003 in the SEER cancer registries [L23]. IRRs for cancers of the oesophagus, stomach, small intestine, colon, and rectum among those with external radiotherapy were 0.61 (95% CI: 0.13, 2.80), 1.42 (95% CI: 0.82, 2.46), 1.50 (95% CI: 0.65, 3.48), 1.44 (95% CI: 1.21, 1.73), and 1.46 (95% CI: 0.97, 2.19), respectively, in comparison with patients who had received surgery only. Corresponding IRRs among patients with brachytherapy were 1.83 (95% CI: 0.54, 6.18), 1.95 (95% CI: 1.10, 3.48), 0.80 (95% CI: 0.25, 2.57), 1.03 (95% CI: 0.82, 1.29), and 1.08 (95% CI: 0.65, 1.78). ERRs per unit dose were not available.

Table 16. Studies identified in the literature review with risk estimates for second primary gastrointestinal cancer

| *Study details* | *Treatment modality, fractionation details, dose type reported, RBE details* | *Number of subjects* | *Second primary gastrointestinal cancer cases (controls, matched, unmatched)* | *Type of first primary cancer* | *Period of first cancer diagnosis* |
|---|---|---|---|---|---|
| International study on patients treated for cancer of the cervix [B33] | Orthovoltage X-rays (200-400 kVp), megavoltage, external beam ($^{60}$Co y-rays), brachytherapy | 150 000 | Stomach: 348 cases and 658 controls<br>Small intestine: 22 cases and 44 controls<br>Colon: 409 cases and 759 controls<br>Rectum: 488 cases and 901 controls | Cervical cancer | 1920 to ≥ 1965 |
| Thyroid cancer patients cohort administered with $^{131}$I in two institutions in France [D7] | Diagnostic or therapeutic administration of $^{131}$I without external beam radiotherapy | 1 771 | 80 including colorectal cancer: 13 | Thyroid cancer | 1950–1989 |
| Pooled cohort study of thyroid cancer patients in France, Italy, and Sweden [R16] | $^{131}$I administration and external beam radiotherapy | 6 841 | 576 including colorectal cancer: 69 | Thyroid cancer | 1934–1995 |
| Cohort study of thyroid cancer patients in France and Italy [R18] | $^{131}$I administration and external beam radiotherapy, doses were estimated by ICRP conversion factors and Dos-EG software | 3 708 | Colorectal cancer: 15 cases and 76 controls | Thyroid cancer | 1933–1998 |
| Five Nordic countries cohort, CACO [S57] | Some orthovoltage beams, some high voltage X- and gamma rays, Dos-EG software | 25 120 | Digestive cancer: 22 cases and 60 controls | Childhood cancer | 1960–1987 |
| Uterine corpus cancer survivors identified from Surveillance, Epidemiology and End Results Program (SEER) [L23] | Brachytherapy or external radiotherapy or combined | 60 949 including 37 534 with any radiotherapy | Oesophagus:<br>Ext. beam therapy group: 2<br>Brachytherapy group: 5<br>Stomach:<br>Ext. beam therapy group: 19<br>Brachytherapy group: 23<br>Small intestine:<br>Ext. beam therapy group: 8<br>Brachytherapy group: 4<br>Colon:<br>Ext. beam therapy group: 178<br>Brachytherapy group: 117<br>Rectum:<br>Ext. beam therapy group: 34<br>Brachytherapy group: 24 | Uterine corpus cancer | 1973–2003 |

| *Study details* | *Treatment modality, fractionation details, dose type reported, RBE details* | *Number of subjects* | *Second primary gastrointestinal cancer cases (controls, matched, unmatched)* | *Type of first primary cancer* | *Period of first cancer diagnosis* |
|---|---|---|---|---|---|
| International nested case-control study within a cohort of breast cancer survivors [M27] | Treated with medial and lateral tangential or direct chest radiotherapy fields (including boost doses), with/without supraclavicular, direct internal mammary chain (IMC), and axillary fields<br>Beam energies included $^{60}$Co gamma rays, orthovoltage X-rays (100–400 kVp), photons (4–18 MV), and electrons (6–20 MeV)<br>Cumulative target doses to the breast and/or peripheral regions were typically 40–50 Gy, using conventional fractionation. | 289 748 | Oesophagus: 252 cases and 488 controls | Breast cancer | 1943–2003 |
| Case-control study in a cohort of childhood cancer survivors at St Jude Children's Research Hospital [N18] | Not available | 13 048 | Colorectal: 19 cases and 220 controls | Childhood cancer including ALL, AML, CML, Hodgkin lymphoma | 1960–2009 |
| Case-control study in British-French and Nordic cohorts of childhood cancer survivors [T27] | Not available | Total: 29 688<br>French-British cohort: 4 568<br>Nordic cohort: 25 120 | Digestive: 58 cases and 167 controls | Childhood cancer | 1946–1985 |
| Childhood Cancer Survivor Study [B39] | Not available | 14 135 | Salivary gland: 23 | Childhoood cancer including leukaemia, HL, cancers of CNS, kidney, soft tissue, and bone, NHL, and neuroblastoma | 1970–1986 |
| International case-control study in a cohort of cervical cancer survivors [K22] | External beam treatments typically were given in 1.5 Gy to 3.0 Gy fractions over 4–6 weeks for a total dose of 30–50 Gy<br>Energies included orthovoltage, 160–400 kVp (47.9%), $^{60}$Co (19.1%), megavoltage photons 4-45 MV (21.3%), combined energies (4.3%), and unknown energy (7.4%) | 53 547 | Stomach: 201 cases and 378 controls | Cervical cancer | 1943–1995 |

| *Study details* | *Treatment modality, fractionation details, dose type reported, RBE details* | *Number of subjects* | *Second primary gastrointestinal cancer cases (controls, matched, unmatched)* | *Type of first primary cancer* | *Period of first cancer diagnosis* |
|---|---|---|---|---|---|
| International case-control study of stomach cancer in a cohort of Hodgkin lymphoma survivors in seven countries [M28] | Patients generally were treated with mantle fields, with or without sub-diaphragmatic fields, with cumulative target doses of 25–45 Gy using conventional fractionation. | 19 882 | Stomach: 89 cases and 190 controls | Hodgkin lymphoma | 1953–2003 |
| International study of Hodgkin lymphoma survivors [D25] | The most commonly administered radiation treatment was the mantle field (52% of patients) with cumulative target doses of 25–45 Gy using conventional fractionation | 19 882 | Pancreas: 36 cases and 70 controls | Hodgkin lymphoma | 1953–2003 |
| International studies of testicular cancer survivors [H19] | Daily target doses were 1.8–2.0 Gy, resulting in cumulative doses typically ranging from 25 Gy to 50 Gy. | 22 269 | Stomach: 92 cases and 180 controls | Testicular cancer | 1959–1987 |
| International studies of testicular cancer survivors [H20] | Patients were generally treated with dog-leg fields (para-aortic and ipsilateral iliac nodes) or para-aortic fields only<br>Daily target doses were 1.8–2.0 Gy resulting in cumulative doses ranging between 25–50 Gy | 23 982 | Pancreas: 80 cases and 145 controls | Testicular cancer | 1947–1991 |
| Pooled analysis of three international studies of Hodgkin lymphoma, testicular cancer and cervical cancer survivors [G7] | Cervical cancer survivors were commonly treated with both external beam therapy and brachytherapy, but the dose contributed by brachytherapy was not included in analyses for comparability with the Hodgkin lymphoma and testicular cancer studies (in which brachytherapy was not used). | Number unknown | Stomach: 327 cases and 678 controls | Hodgkin lymphoma<br>Testicular cancer<br>Cervical cancer | 1943–1995<br>Mean: 1969.8,<br>Median: 1970 |
| French Childhood Cancer Survivors Study (CCSS), five-year survivors children treated for first primary cancer [A22] | Not available | 7 032 | Colorectal: 36 cases (21 colon, 16 rectum) and 140 controls | Childhood cancer: Hodgkin lymphoma; Non-Hodgkin lymphoma; Soft-tissue sarcoma; Nephroblastoma; Central nervous system/Neuroblastoma; Other cancers (gonadal, thyroid tumour, and others) | 1947–1994 |

336. The risk of second primary stomach cancer was reported in an international case-control study (201 cases and 378 controls) within a cohort of 53,547 cervical cancer survivors in Canada, Denmark, Finland, Sweden, and the United States [K22]. The analysis showed that the estimate of the EOR per unit dose for second primary stomach cancer was 0.11 (95% CI: 0.001, 0.48) $Gy^{-1}$.

337. An international case-control study (89 cases and 190 controls) of stomach cancer was conducted in a cohort of 19,882 Hodgkin lymphoma survivors in seven countries [M28]. The EOR per unit dose for second primary stomach cancer was 0.09 (95% CI: 0.04, 0.21) $Gy^{-1}$.

338. Allodji et al. reported the risks of second primary colorectal cancer in 36 patients, including 21 colon and 15 rectum cancers, after treatment for childhood cancer using a nested case-control design with data from French CCSS [A22]. Using a mean dose within the colorectal cancer site and adjusting for chemotherapy agents and regimens, the EOR per unit dose was 0.13 (97.5% CI: 0.02, 0.48) $Gy^{-1}$ for colorectal cancer, 0.20 (97.5% CI: 0.02, 1.36) $Gy^{-1}$ for colon cancer, and 0.08 (97.5% CI: –0.02, 0.75) $Gy^{-1}$ for rectal cancer.

339. The risk of second primary salivary gland cancer for a group of 23 patients in the North American CCSS was reported by Boukheris et al. [B39]. Radiation doses to the salivary glands were estimated from medical records and a cohort design with Poisson regression analysis was used to assess risks with respect to radiation dose, adjusting for other potential confounders. The overall EAR for salivary gland cancer was 9.8 per $10^5$ PY, and the ERR per unit dose to the salivary gland was 0.36 (95% CI: 0.06, 2.5) $Gy^{-1}$.

340. In a study conducted by de Vathaire et al. of 1,771 patients treated for thyroid cancer at two French hospitals, 1,497 of whom had been treated with $^{131}I$ and none had external radiotherapy, 13 patients were reported with second primary colorectal cancer [D7]. The ERR following administration of $^{131}I$ among the 1,275 patients followed five years or more was 0.34 (90% CI: 0.05, 1.09) $GBq^{-1}$ for digestive cancer (n=12) and 0.47 (90% CI: 0.1, 1.6) $GBq^{-1}$ for colorectal cancer (n=11). The authors reported that the elevated risks were probably caused by the accumulation of $^{131}I$ in the colon lumen.

341. An international cohort study of 6,841 thyroid cancer patients in France, Italy, and Sweden was conducted by Rubino et al. [R16], and risks of second primary cancers (n=576) including colorectal cancer (n=69) were reported. The ERR following administration of $^{131}I$ was 0.10 (95% CI: 0.08, 0.27) $GBq^{-1}$ for colorectal cancer among all patients, and it was 0.02 (95% CI: NA, 0.20) $GBq^{-1}$ and 0.15 (95% CI: 0.02, 0.38) $GBq^{-1}$ among patients with and without external radiotherapy.

342. A case-control study of colorectal cancer was conducted in an international cohort of 3,708 thyroid cancer patients in France and Italy [R18]. Data on 15 cases and 76 matched controls were analysed and ORs for second primary colorectal cancer were 1.8 (95% CI: 0.2, 23) and 9.4 (95% CI: 0.4, 214) for dose categories of 0–0.1 Gy and >0.1 Gy in comparison with no radiotherapy group.

343. A hospital-based case-control study (19 cases and 220 controls) was conducted in a cohort of 13,048 childhood cancer survivors, and the risk of second primary colorectal cancer was reported [N18]. Logistic regression analyses showed ORs for secondary colorectal cancer of 1.7 (95% CI: 1.2, 2.5) and 1.7 (95% CI: 1.1, 2.9) for every 10 Gy dose increment compared with two different control groups, which correspond to EOR estimates of 1.3 (95% CI: −0.5, 3.1) and 0.8 (95% CI: −0.6, 2.2).

344. For the same cohort of testicular cancer survivors included in the study by Hauptmann et al. [H20], the risk of second primary pancreas cancer was also analysed based on 80 pancreatic cancer cases and 145 matched controls. The EOR per unit dose for pancreatic cancer was 0.12 (95% CI: 0.03, 0.42) $Gy^{-1}$.

345. Dores et al. conducted an international case-control study within a cohort of 19,882 Hodgkin lymphoma survivors diagnosed between 1953 and 2003 including 36 cases of second primary pancreatic

cancer and 70 matched controls [D25]. The EOR per unit dose for pancreatic cancer was 0.098 (95% CI: 0.015, 0.42) $Gy^{-1}$ with respect to the mean unit radiation dose to the part of the pancreas (head, body, tail) where the tumour was located for cases and to the corresponding anatomical site in matched controls.

346. Morton, et al. conducted an international nested case-control study (252 cases and 488 controls) of oesophagus cancer within a cohort of 289,748 breast cancer survivors in Canada, Denmark, Finland, Sweden, and the United States [M27]. The risk of second primary oesophagus cancer increased with increasing radiation dose to the tumour location ($p$ for trend <0.001), with doses of more than 35 Gy associated with an OR of 8.3 (95% CI: 2.7, 28). The ERR per unit dose was not available, but risks for specific dose categories were reported.

347. Svahn-Tapper et al. conducted a case-control study of digestive cancer (22 cases and 60 controls) nested in a cohort of 25,120 childhood cancer survivors in Denmark, Finland, Iceland, Norway, and Sweden, and analysed risks of second primary digestive cancer [S57]. The ERR per unit dose for second primary cancer of digestive tract was 0.1 (95% CI: 0.0, 0.8) $Gy^{-1}$.

348. The risk of second digestive cancer was also reported in a case-control study (58 cases and 167 controls) in France, the United Kingdom, and five Nordic cohorts of 29,688 childhood cancer survivors [T27]. The EOR per unit dose for digestive cancers was 0.13 (95% CI: 0.05, 0.32) $Gy^{-1}$ for both sexes, with no significant difference between men (0.15 (95% CI: 0.05, 0.35) $Gy^{-1}$) and women (0.18 (95% CI: −0.01, 0.44) $Gy^{-1}$).

### (d) *Results of the meta-analysis of risks of second primary gastrointestinal cancer*

349. The primary random effects meta-analysis model pooled the log RR from the results of 19 non overlapping individual studies of risks of second primary gastrointestinal cancer. The meta-analysis was only based on data for ERRs since information on EAR per unit dose in the studies reviewed was insufficient. The pooled ERR per unit dose estimated in the primary analysis was 0.04 (95% CI: 0.01, 0.07). The primary analysis also estimated a value for $I^2$ of 47.7% (95% CI: 10.8, 69.3), indicating that this percentage of the variability in the estimates of the effect size of radiotherapy is due to the differences between studies. The Cochran's Q test ($p$=0.01) indicated statistically significant heterogeneity between the individual studies. ERRs of second primary gastrointestinal cancer per unit dose estimated in the primary analysis are shown figure XXII.

350. A sensitivity analysis pooling the ERRs from the same 19 studies on the original scale resulted in only a slightly different pooled estimate of the ERR per unit dose of 0.03 (95% CI: 0.01, 0.06) $Gy^{-1}$. All subsequent results for the pooled estimates of the ERR per unit dose in further sensitivity and subgroup analyses were based on pooling the log RR. In leave-one-out diagnostics, the pooled ERRs per unit dose for the 18 remaining studies varied between 0.02 and 0.05 $Gy^{-1}$. The highest ERR per unit dose of 0.05 (95% CI: 0.02, 0.09) $Gy^{-1}$ was obtained by leaving out the data from the study on rectal cancer by Lönn et al. [L23], since this study was unusual in having an unexposed surgery-only comparison group. In the sensitivity analysis that applied the linear uncorrelated model relevant to studies the reported only risks for specific dose categories, the pooled estimate of the ERR per unit dose was 0.05 (95% CI: 0.02, 0.08) $Gy^{-1}$, $I^2$ was 58.3% (95% CI: 30.7, 74.9) and the p-value for the Cochran's Q test was 0.001. If all of the 21 study results were included in the meta-analysis without eliminating studies with potentially overlapping data, the ERR per unit dose pooled estimate was 0.05 (95% CI: 0.02, 0.08) $Gy^{-1}$.

Figure XXII. Risks of gastrointestinal cancers per unit dose estimated in the primary analysis

The CIs were derived from the standard error of the point estimate based on the assumption of a log-normal distribution and may not coincide with the CI as originally published

| STUDY | ERR PER UNIT DOSE (95% CI) $Gy^{-1}$ | WEIGHT |
|---|---|---|
| Boice et al. (Stomach) [B35] | 0.11 (-0.42, 1.11) | 0.2% |
| Boice et al. (Colon) [B35] | -0.00 (-0.02, 0.01) | 16.8% |
| Boice et al. (Rectum) [B35] | 0.01 (-0.01, 0.04) | 15.5% |
| Boice et al. (Pancreas) [B35] | -0.07 (-0.54, 0.89) | 0.2% |
| Rubino et al. [R18] | 45.95 (0.02, 2 150.50) | 0.0% |
| Svahn-Tapper et al. [S57] | 0.10 (-0.18, 0.48) | 0.9% |
| Lönn et al. (Stomach) [L23] | 0.12 (-0.09, 0.38) | 1.7% |
| Lönn et al. (Small intestine) [L23] | 0.01 (0.02, 0.05) | 13.9% |
| Lönn et al. (Rectum) [L23] | 0.01 (-0.00, 0.02) | 16.9% |
| Lönn et al. (Pancreas) [L23] | -0.06 (-0.28, 0.23) | 1.0% |
| Morton et al. [M27] | 0.09 (0.03, 0.15) | 10.6% |
| Nottage et al. [N18] | 1.30 (-0.20, 5.59) | 0.1% |
| Tukenova et al. [T27] | 0.13 (0.01, 0.27) | 4.5% |
| Boukheris et al. [B41] | 0.36 (-0.25, 1.47) | 0.2% |
| Kleinerman et al. [K22] | 0.11 (-0.09, 0.35) | 1.9% |
| Dores et al. [D25] | 0.10 (-0.07, 0.30) | 2.5% |
| Hauptmann et al. [H20] | 0.12 (-0.05, 0.32) | 2.7% |
| Gilbert et al. [G7] | 0.09 (0.01, 0.17) | 8.0% |
| Allodji et al. [A22] | 0.13 (-0.0.4, 0.33) | 2.6% |
| **Random effects model** | **0.04 (0.01, 0.07)** | **100.0%** |
| **Prediction interval** | **(-0.04, 0.13)** | |

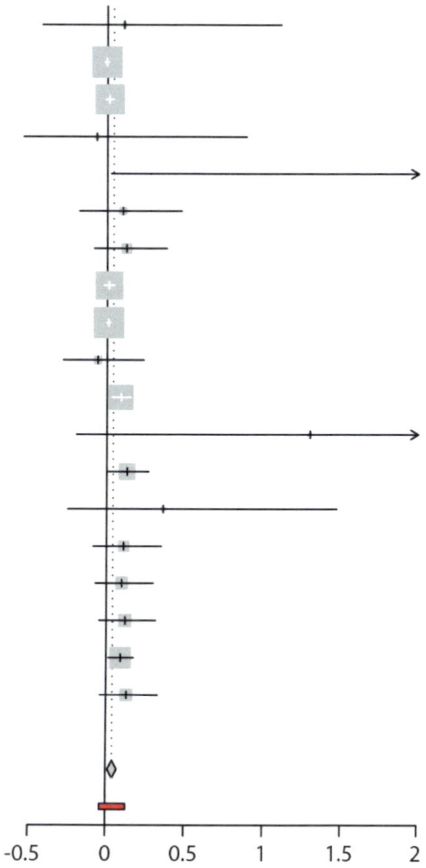


Heterogeneity: $I^2 = 48\%, p = 0.01$

351. Ten studies contained direct estimates of ERR per unit dose. Based on these studies alone, the pooled estimate of the ERR per unit dose was 0.10 (95% CI: 0.06, 0.14) $Gy^{-1}$. Risks in four publications related only to exposures in childhood or early adolescence for which the pooled estimate of the ERR per unit dose was 0.13 (95% CI: 0.04, 0.24) $Gy^{-1}$. Results of nine studies were for risks only related to exposures in adulthood for which the pooled estimate of the ERR per unit dose was 0.01 (95% CI: −0.00, 0.02) $Gy^{-1}$. Results of four studies were related to exposed patients without chemotherapy for which the pooled estimate of the ERR per unit dose was 0.00 (95% CI: −0.02, 0.02) $Gy^{-1}$. Results of four studies were related to exposed patients without chemotherapy for which the pooled estimate of the ERR per unit dose was 0.00 (95% CI: −0.02, 0.02) Gy-1, while four study published risks related to exposures adjusted for chemotherapy with a pooled estimate ERR per unit dose of 0.11 (95% CI: 0.04, 0.17) $Gy^{-1}$. Risks in twelve publications were related pertained to only external exposures and the pooled estimate of the ERR per unit dose was 0.06 (95% CI: 0.01, 0.10) $Gy^{-1}$. Based on the subgroup analysis of the results of the 15 studies without external comparisons of baseline rates, the pooled estimate of the ERR per unit dose was also 0.06 (95% CI: 0.02, 0.11) $Gy^{-1}$. The results of the sensitivity analysis are summarized in figure XXIII. Results were not reported if fewer than three studies contributed to the meta-analysis.

Figure XXIII. Comparison of risks per unit dose estimated in the primary analysis and in the sensitivity analyses for second primary gastrointestinal cancer

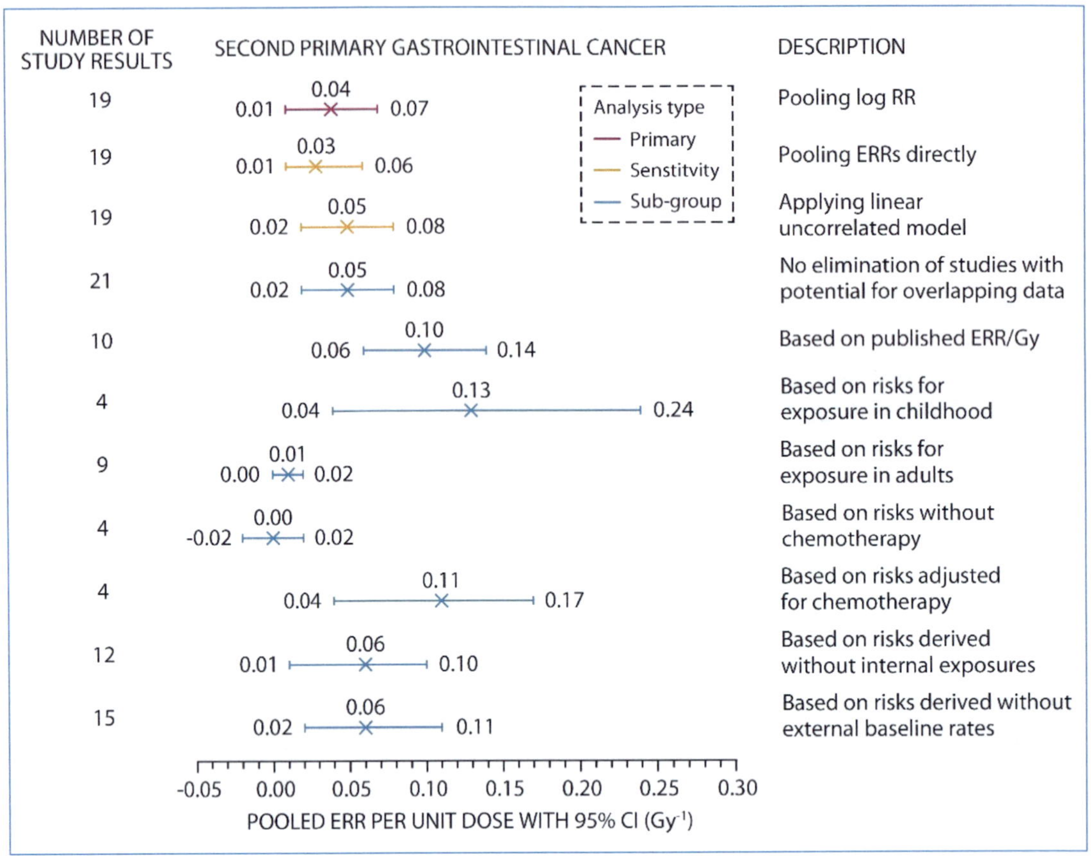


## (e) *Summary*

352. Nineteen studies identified in the literature review provided information on risks of second primary gastrointestinal cancer in relevant organs after radiation treatment for first cancers. This cancer type covers a broad range of organs, from salivary glands to rectum, with different radiosensitivities and magnitudes of risk, which could affect the meta-analysis result. The studies described had either cohort or nested case-control designs, and were conducted in countries in North America or Europe. It was not always possible to clearly distinguish the follow-up period from cancer diagnosis, or period of radiation treatment or age at first radiation treatment. Therefore, it was not feasible to assess effects modification of these factors on risks of radiation-induced second gastrointestinal cancer. Most of the studies reported results based on patients treated with radiation and chemotherapy together, but all studies except one reporting risks per unit dose provided risks adjusted for chemotherapy.

353. Several studies reported cumulative incidence rates of second primary gastrointestinal cancer after first cancer treated with radiotherapy. The cumulative incidence rates varied according to cancer types, cohorts, and other factors, ranging generally from less than one to a few percentages at several decades after the first cancer diagnosis or treatment.

354. The mean or median latency period of second primary gastrointestinal cancer from the diagnosis or treatment of the first primary cancer ranged from about 12 to 27 years depending on the types of second

cancer and populations. Although most studies reported mean or median latency periods of around 20 years, some studies suggested that the shortest latency period was less than 10 years.

355. Although information about the shape of dose-response relationship of second primary gastrointestinal cancer is limited, available data in the studies included in this evaluation suggest that this relationship is linear and increases with the dose, confirming the findings of the UNSCEAR 2000 and 2006 reports [U3, U4].

## 6. Thyroid cancer

### (a) *Studies with information on cumulative risks*

356. Information on cumulative incidence of second primary thyroid cancer after radiotherapy is reported in the studies by de Vathaire et al. and Green et al. [G21]. The information from these studies is summarized in table 17. The table shows that, at time since primary cancer of 30 years, the cumulative incidence rate ranged from 1.5% to 11.9%, although the number of second primary thyroid cancers from which this range was estimated was very low. Women generally have higher incidence rates for thyroid cancer, and therefore sex distribution can strongly affect the reported estimates of cumulative incidence.

Table 17. Studies with information on cumulative incidence of second primary thyroid cancer after radiotherapy

| *Information* | *Study reference* | |
|---|---|---|
| | *De Vathaire et al. [D8]* | *Green et al. [G21]* |
| Country of study | France and the United Kingdom | United States |
| Type of primary cancer | Childhood Cancer Survivors Study | Hodgkin lymphoma |
| Number of survivors | 4 096 | 182 |
| Age at treatment (years) | <15 | <20 |
| Diagnosis time frame | before 1986 | 1960–1989 |
| Second primary cancer incidences | 14 | 6 |
| Cumulative incidence (%, 95% CI or ±SD) | 1.5% (95% CI: 0.6, 2.3) | 11.89 (±6.05) |
| Time frame[a] (years) | 30 | 30 |

[a] Time since primary cancer (unless otherwise specified).

### (b) *Studies with information on latency periods between treatment for first primary cancer and second primary thyroid cancer occurrence*

357. Studies that reported latency periods between the radiation treatment for first primary cancer and the occurrence of second primary thyroid cancer are summarized in table 18. The table shows that, in general, the shortest latency period for second primary thyroid cancer reported in the literature was three years and that latency period extended beyond 30 years after treatment.

Table 18. Studies with information on latency periods between radiation treatment of first primary cancer and occurrence of second primary thyroid cancer

| *Study reference* | *First primary cancer* | *Latency period (years)* | |
|---|---|---|---|
| | | *Median* | *Range* |
| Boice et al. [B33] | Cervical cancer | | 5–>30 |
| Hallquist et al. [H8] | Malignant diseases | 13 | |
| de Vathaire et al. [D8] | Childhood cancer | | 15–19[a] |
| Green et al. [G21] | Hodgkin lymphoma | | 7.79–23.5 |
| Bhatia et al. [B23] | Childhood acute lymphoblastic leukaemia | 9.7 | 5.5–11.8 |
| Acharya et al. [A1] | Childhood cancer | 13 | 5.5–11.9 |
| Rubino et al. [R15] | Breast cancer | 9 | 3–27 |
| Ronckers et al. [R11] | Childhood cancer | | 6–29 |
| Finke et al. [F7] | Childhood cancer | 9.3 | 4.0–17.8 |

[a] Peak incidence.

## (c) *Studies with risk estimates with respect to thyroid doses*

358. Studies identified in the literature review that reported risks of second primary thyroid cancer with respect to the relevant organ or tissue dose are described in this section. A summary of the main features of these studies is given table 19, while full details are available in electronic attachment C. Some of these studies were found to have overlapping since they analysed data for the same cohorts. To prevent a biased meta-estimator of the combined risks in the meta-analysis, references with overlapping data are identified in the paragraphs below.

Table 19. Summary of studies identified in the literature review with risk estimates for second primary thyroid cancer

| *Study details* | *Treatment modality, fractionation details, dose type reported, RBE details* | *Number of subjects* | *Second primary thyroid cancer cases (controls, matched, unmatched)* | *Type of first primary cancer* | *Period of first cancer diagnosis* |
|---|---|---|---|---|---|
| International study on patients treated for cancer of the cervix [B33] | Orthovoltage X-rays (200–400 kVp), megavoltage, external beam ($^{60}Co$), brachytherapy | 150 000 | 43 cases and 81 matched controls | Cervical cancer | 1920 to after 1965 |
| U.S. Late Effects Study Group cohort and CACO study [T26] | Mainly orthovoltage radiation, also higher energy megavoltage units | 9 170 | 23 cases and 89 matched controls | Childhood cancer | 1936–1979 |
| Pooled analysis of seven cohorts, two on SPC, i.e., primary cancer (cervical cancer and childhood cancer)<br>Late Effects Study Group [R10] | Orthovoltage X-rays, megavoltage X-rays, radium implants, mean thyroid dose | | 43 cases and 81 controls<br>22 cases and 82 controls | Cervical cancer<br>Childhood cancer | |
| British and French Childhood Cancer Survivor Study, 1942–1992 [D8] | Orthovoltage, $^{60}Co$, high-energy X-rays, electrons | 4 400 | 14 cases | Childhood cancer | 1942–1985 |
| French cohort [A4] | Cobalt or electrons, mean: 27 fractions, range: 12–52 Dos-EG program for individual doses | 7 711 | 8 cases and 192 matched controls | Breast cancer | 1954-1983 |
| French Gustave Roussy Institute Breast Cancer cohort, CACO [R17] | External beam radiotherapy | 7 711 | 8 cases | Breast cancer | 1954–1983 |
| British and French Childhood Cancer Survivor Study [R15] | Orthovoltage X-rays (200–250 kV), $^{60}Co$ rays, megavoltage X-rays (4–25 MV) and electrons (6–22 MV). | 544 | 5 cases | Neuroblastoma in childhood | 1948– 1986 |
| North America Childhood Cancer Survivor Study, nested CACO [S40] | No details given | 14 054 five-year survivors | 69 cases and 265 matched controls | Cancer during childhood involving RT to the thorax or head and neck region | 1970–1986 |
| North America Childhood Cancer Survivor Study, CACO study [R11] | No details given | 14 054 | 69 cases and 265 matched controls | Childhood cancer | 1970–1986 |
| Five Nordic countries cohort, CACO [S57] | Some orthovoltage beams, some high voltage X- and gamma rays, software package, Dos-EG | 25 120 | 14 cases and 41 controls | Childhood cancer | 1960–1987 |
| North America Childhood Cancer Survivor Study, cohort study [C10] | CNS (including spinal) radiation | 3 579 ALL survivors and 3 846 siblings | 13 cases | ALL in childhood | 1970–1986 |

| *Study details* | *Treatment modality, fractionation details, dose type reported, RBE details* | *Number of subjects* | *Second primary thyroid cancer cases (controls, matched, unmatched)* | *Type of first primary cancer* | *Period of first cancer diagnosis* |
|---|---|---|---|---|---|
| North America Childhood Cancer Survivor Study [B27] | A wide range of treatment modalities | 12 547 | 119 cases | Childhood cancer: Leukemia; Hodgkin lymphoma; Non-Hodgkin lymphoma; Central nervous system cancer; Soft-tissue sarcoma; Kidney cancer; Bone cancer Neuroblastoma | 1970–1986 |
| Surveillance, Epidemiology and End Results Program [L23] | Brachytherapy or external radiotherapy or these combined | 60 949 | 25 cases | Uterine corpus cancer | 1973–2003 |
| Pooled analysis (2 cohorts + 2 CACO) including those already in this table [B27, D8, S57, T26, V11] | Pooled study–mixed | 16 757 | 187 cases | Childhood cancer | 1942-1986 |
| British and French Childhood Cancer Survivor Study, 1942–2014 [D10] | Brachytherapy, external radiotherapy | 4 338 | 55 cases | Childhood cancer | 1942–1985 |
| German Childhood Cancer Registry, CACO [F7] | No details given | 33 809 | 16 cases and 34 matched controls | Childhood cancer | 1980–2000 |

359. Adjadj et al. [A4] reported risks of developing thyroid carcinoma after treatment for breast cancer in a case-control study nested in a French cohort of 7,711 women treated between 1954 and 1983. The analysis was based on eight cases and 192 matched controls with median absorbed thyroid doses of 6.6 Gy for cases and 9.4 Gy for controls. An ERR per unit thyroid dose for second primary thyroid cancer of 0.012 (95% CI: −0.01, 0.07) $Gy^{-1}$ ($p$=0.8) was reported.

360. Boice et al. [B33] reported risks of second primary thyroid cancer in an international study on patients treated for cancer of the cervix. Thyroid doses were reconstructed for each patient on the basis of the original radiotherapy records and the mean thyroid dose was 0.11 Gy. The overall RR was 2.35 (90% CI: 0.6, 8.7) and the RR at 1 Gy was 13.30 (90% CI: 0.00, 77.0).

361. In an ongoing series of papers from the British and French CCSS on patients treated during the period between 1942 and 1985, de Vathaire et al. [D8] reported risks from 4,096 childhood cancers with a mean follow up period of 15 years. There were 14 thyroid cancers reported. The mean thyroid dose for the survivors treated with radiotherapy was 0.5 Gy for the dose category 0–<5 and 23 Gy for the dose category >5 Gy. Risks in four dose categories were reported but no overall risk per unit dose was given. These results were updated when de Vathaire et al. [D10] carried out a study of the risks for 4,338 childhood cancer survivors with a mean follow up period a mean of 27 years. There were 55 thyroid cancers reported in the study. The mean thyroid dose for the childhood cancer survivors treated with radiotherapy was 6.4 Gy, with a median of 0.7 Gy and a range of <0.001–72 Gy. The ERR per unit dose was 4.9 (95% CI: 1.1, 20.7) $Gy^{-1}$ and a non-linear model for the dose-response relationship was preferred.

362. Rubino et al. [R15] analysed data from eight British and French treatment centres on 544 patients diagnosed with neuroblastoma before age 16 years between 1948 and 1986, and surviving at least five years to investigate the development of second primary cancer. The follow-up finished at the end of 1991 or 1990 in British and French centres, respectively. The average dose to the thyroid received by the five patients who developed a second primary thyroid cancer was 6.7 Gy. The corresponding ERR per unit thyroid dose was reported to be 0.50 (95% CI: <0, 16) $Gy^{-1}$.

363. In view of the potential for data overlap between the studies by de Vathaire et al. [D8] de Vathaire et al. [D10] and Rubino et al. [R15], it was decided that only the risks from the study by de Vathaire et al. [D10] could be used in the primary meta-analysis, because of the highest number of cases and the most recent follow-up in it.

364. In an ongoing series of papers from the North America CCSS, results from a case-control study were reported by Ronckers et al. [R11] on 69 thyroid cancer cases and 265 controls nested in the cohort of 14,054 survivors. An evaluation of the shape of the dose-response relationship was provided. A model with a linear term in thyroid dose and a negative exponential in dose squared gave the best parsimonious fit to the data with an ERR per unit dose of 1.3 (95% CI: 0.4, 4.1) $Gy^{-1}$ at doses below 6 Gy followed by a relative decrease in ERR [S40].

365. Chow et al. [C10] also analysed this North America CCSS cohort but restricted the study to data on childhood acute lymphoblastic leukaemia survivors. Although the primary outcomes of interest were hypothyroidism and hyperthyroidism, a SIR of 4.8 (95% CI: 2.6, 8.2) for 13 thyroid cancers was reported and the mean thyroid dose was 13.4 Gy (range: 0.1–21.4).

366. Bhatti et al. [B27] provided further analysis of the North America CCSS based on 12,547 childhood cancer survivors diagnosed from 1970 to 1986 and followed up to 2005. The risk models were calculated based on 119 second primary thyroid cancers and thyroid doses, D. The preferred risk model was reported to be $ERR = 1.381De^{-0.00164\,D}$, linear up to 20 Gy but CIs were not given for the fitting parameters. At thyroid doses greater than 20 Gy, a downturn in the dose-response relationship was observed.

367. In view of the data overlap between the studies by Ronckers et al. [R11], Chow et al. [C10] and Bhatti et al. [B27] it was decided that only the data on risks from the study by Bhatti et al. [B27] would be used in the primary meta-analysis, because of the highest number of cases in it.

368. Svahn-Tapper et al. [S57] published results from a case-control study nested in the cohort of patients in five Nordic countries less than 20 years old at first cancer diagnosis between 1960 and 1987, with second primary cancers diagnosed up to 1991. AERR per unit thyroid dose of 1.7 (95% CI: 0.1, 22.1) $Gy^{-1}$ was reported for the 14 second primary thyroid cancers and the 41 controls.

369. Finke et al. [F7] published results from a case control study using data from the German Childhood Cancer Registry on patients aged 14 years or younger at the time of treatment for cancer. Neck doses were used in the analysis and mean neck doses of 30.3 Gy (range: 25–35 Gy) for the cases and 32.5 Gy (range: 20–40 Gy) for the controls were reported. Based on the analysis of the 17 second primary thyroid cancers and the 32 controls, the OR for second primary thyroid cancer per unit neck dose was 1.051 (95% CI: 0.984, 1.123) $Gy^{-1}$.

## (d) *Results of the meta-analysis of risks of second primary thyroid cancer*

370. Several studies have reported a non-linear dose-response relationship for thyroid cancer and so this was considered in the analysis by considering risk for specific dose categories under a specific dose cut-off. The primary random effects meta-analysis model applied a dose cut-off of 15 Gy, under which risk for specific dose categories were considered, and pooled the log RR from nine non-overlapping individual studies of risks of second primary thyroid cancer. The meta-analysis was only based on data for ERRs since information on EAR per unit dose in the studies reviewed was insufficient. The pooled ERR per unit dose estimated in the primary analysis was 0.68 (95% CI: 0.09, 1.60). The primary analysis also estimated a value for $I^2$ of 70.1% (95% CI: 40.5, 85.0), indicating that this percentage of the variability in the estimates of the effect size of radiotherapy is due to the differences between studies. The Cochran's Q test ($p$=0.001) indicated a statistically significant heterogeneity between the individual studies. ERRs of second primary thyroid cancer per unit dose estimated in the primary analysis are shown in figure XXIV.

371. A sensitivity analysis pooling the ERRs from the same nine studies on the original scale resulted in a much lower pooled estimate of the ERR per unit dose of 0.02 (95% CI: –0.01, 0.06) $Gy^{-1}$. The reason for this discrepancy lies in the highly asymmetrical CIs, especially in two of the nine individual studies contributing to the pooled estimate risk. Applying Wald-type statistical assumptions for this sensitivity analysis (i.e., assuming that the CIs are symmetrical about the nine individual study central risk estimates) causes the meta-analysis weightings to be very different than the weightings obtained from applying the log-scale approach and may cause biased and inappropriate results. Therefore, the result from the primary analysis obtained from pooling risks on the log-scale is more appropriate. All subsequent results for the pooled estimate of the ERR per unit dose in further sensitivity and subgroup analyses are based on pooling the log RR. In leave-one-out diagnostics, the pooled estimate of the ERRs per unit dose for the eight remaining studies varied between 0.30 and 1.12 $Gy^{-1}$. Omitting the Lönn et al. [L23] study, which was unusual in having an unexposed surgery-only comparison group resulted in a risk of 0.30 (95% CI: –0.01, 0.70) $Gy^{-1}$. In the sensitivity analysis that applied the linear uncorrelated model relevant to studies only publishing risk for specific dose categories, the pooled estimate of the ERR per unit dose was 0.78 (95% CI: 0.11, 1.88) $Gy^{-1}$; $I^2$ was 72.4% (95% CI: 45.9, 86.0), and the $p$-value for the Cochran's Q test was 0.0003. If the risks from all of the 15 studies were included in the meta-analysis without eliminating studies with potentially overlapping data, the pooled estimate of the ERR per unit dose would be 0.53 (95% CI: 0.21, 0.93).

Figure XXIV. Risks of second primary thyroid cancer per unit dose to the thyroid estimated in the primary analysis

The CIs were derived from the standard error of the point estimate based on the assumption of a log-normal distribution and may not coincide with the CI as originally published

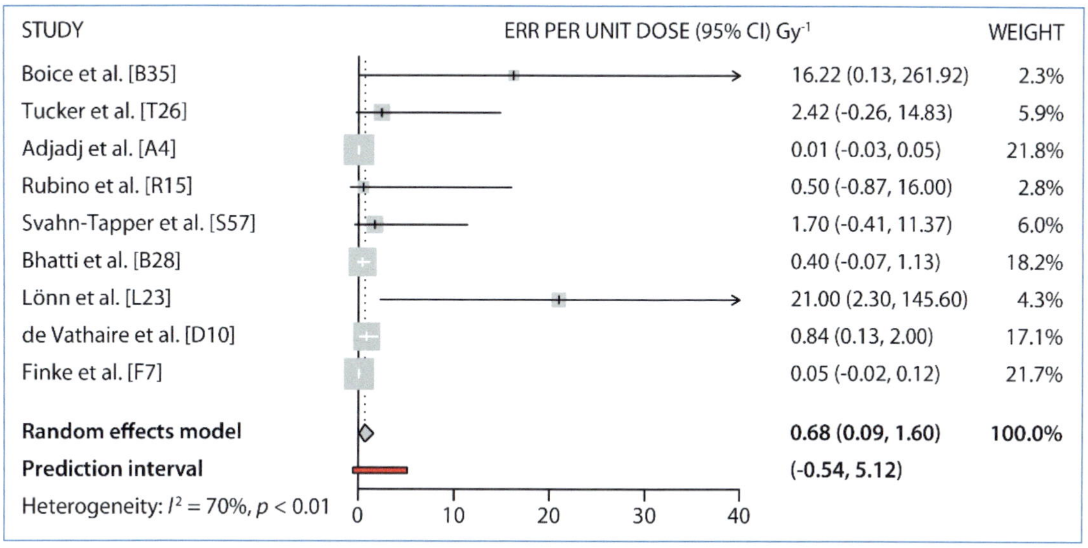


372. Sensitivity analyses were carried out by applying a dose cut-off of either 5 Gy or 20 Gy under which dose category specific RRs were considered in the analysis. For an analysis that pooled the log RR from nine non-overlapping individual studies the pooled estimates of the ERR per unit dose were 1.20 (95% CI: 0.18, 3.08) $Gy^{-1}$ or 0.43 (95% CI: 0.04, 0.99) $Gy^{-1}$, respectively. Based on the subgroup analysis of the risks in the eight studies without external comparisons of baseline rates, the pooled estimate of the ERR per unit dose was 0.3 (95% CI: –0.01, 0.70) $Gy^{-1}$. These results are summarized in figure XXV.

373. Three studies published direct estimates of the ERR per unit dose. Based on these risks alone, the pooled estimate of the ERR per unit dose was 0.14 (95% CI: –0.32, 0.91) $Gy^{-1}$. Six studies, without potentially overlapping data, published risks related only to exposures in childhood or early adolescence and the pooled estimate of the ERR per unit dose for these studies was 0.42 (95% CI: 0.01, 0.98) $Gy^{-1}$. Three studies published risks related only to exposures in adulthood and the pooled estimate ERR per unit dose for these studies was 4.59 (95% CI: –0.36, 48.15). Only two studies published risks related to exposed patients without chemotherapy. Therefore, no meta-analysis was performed for these subgroups. Five studies published risks related to exposures adjusted for chemotherapy and the pooled estimate of the ERR per unit dose was 0.42 (95% CI: 0.01, 1.01) $Gy^{-1}$. Seven studies published risks related to external exposures only and the pooled estimate of the ERR per unit dose for these studies was 0.80 (95% CI: 0.10, 1.95) $Gy^{-1}$.

Figure XXV. Comparison of risks per unit dose estimated in the primary analysis and in the sensitivity analyses for second primary thyroid cancer

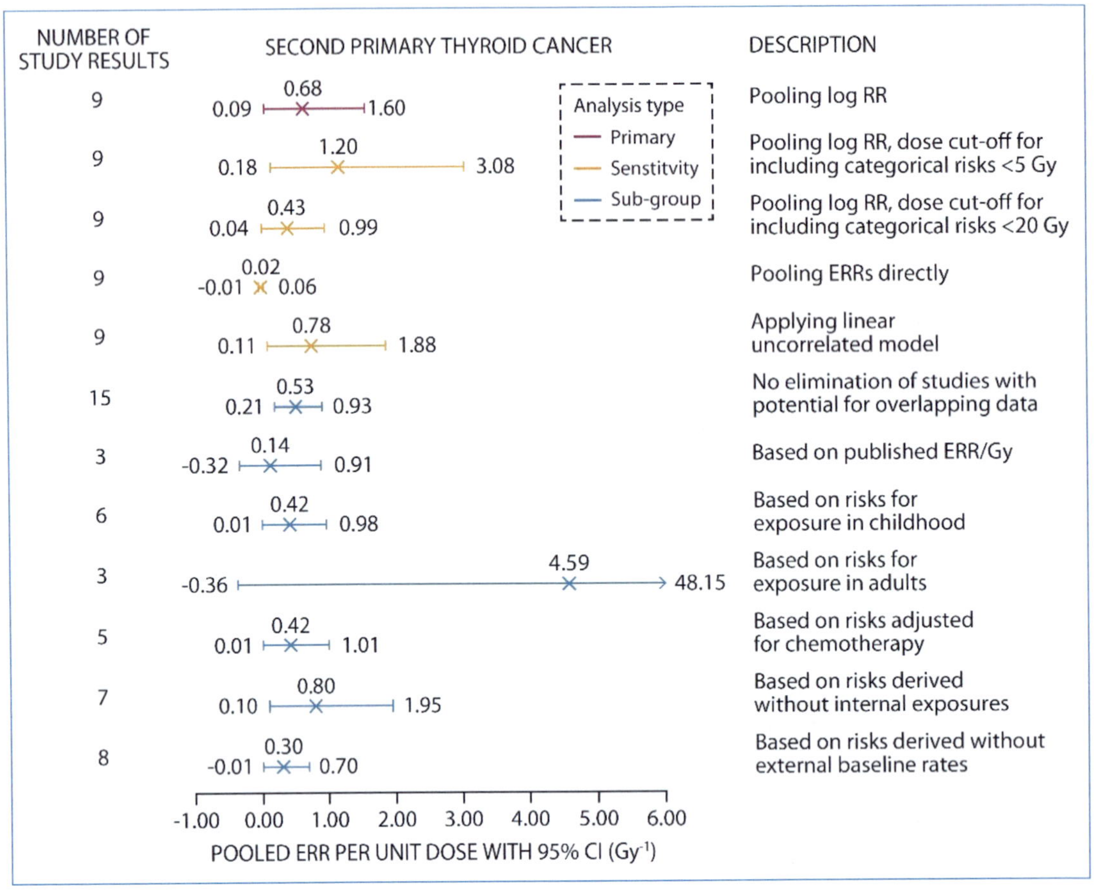


## (e) *Summary*

374. Fifteen studies included in the literature review provided information on risks of second primary thyroid cancer per unit thyroid dose after radiation treatment for first cancers. The studies had either cohort or nested case-control designs and were conducted in countries in North America and Europe. It was not always possible to clearly distinguish the follow-up period from cancer diagnosis period, the period of radiation treatment or the age at first radiation treatment. Therefore, it was not considered feasible to extract a consistent set of information on these factors or to draw conclusions based on risk effect modification by these factors. Most studies reported risks per unit thyroid dose based on patients treated with radiation and chemotherapy together, but the risks were adjusted for chemotherapy.

375. Only two studies were found that reported cumulative incidence of second primary thyroid cancer after radiotherapy, ranging from 1.5% to 11.9% at 30 years after the first childhood cancer, although the number of second primary thyroid cancers included in these studies was very low.

376. The median latency period of second primary thyroid cancer from the treatment of the first primary cancer ranged from about 9 to 19 years, with most studies reporting median latencies of around 10 years. The shortest latency period reported was three years.

377. Although information about the shape of dose-response relationship of second primary thyroid cancer is limited, available data in the studies included in this evaluation indicate that this relationship is linear and increases with the dose, with the risk levelling off at high doses above approximately 10 Gy.

## 7. Brain tumour

378. Brain tumour in this annex refers in general to both brain and CNS tumours, despite some CNS tumours not being located within the brain. The most common types are glioma and meningioma. The gliomas are malignant and most meningiomas are benign. Depending on the tumour location, benign and malignant tumours of the brain can have similar symptoms and outcomes and therefore, they are usually not distinguished in many tumour registries and epidemiological studies [U4]. Data on benign brain tumours are more limited because, historically, collection of data for tumour registries has been limited to malignant tumours and many countries have not collected information on benign brain tumours or have not collected it in a standardized manner, although some Nordic countries have been recording benign brain tumours since the 1950s [M10].

### (a) *Studies with information on cumulative risks*

379. Studies that reported information on cumulative incidence of second primary brain cancer after radiotherapy are summarized in table 20. The table shows that the cumulative incidence is less than 2% at time since first primary cancer of 10 years and increases with time since exposure, reaching a value of just over 9.1% at 40 years.

Table 20. Summary of information on cumulative incidence of second primary brain tumour after radiotherapy

ALL: Acute lymphoblastic leukaemia; CNS: Central nervous system

| *Study* | *Country of study* | *Type of primary cancer* | *Number of survivors* | *Age at treatment (years)* | *Diagnosis time frame* | *Second primary cancer incidence* | *Cumulative incidence (%, 95% CI or ±SD)* | *Time frame*[a] *(years)* |
|---|---|---|---|---|---|---|---|---|
| Brada et al. [B40] | United Kingdom | Pituitary adenoma | 334 | <79 | 1962–1986 | 5 | 1.3 (0.4, 3.9)<br>1.9 (0.7, 5.0) | 10<br>20 |
| Rosso et al. [R13] | Italy | Childhood ALL | 1 664 | <18 | 1960–1984 | 9 | 0.1 (0, 0.2)<br>1.9 (0.5, 3.2) | 5<br>10 |
| Bhatia et al. [B23] | Canada and the United States | Childhood ALL | 8 831 | <21 | 1983–1995 | 19 | 0.47 (0.2, 0.6) (63% exposed)<br>0.90 (0.4, 1.4) (63% exposed) | 10<br>15 |
| Armstrong et al. [A27] | Canada and the United States | Childhood CNS malignancies | 1 877 | <21 | 1970–1986 | 79 | 1.0 (0, 2.3) (unexposed)<br>5.2 (2.1, 8.3) (<50 Gy)<br>7.1 (4.5, 9.6) (≥50 Gy) | 25 |
| Taylor et al. [T6] | United Kingdom | Childhood cancer | 17 980 | <15 | 1940–1991 | 247 | Overall: 3.6 (2.9, 4.3)<br>For first primary CNS tumours<br>Unexposed: 1.4 (0.6,2.8)<br>Overall exposed: 9.1 (7.9, 11.7)<br>Meningiomas: 6.3 (4.5, 8.5)<br>Gliomas: 2.4 (1.3, 4.1) | 40 |

*a* Time since primary cancer (unless otherwise specified).

## (b) *Studies with information on latency periods between treatment for first primary cancer and occurrence of second primary brain cancer*

380. Studies that reported latency periods between radiotherapy for first primary cancer and the occurrence of second primary brain tumour [B40, L20, R13] and studies with information on the latency period for most first primary cancers treated with radiation [A27, B23, N10, S57, T6] are summarized in table 21. The table shows that the shortest latency period for second primary brain tumour reported in the studies reviewed was five years or less and that the latency period extended to 52 years after treatment.

Table 21. Studies with information on latency periods between radiation treatment of first primary cancer and occurrence of second primary brain tumour

ALL: Acute lymphoblastic leukaemia; CNS: Central nervous system

| *Study reference* | *First primary cancer* | *Second primary cancer* | *Latency period (years)* | | |
|---|---|---|---|---|---|
| | | | *Mean* | *Median* | *Range* |
| Brada et al. [B40] | Pituitary adenoma | | 6, 7, 7, 10, 21 (5 cases total) | | |
| Rosso et al. [R13] | Childhood ALL | | | | 1.1–8.8 |
| Little et al. [L20] | Childhood cancer | | 15.8 | | 3.5–35.4 |
| Bhatia et al. [B23] | Childhood ALL | | | 7.1 [a] | 3.9–13.0 [a] |
| Neglia et al. [N10] | Childhood cancer | Glioma | | 9 [a] | 5–28 |
| | | Meningioma | | 17 [a] | |
| Svahn-Tapper et al. [S57] | Childhood and adolescence cancer | | 8.8 [a] | | |
| Armstrong et al. [A27] | Childhood CNS malignancies | Malignant tumour | | 14.0 [a] | |
| | | Benign tumour | | 18.8 [a] | |
| Taylor et al. [T6] | Childhood cancer | Glioma | 17.4 [a] | | 5–52 [a, b] |
| | | Meningioma | 23.1 [a] | | |

[a] Not all after first primary cancer with radiotherapy.

[b] The second primary brain tumour diagnosed within the first five years after first primary cancer were excluded because of the cohort composed of five-year survivors of first primary cancer.

## (c) *Studies with risk estimates with respect to brain doses*

381. Studies identified in the literature review that reported risks of second primary brain tumour with respect to the dose to the brain are described in this section. A summary of the main features of these studies is given table 22, while full details are available in electronic attachment C. Some of these studies were found to have overlapping since they analysed data for the same cohorts. To prevent a biased meta estimator of the combined risks in the meta-analysis, references with overlapping data are identified in the paragraphs below.

382. Bhatia et al. [L20] provided information on the United States-Canada cohort of 8,831 children diagnosed with acute lymphoblastic leukaemia treated on one of the 12 Children's Cancer Group

therapeutic protocols between 1983 and 1995 and followed up until 1999 for a total of 54,883 PY. Nineteen cases of second brain tumour were reported with a median age of 3.8 years at diagnosis of ALL. The SIR for malignant brain tumour was 10.1 (95% CI: 5.9, 16.2; n=17, excluding 2 meningiomas). The RR of second brain tumour after ALL with radiation to the craniospinal axis was 2.4 (95% CI: 1.1, 5.2), compared with that of patients who received no radiation, and increased with radiation dose (RR=2.1 (95% CI: 0.7, 6.3) for a dose of 18 Gy; RR=4.2 (95% CI: 0.5, 37.7) for a dose of 24 Gy).

383. Svahn-Tapper et al. reported findings of the second primary cancer case-control study nested within Nordic cohort of 25,120 cancers diagnosed before age 20 during the period between 1960 and 1987, and followed up until 1991 [T6]. In this study individual site-specific doses were reconstructed. For brain tumour, 48 cases were identified and matched with 143 controls. The RRs reported were 0.7 (95% CI: 0.2, 2.7), 1.5 (95% CI: 0.3, 8.4), 3.4 (95% CI: 1.1, 10.2) and 10.0 (95% CI: 1.9, 52.8) for dose categories of 0–1, 1–5, 5–30 and >30 Gy, respectively (dose to the reference group was 0 Gy). These results indicated that the risk of second brain tumour increased with radiation dose to first cancer, significantly with a dose above 5 Gy. The estimate for the additive ERR per unit dose was 0.1 (95% CI: 0.0, 0.5) $Gy^{-1}$.

384. There is useful information on the dose-response relationship for types of second brain tumour in several studies. Little et al. carried out a case-control study of second brain tumours nested within British-French cohort of 4,199 survivors of childhood solid tumour, with average follow-up of 15.1 years (63,309 PY) [L20]. The individual whole brain dose was estimated using radiotherapy doses to 10 areas of the brain; the mean dose was 6.2 Gy (range: 0.0–82.7 Gy). There were 22 cases of second brain tumour, including 12 malignant and 10 benign or unspecified. For all brain tumours, the estimate of the ERR per unit dose was 0.19 (95% CI: 0.03, 0.85) $Gy^{-1}$ with a linear model, and 0.05 (95% CI: 0.02, 0.09) $Gy^{-1}$ with an exponential model. The dose-response relationship remained statistically significant after adjustment for chemotherapy. The radiation-related risk of benign/unspecified brain tumour was significantly higher than that of malignant brain tumour ($p\leq0.05$); there was no significant change of malignant brain tumour risk with radiation dose ($p>0.2$).

385. Taylor et al. [T6] examined the risk of second brain tumour among 17,980 patients in the British CCSS with at least five-year survival diagnosed before age 15 between 1940 and 1991, and followed to 2002 with mean follow-up period of 17.3 years (310,816 PY). In the study 247 cases of second brain tumour were reported, including 137 meningiomas, 73 gliomas and 37 others. The SIR for glioma was higher after radiotherapy at 14.3 (95% CI: 10.9, 18.7) compared with that for treatment without radiotherapy at 6.1 (95% CI: 3.1, 11.0). Within the case-control study, the ERR per unit dose was 5.1 (95% CI: 0.7, 107.7) $Gy^{-1}$ for meningioma and 0.079 (95% CI: 0.021, 0.229) $Gy^{-1}$ for glioma/PNET. In both cases the ERR increased linearly with increased cumulative exposure to radiation ($p<0.001$). For meningioma and glioma/PNET, the RRs associated with an exposure of at least 40 Gy were 479 times and 4 times higher than those experienced in unexposed tissue, respectively [L20].

Table 22. Summary of studies identified in the literature review with risk estimates for second primary brain cancer

| *Study details (study period)* | *Treatment modality, fractionation details, dose type reported, RBE details* | *Number of subjects* | *Second primary brain/CNS tumour cases (controls, matched, unmatched)* | *Type of first primary cancer* | *Period of first cancer diagnosis* |
|---|---|---|---|---|---|
| Swedish cohort of infants treated for skin haemangioma, 1958–1989 [L19] | $^{226}Ra$ applicators or implants brachytherapy, 81% use gamma ray, 17% use beta ray | 11 807 | 34 | Infants skin haemangioma | 1930–1965 |
| French-British childhood solid tumour survivors cohort, nested CACO [L20] | EBRT with fractions | 4 199 | 22 (1 case matched, 15 controls) | Childhood solid tumours | |
| United States–Canada childhood acute lymphoblastic leukemia survivors cohort, 1983–1999 [B23] | Children's Cancer Group (CCG) therapeutic protocols | 8 831 | 19 | Childhood acute lymphoblastic leukaemia | 1983–1995 |
| United States–Canada Childhood Cancer Survivor Study (CCSS) cohort, nested CACO, 1975–2001 [N10] | EBRT | 14 361 | 116 (464 controls matched) | Childhood cancer | 1970–1986 |
| Five Nordic countries childhood cancer survivors cohort, nested CACO, 1960–1991[S57] | EBRT | 25 120 | 48 (143 controls matched) | Childhood cancer | 1960–1987 |
| British Childhood Cancer Survivor Study (BCCSS) cohort, nested CACO, 1945–2002 [T6] | EBRT | 17 980 | 247 (1:1 controls matched) | Childhood cancer | 1940–1991 |

386. Neglia et al. [N10] performed a case-control study of second brain tumour nested within the United States-Canada CCSS cohort of 14,361 five-year survivors of cancer diagnosed before age of 21 years during the period 1970–1986, followed to 2001. The study included 116 cases of second brain tumour and each case was matched with four controls (n=464). Tumour site-specific radiation dosimetry was performed. The ORs for meningioma were higher across all categories of radiation dose than those for glioma. The ERR per unit dose estimated using a linear dose-response relationship was 0.69 (95% CI: 0.25, 2.23) $Gy^{-1}$ for all brain tumours, 1.06 (95% CI: 0.21, 8.15) $Gy^{-1}$ for meningioma (n=66) and 0.33 (95% CI: 0.07, 1.71) $Gy^{-1}$ for glioma (n=40). The ERR per unit dose for glioma was highest among children exposed at less than 5 years of age. After adjustment for radiation dose, neither original cancer diagnosis nor chemotherapy was associated with risk.

387. Schonfeld et al. [S24] compared radiation dose reconstruction methods using the second primary cancer case-control study nested within the United States-Canada CCSS between 1970 and 1986. The EOR was estimated from conditional logistic regression, adjusting for first cancer type. For second brain tumour, 64 cases were matched with 244 controls. The EOR per unit dose reported in the paper was 1.25 (95% CI: 0.33, 6.33) $Gy^{-1}$ and 1.20 (95% CI: 0.31, 6.14) $Gy^{-1}$, depending on whether tumour location-specific doses or body-region doses were used. It is likely that data from this study overlap with those from the study by Neglia et al. [N10])

## (d) *Results of the meta-analysis of risks of second primary brain tumour*

388. The primary random effects meta-analysis model pooled the log RR from six non-overlapping individual studies on risks of second primary brain tumour. The meta-analysis was only based on data for ERRs since information on EAR per unit dose in the studies reviewed was insufficient. The pooled ERR per unit dose to the brain estimated in the primary analysis was 0.11 (95% CI: 0.03, 0.19) $Gy^{-1}$. The primary analysis also estimated a value for $I^2$ of 21.5% (95% CI: 0.0, 65.9), indicating that this percentage of the variability in the estimates of the effect size of radiotherapy is due to the differences between studies. The Cochran's Q test ($p$=0.27) indicated that there was no statistically significant heterogeneity between the individual studies. ERRs of second primary brain tumour per unit dose to the brain estimated in the primary analysis are shown in figure XXVI.

Figure XXVI. Risks of second primary brain tumour per unit dose to the brain estimated in the primary analysis

The CIs were derived from the standard error of the point estimate based on the assumption of a log-normal distribution and may not coincide with the CI as originally published

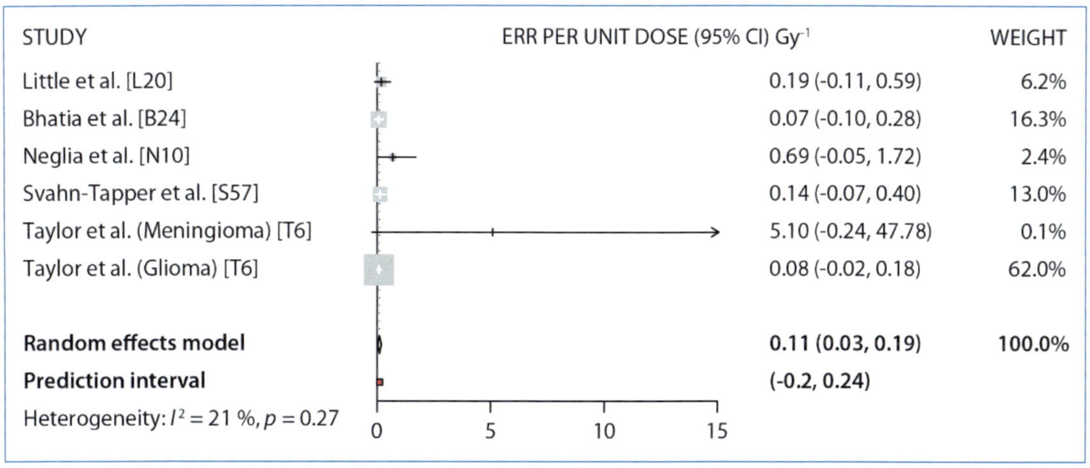


389. A sensitivity analysis pooling the ERRs from the same six studies on the original scale resulted in a lower but statistically compatible pooled estimate of the ERR per unit dose of 0.09 (95% CI: 0.01, 0.18) $Gy^{-1}$. All subsequent results for the pooled estimate of the ERRs per unit dose in the sensitivity analysis were based on pooling the log RR. Leave one-study-out analysis of the risks from the remaining five studies generated varying main central pooled ERRs per unit dose between 0.09 and 0.15 $Gy^{-1}$. In the sensitivity analysis that applied the linear uncorrelated model relevant to studies that reported only risks for specific dose categories, the pooled estimate of the ERR per unit dose was 0.11 (95% CI: 0.03, 0.20) $Gy^{-1}$, $I^2$ was 20.4 (95% CI: 0.0, 65.0), and the p-value for the Cochran's Q test was 0.28. If all seven studies were included in the main meta-analysis without eliminating studies with potentially overlapping data, the pooled estimate of the ERR per unit dose was 0.11 (95% CI: 0.03, 0.20) $Gy^{-1}$.

390. The estimates of the ERR per unit dose could be extracted from five published studies and on the basis of these risks, the pooled estimate of the ERR per unit dose was 0.11 (95% CI: 0.03, 0.21) $Gy^{-1}$. The risks in all six studies in the primary meta-analysis without overlapping data related to exposures in childhood or early adolescence were adjusted for chemotherapy, were based on external exposures only and did not base risk estimation on external rates. Therefore, subgroup analyses for these factors were redundant. There was not enough information to base subgroup analyses on subsets of studies with risks reported in the literature that were related to exposures in adulthood or without chemotherapy. These results of the sensitivity analysis are summarized in figure XXVII. Results were not reported if fewer than three studies contributed to the meta-analysis.

Figure XXVII. Comparison of risks per unit dose estimated in the primary analysis and in the sensitivity analyses for second primary brain tumour

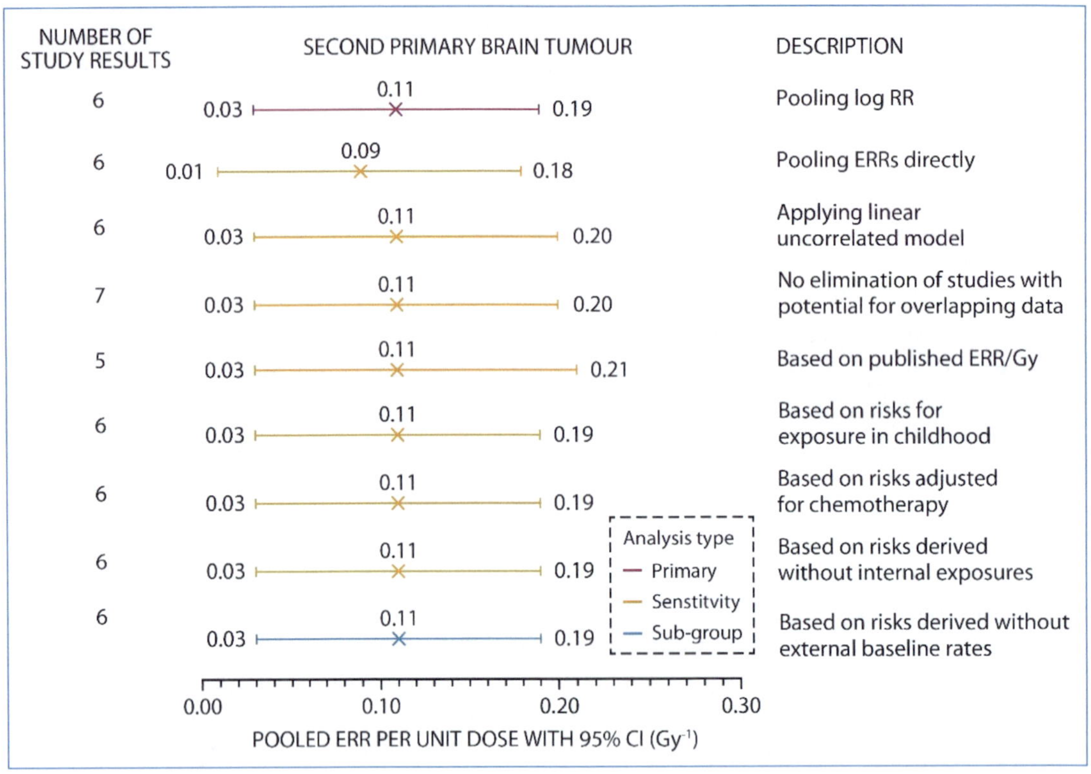


## (e) *Summary*

391. Cases of second primary brain tumour after radiotherapy were mainly observed among survivors of childhood or early adolescence cancer, especially among patients of acute lymphoblastic leukaemia or brain tumour treated with cranial irradiation. The cumulative incidence of second primary brain tumour after first cancer treated with radiotherapy reported in several studies was higher than that treated with other modalities. In relevant studies on dose-response relationship, the risk of second primary brain tumour after radiotherapy increased with brain dose, and the meta-analyses were based on results of studies that assumed a linear dose-response relationship.

## 8. Other cancer sites

392. Risks of second primary cancers after radiotherapy in other sites not considered in this chapter have been also investigated in several epidemiological studies. In particular, an increased risk of different types of skin cancer (e.g. non-melanoma skin cancer including BCC and squamous cell carcinoma) has been observed in many studies. Several studies have also assessed the risks of second kidney cancer after radiotherapy while findings on the association between radiation exposure and kidney cancer are less clear than for skin cancers and other types of cancer. This section provides a brief overview of research showing quantitative information on risks of second primary skin or kidney cancer.

### (a) *Studies with information on cumulative risks*

393. Risks of second primary cancer were investigated by randomized clinical trial design for 1,494 adult patients of acute lymphoblastic leukaemia who were enrolled between 1987 and 2002, and received chemotherapy and/or radiotherapy [T3]. Secondary or concomitant neoplasms were observed in 23 patients including five skin tumours (four BCCs and one melanoma) by February 2005. The occurrence of second primary cancer was not associated with the use of any specific cytotoxic agent. However, the risk of skin tumour increased with radiation dose and transplantation ($p$=0.01).

394. A cohort of 446 paediatric tumour patients treated with radiation, at or below the age of 17 years, between 1954 and 1980 at the former University of Minnesota hospital was followed up to the end of 1995 [G11]. Thirty-seven patients (8.3% of the cohort) developed second primary cancer including eight skin cancers (five BCCs, two melanomas, and one dermatosarcoma). The cumulative risk of developing skin cancer at 30 years was estimated at 4% (95% CI: 2, 8). The SIR of skin cancer excluding meningiomas and BCCs was 6.3 (95% CI: 1.3, 17.6), showing a statistical and significant increase over the baseline rate for general population from the SEER registry.

395. Risks of second primary cancer were evaluated in a cohort of survivors of Wilms' tumour by a population-based British CCSS [T5]. The total follow-up from five-year survival was 27,841 PY, with a mean follow-up period of 19.3 years. There were 81 second primary cancers observed among the survivors, including 52 solid tumours other than non-melanoma skin cancer, 26 BCCs, and three acute myeloid leukaemia. The cumulative incidences of BCCs as a second primary cancer by age 30, 40, and 50 were 1.0% (0.5–1.9), 5.4% (3.4–8.0), and 6.0% (3.8–9.0), respectively. Cumulative incidence of BCCs reported by years of follow-up were 0.07% (0.01–0.4), 1.0% (0.5–1.9), 4.5% (2.9–6.7), and 5.7% (3.6–8.6) at 10, 20, 30 and 40 years, respectively.

396. In a cohort of 1,368 five-year survivors of childhood cancer treated between 1966 and 1996, the risk of second primary cancer was investigated up to 2002 with a median follow-up time of 16.8 years [C4]. A total of 18 survivors developed BCC as a second primary cancer, and the corresponding SIR was 8.95 (95% CI: 5.30,14.1). The highest absolute excess risk was observed for BCC (0.94 per 1,000 PY). Radiation was a strong risk factor for BCC and meningioma, with all meningiomas and 94% of the BCCs occurring after radiotherapy.

397. An analysis of the second primary cancer risk performed for 14,359 five-year survivors within the CCSS treated during the period 1970–1986 [F16] reported that non-melanoma skin cancer (n=1,574) was the most common type as a second primary cancer out of 2,703 second primary cancer cases with 325,119 PY of follow-up and a mean follow-up period of 22.7 years. The cumulative incidence rate of non-melanoma skin cancer was 9.1% (95% CI: 8.1, 10.1). For melanoma, the SIR was 3.3 (95% CI: 2.4, 4.5) on the basis of 48 observed skin cancer cases with a RR of 4.1 (95% CI: 2.7, 6.7).

398. Armstrong et al. reported long-term outcomes including mortality and subsequent neoplasms among five-year adult survivors of CNS malignancies diagnosed between 1970 and 1986 within the CCSS [A27]. The 1,877 survivors evaluated for the occurrence of subsequent neoplasms accrued more than 27,550 PY with a median follow-up period of 19.6 years (range: 5.1, 34.6). There were 255 confirmed subsequent neoplasms in 151 survivors including 112 non-melanoma skin cancers and five melanomas. The cumulative incidence rate of non-melanoma skin cancer was 0.2 (95% CI: 0.0, 0.3), 0.4 (0.1, 0.7), 1.0 (0.5, 1.5), and 2.9 (1.8, 4.0) at 10, 15, 20, and 25 years since initial diagnosis, respectively.

399. Using data from the CCSS cohort, the cumulative incidence rate of second first subsequent melanoma at 35 years from initial cancer diagnosis was 0.55% (95% CI: 0.37, 0.73) [P1]. However, data on the precise location of the melanoma was not sufficient to allow for the estimation of radiotherapy field location and its association with the development of subsequent melanoma.

400. Long-term risks of second primary cancers in 1,253 patients diagnosed with Hodgkin disease before age 40 between 1966 and 1986 were assessed by van Leeuwen et al. [V5]. A total of 137 patients developed second primary cancers with a median follow-up period of 14.1 years, among which seven were melanomas.

401. Therapy-related risk factors for the development of melanoma after childhood cancer were analysed by cohort and nested case-control design for 29,521 three-year survivors of cancer patients younger than 20 years old [G23]. The overall cumulative incidence of melanoma were 0.02% ± 0.02% 0.09% ± 0.05% and 0.4% ± 0.19% at 10 years, 20 years and 30 years after the diagnosis of the first cancer, respectively.

## (b) *Studies with information on latency periods between treatment for first primary cancer and occurrence of second primary cancer*

402. Gold et al. reported a median value of 17 years for the latency period of skin cancer after initial radiotherapy in a cohort study of 446 paediatric tumour patients [G11]. A median time of 18.9 years (range: 5.6–35.2 years) was reported for occurrence of melanoma as a second primary cancer in the CCSS cohort [F16]. In the extended follow-up, the median latency period for second primary melanoma was 21.0 years (range: 5.6–35.4 years) [P1].

## (c) *Studies with risk estimates with respect to doses*

403. Several publications have provided risk estimates of skin cancer with respect to radiation doses. Results of a case-control study nested in a cohort of 25,120 childhood cancer survivors in Denmark, Finland, Iceland, Norway, and Sweden were reported on skin cancer (19 cases and 54 controls) [S57]. The authors mentioned no significant risk factor could be found for skin cancer; the OR for group in the dose category of 0–1 Gy was 1.1 (95% CI: 0.2, 6.3) compared with the reference group (dose of 0 Gy), and the corresponding ERR per unit dose was 0.2 (95% CI: 0.0, 1.7) $Gy^{-1}$.

404. Risks of melanoma following adulthood cancer were evaluated in a case-control study (57 cases and 171 controls) [D30]. The ORs for dose categories of 0–1, 1–5, 5–20 and >20 Gy adjusted for family history, chemotherapy, and hormone therapy were 2.8 (95% CI: 0.9, 8.7), 1.4 (95% CI: 0.2, 8.2), 1.8 (95% CI: 0.4, 8.5), 0.8 (95% CI: 0.1, 4.8), respectively. No risk of melanoma was associated with radiotherapy (OR for dose category of 1 Gy=1.01 (95% CI: 0.96, 1.07). The ERR per unit dose to the skin was estimated to be 0.014 (95% CI: –0.03, 0.07) $Gy^{-1}$.

405. A nested case-control study by Guerin et al. [G23] showed that the risk of melanoma increased after radiotherapy as a function of the local radiation dose. The ORs for dose categories of 0–1, 1–15 and >15 Gy were 1.4 (95% CI: 0.28, 7.0), 3.2 (95% CI: 0.37, 27), and 13 (95% CI: 0.94, 174), respectively, in comparison with no radiotherapy group.

406. Radiation-related risk of BCC was assessed by case-control design using data on 199 cases and 597 controls from the CCSS [W11]. The OR for patients who received doses of 35 Gy or more to the skin site versus patients who were not treated with radiotherapy was 39.8 (95% CI: 8.6, 185). Results were consistent with a linear dose-response relationship; the ERR per unit dose was estimated at 1.09 (95% CI: 0.49, 2.64) $Gy^{-1}$. Radiation doses to the skin of more than 1 Gy were associated with an increased risk of BCC.

407. Information on estimates of risk for kidney cancer is limited. The international study of almost 150,000 cervix cancer patients provided an estimate of RR at 1 Gy of 1.71 (90% CI: 1.03, 3.24), and an

estimate of the EAR per 10,000 PY Gy of 1.10 (90% CI: 0.05, 3.50) [B33]. There was only a suggestion of increasing risk with increasing dose to the kidney among 10-year survivors ($p$=0.17). Dose category specific RRs were also provided in this study; the RRs for dose categories of 0–1, 1–2, 2–3 and >3 Gy were 0.78 (90% CI: 0.3, 1.9), 1.34 (90% CI: 0.7, 2.7), 1.10 (90% CI: 0.5, 2.1), and 0.99 (90% CI: 0.4, 2.2), respectively.

408. Risks of second kidney cancer after childhood cancer were evaluated by de Vathaire, et al. [D11] in a cohort of 4,350 five-year cancer survivors who had been treated for cancer as children in France and the United Kingdom. Compared to the general population, the incidence of kidney cancer was 5.7-fold higher (95% CI: 1.4, 14.7) if radiotherapy was not performed or if the absorbed dose of less than 1 Gy was delivered to the kidney, but 66.3-fold higher (95% CI: 23.8, 142.5) if the dose was 10 to 19 Gy, and 14.5-fold higher (95% CI: 0.8, 63.9) for larger radiation doses. The estimated ERR per unit dose was 0.20 $Gy^{-1}$ (CIs not available) using a simple linear ERR model, while the linear term of the linear exponential model was 1.1 (95% CI: 0.1, 9.8) $Gy^{-1}$.

## 9. All solid cancers combined

409. The risk of second primary cancer after radiotherapy for all solid cancers combined has also been evaluated by studies of various cohorts. In particular, patients treated with radiotherapy for Hodgkin lymphoma mainly during childhood or adolescence have often shown an increased risk of second primary solid cancer. A study by O'Brien et al. [O1] investigated second primary cancers in 110 survivors of paediatric Hodgkin lymphoma treated with low-dose radiation and chemotherapy at Stanford University in the United States. Eighteen patients developed one or more second primary cancers with a median follow-up of 20.6 years. Treatment included six cycles of chemotherapy and radiotherapy with doses in the range of 15 to 25.5 Gy, followed by additional doses of 10 Gy for selected patients. The SIR of second solid cancer combined was 19.8 (95% CI: 11.5, 31.7) based on 17 cases. Another study by Ng et al. [N13] reported a long-term risk of second primary cancer after Hodgkin lymphoma treated with radiotherapy and with or without chemotherapy. A total of 131 second solid cancers were reported among 1,319 patients in the United States with a median follow-up of 12 years. Overall SIR of solid cancer was 3.5 (95% CI: 3.0, 4.2), and SIRs increased with increasing follow-up period: 0.7 (95% CI: 0.3, 1.4), 2.0 (95% CI: 1.2, 3.1), 3.5 (95% CI: 2.3, 4.9), 7.7 (95% CI: 5.7, 10.3), and 9.2 (95% CI: 6.5, 12.4) for 0–5, 5–10, 10–15, 15–20, and 20+ years of follow-up, respectively.

410. A study of second primary cancers among 1,253 survivors of Hodgkin lymphoma diagnosed before age 40 in the Kingdom of the Netherlands [V5] reported that 26.8% and 6.0% of the survivors received either radiotherapy or chemotherapy only, while the rest received a combination of both treatments. A total of 144 invasive second primary cancers in 137 patients were observed including 106 solid tumours with a median follow-up of 14.1 years. The 25-year cumulative risk of all solid tumours was 23.3% (95% CI: 18.9, 28.6), and a significantly elevated SIR of 6.1 (95% CI: 5.0, 7.4) was reported. The SIR for all solid tumours decreased with age at diagnosis for Hodgkin lymphoma: 13.9 (95% CI: 9.0, 20.6), 6.5 (95% CI: 4.7, 8.8), and 4.2 (95% CI: 3.0, 5.8) for age at diagnosis of $\leq$20, 21–30, and 31–39 years, respectively. Results of an international study of 5,925 Hodgkin lymphoma survivors diagnosed before age 21 and identified in 16 cancer registries in Europe and North America, were reported by Metayer et al. [M19]. A total of 195 second primary cancers were reported including 157 solid cancers with a SIR of 7.0 (95% CI: 5.9, 8.2) and a mean follow-up of 10.5 years. The cumulative risk of developing any solid cancer was 6.5% (95% CI: 5.3, 7.7) and 11.7% (95% CI: 9.7, 14.1) at 20 and 25 years after Hodgkin lymphoma diagnosis. Increased risks persisted for 25 years after the Hodgkin lymphoma diagnosis (SIR=4.6 (95% CI: 2.6, 7.6)).

411. Increased risks of solid cancer after radiotherapy in childhood or adolescence were observed among survivors of other primary tumour types. Around 748 non-Hodgkin lymphoma patients were followed up for second primary cancer in the European Organization for Research on Treatment of Cancer trials [M32]. All patients received chemotherapy; 65% of them received involved-field radiotherapy, and 14% high-dose radiotherapy. A total of 37 solid cancers were reported with a median follow-up of 9.4 years, and the corresponding SIR was 1.0 (95% CI: 0.7, 1.3). The risk of second solid cancer appeared to be clearly age-related with SIRs of 3.6 (95% CI: 2.0, 6.0) for patients aged less than 45 years at time of treatment for non-Hodgkin lymphoma and 0.4 (95% CI: 0.2, 0.6) for patients aged 45 years or older.

412. Solid cancer risks among survivors of non-Hodgkin lymphoma were also reported in a study by Bluhm et al. [B30]. Mortality and incidence rates of second primary cancers were evaluated among 1,082 five-year survivors of non-Hodgkin lymphoma in the CCSS. Overall, 65% received a combination of chemotherapy and radiotherapy, and 30% received chemotherapy only. There were 27 solid second primary cancers (SIR=3.9 (95% CI: 2.6, 5.7)). In a multivariate model including all of these factors, only the female sex (RR=2.8 (95% CI:1.5, 5.1)) and cardiac radiotherapy exposure (RR=2.1 (95% CI: 1.1, 3.9)) showed significant associations. In a Poisson regression analysis, shielded or unshielded breast exposure to radiotherapy was associated with a 7.5-fold increase in risk (95% CI: 2.6, 26.7) of second solid tumour. Exposure to the thyroid was associated with a more than 2.6-fold increase in risk (95% CI: 1.0, 7.1).

413. Risks of second solid cancer were evaluated among 668 retinoblastoma patients in the Kingdom of the Netherlands [M7]. The majority of the cohort members (n=392) were treated with surgery alone, while and 240 patients received radiotherapy. After a median follow-up of 21.9 years, a total of 74 second primary cancers were diagnosed, 12 in non-hereditary patients and 62 in hereditary patients. Most of the hereditary patients (70%) were treated with radiotherapy, whereas only 8% of the non-hereditary patients were treated with radiotherapy. The SIRs for solid cancers were 5.06 (95% CI: 2.52, 9.05) among hereditary retinoblastoma patients and 1.01 (95% CI: 0.33, 2.36) among non-hereditary patients. EBRT and combination of EBRT and chemotherapy were associated with increased risks of solid cancer; SIRs were 4.89 (95% CI: 1.79, 10.6) and 12.7 (95% CI: 3.45, 32.4), respectively.

414. Increased risk of second solid cancer was also reported in the studies of childhood cancer survivors as a whole in Europe and North America. In a cohort study of 4,400 survivors of childhood cancer in France and the United Kingdom [D9], 31 patients developed a solid second primary cancer among the 1,045 treated with radiotherapy alone, compared to 5.6 expected (RR=5.6 (95% CI: 3.8, 7.8)) while three second primary solid cancers occurred among the 406 patients treated with surgery alone, compared to 2.0 expected second primary cancer (RR=1.5 (95% CI: 0.6, 4)).

415. In a case-control study nested in the childhood cancer survivor study in the Kingdom of the Netherlands, Cardous-Ubbink et al. [C4] showed that radiotherapy alone and combined modality therapy including recurrence treatment were associated with the highest SIRs of solid cancers: 16.5 (95% CI: 7.56, 31.4) and 20.1 (95% CI: 9.18, 38.1), respectively. In a Cox regression analysis, radiotherapy was significantly associated with solid cancer risk (HR=3.39 (95% CI: 1.68, 6.84)), in contrast to chemotherapy, for which an HR of 0.72 (95% CI: 0.37, 1.40) was estimated.

416. In a study of 25,965 two-month survivors of childhood cancer identified from the SEER registries, childhood cancer patients whose initial treatment included radiotherapy or chemotherapy or both were shown to be at higher risk of developing a new solid cancer than those not treated [I9]. The adjusted RR for second solid cancer was significantly elevated for patients treated with both radiotherapy without chemotherapy (RR=2.8 (95% CI: 1.8, 4.3)) and radiotherapy with chemotherapy (RR=3.2 (95% CI: 2.1, 4.9)), compared with patients who had received neither radiotherapy or chemotherapy.

417. A case-control study nested in a cohort of 33,809 childhood cancer survivors in Germany also showed a significantly elevated risk of solid cancer risk [K1]. The RR for solid cancer of patients treated with radiotherapy was 4.46 (95% CI: 2.51, 7.95) on the basis of 167 cases. This estimate of RR remained unchanged from the estimate without adjustment for chemotherapy.

418. Several studies have also evaluated the risk of all second solid cancers combined in populations treated with radiotherapy for tumours in adulthood. One of the largest relevant studies is that of Chaturvedi et al. [C8], who investigated second primary cancers among 104,760 one-year survivors of cervical cancer from 13 cancer registries in Denmark, Finland, Norway, Sweden, and the United States. A total of 12,496 second primary cancers were reported including 11,720 solid cancers, with a mean follow-up of 12.2 years.

419. The SIRs for all solid cancer combined for patients with and without radiotherapy were 1.36 (95% CI: 1.33, 1.39) and 1.06 (95% CI: 1.02, 1.11). The SIRs were elevated to a statistically significant extent for cancers at heavily (SIR=1.59 (95% CI: 1.54, 1.66)), moderately (SIR=1.30 (95% CI: 1.22, 1.39)), and lightly (SIR=1.21 (95% CI: 1.16, 1.26)) irradiated sites.

420. Another large study by Lönn et al. [L23] investigated second primary cancers among 60,949 one-year survivors of uterine corpus cancer treated with different radiotherapy modality such as brachytherapy and EBRT. A total of 7,428 second primary cancers were observed including 6,538 solid cancers. The RRs of solid cancers among patients with combination of radiotherapy, EBRT, and brachytherapy were 1.24 (95% CI: 1.13, 1.35), 1.14 (1.06, 1.21), and 1.06 (0.98, 1.14), respectively, compared with patients with surgery only. Increased risks following combination radiotherapy were found for second solid cancers occurring at heavily (RR=1.93 (95% CI: 1.60, 2.32)) and moderately irradiated sites (RR=1.55 (1.17, 2.06)).

421. In a study of Rubino et al. [R16], 6,841 thyroid cancer patients treated with $^{131}$I or without and external beam therapy in France, Italy, and Sweden were followed up in an analysis of second primary cancers. A linear dose-response relationship without any significant quadratic effect best fitted the data on the solid tumours ($\chi^2$=0.46; $p$=0.5) and an ERR of 0.035 (95% CI: 0.009, 0.069) $GBq^{-1}$ following administration of $^{131}$I was estimated. The ERR for patients who were treated with external beam therapy was similar to the ERR for patients who did not receive external radiotherapy.

422. As discussed above, there is strong epidemiological evidence for an increased risk of second primary combined solid cancers among patients treated with radiotherapy. In general, the radiotherapy related risk (i.e. RR or SIR) seems to be higher following childhood exposure than adulthood exposure. A difficulty arises in assessing risk for all solid cancers combined because radiation is delivered to partial sections of the body of patients rather than to the whole body. It was therefore decided to exclude data from studies who reported risks of all solid cancer combined risk per unit proxy or reference dose (e.g., colon dose) from the meta-analysis because of dose heterogeneity in each study. Additionally, there were only a few available studies on the dose-response relationship for all solid cancers combined. However, for the purpose of assessing the risk of solid cancer combined but with a high degree of dose homogeneity to the organs at risk, it was decided to combine the risks used in each of the six site-specific meta-analyses and the results of this analysis are reported in the next subsection.

## (a) *Results of the meta-analysis for six cancer sites*

423. A The primary random effects meta-analysis model pooled the log RR from 63 non-overlapping individual studies used in the meta-analyses for the six second primary solid cancer sites considered (sarcoma, breast, lung, gastrointestinal, thyroid, and brain). The pooled ERR per unit dose estimated in

the primary analysis was 0.08 (95% CI: 0.06, 0.11). The primary analysis also estimated a value for $I^2$ of 59.3% (95% CI: 46.3, 69.1), indicating that this percentage of the variability in the estimates effect size of radiotherapy is due to the differences between studies. The Cochran's Q test ($p$<0.0001) indicated that there was a statistically significant heterogeneity between the individual studies. ERRs of second primary cancers for the six cancer sites estimated in the primary analysis are shown in figure XXVIII. The CIs were reconstructed from the standard errors that were used to calculate the study weightings applied in the meta-analysis, and for this reason, may differ slightly from those published in the original papers. Leave one-study-out diagnostics caused the resulting 63 main central pooled ERRs per unit dose for the 62 remaining studies to vary between 0.07 and 0.09.

Figure XXVIII. Risks per unit organ dose for the six second primary solid cancer sites estimated in the primary analysis

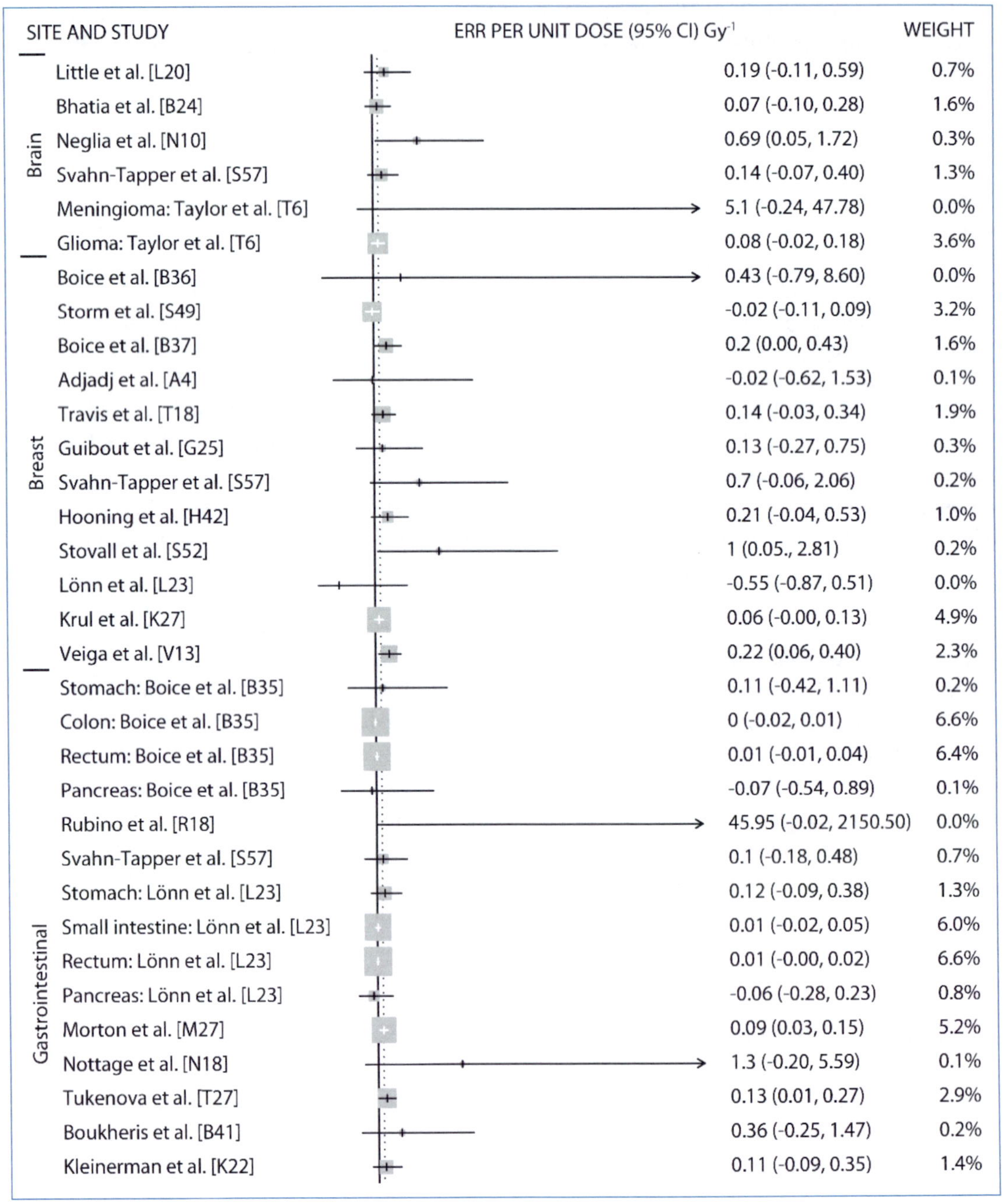

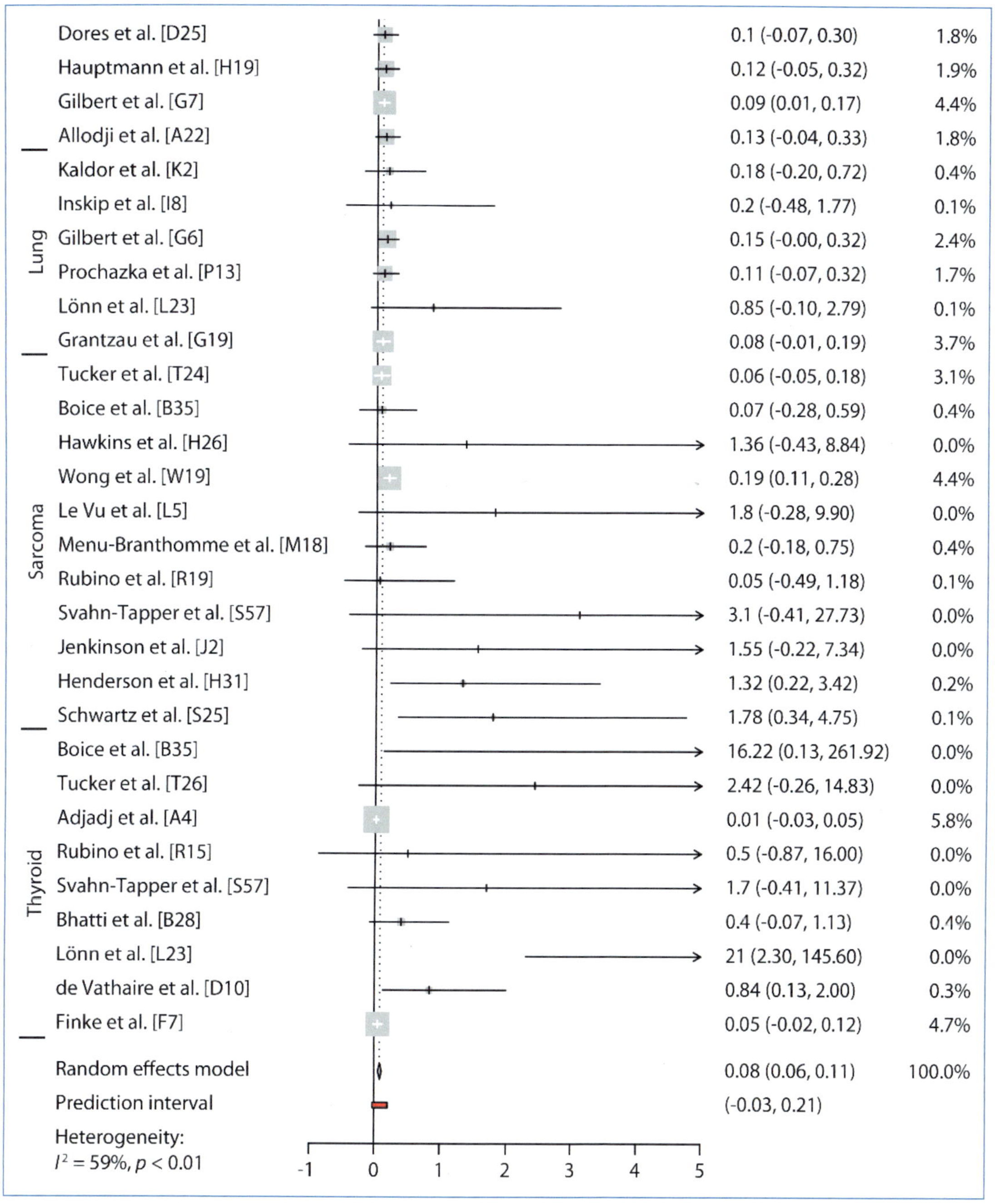


## C. Risk prediction models and calculations of assigned share (AS)

424. Evidence suggests that second primary cancer risk after radiotherapy is affected by the following factors:

*(a)* Age of the patient at the time when radiotherapy was applied in the treatment of the first primary cancer;

*(b)* Attained age of the patient in the follow-up after radiotherapy;

*(c)* Sex of the patient;

*(d)* Type of organ or tissue receiving the radiation dose and within which the second primary cancer was found;

*(e)* Magnitude of the organ dose and the size of the irradiated area;

*(f)* Dose rate to the organ of interest;

*(g)* Genetic risk factors (cf. [N14] for a review of the relevant papers); and

*(h)* Possible influence of chemotherapy applied in combination with radiotherapy.

425. General prediction models of radiation risk for outcomes of interest, such as cancer, are often based on lifetime risks, based on age-specific cancer incidence and mortality rates and all-cause mortality rates for the population of interest, along with detailed dose-response models for radiation-induced cancer risk with effect modifiers such as age at exposure, attained age, and sex.

426. However, for the cohort of cancer survivors treated with radiotherapy, the development of risk prediction models for the incidence of a second primary cancer related to the radiation treatment is more complicated. Ideally, these models need to consider at least the effect modifiers *(a)* to *(d)* listed above, but these are associated with more uncertainties and unknown factors than in risk prediction models for the general population [S16]. Uncertainties include the effect of the first primary cancer type on life expectancy and the effect of the risk factors for the first primary cancer, such as genetic predisposition, smoking levels, body mass index, dietary quality, and physical activity levels. Other uncertainties and unknown factors include the lack of data on age-specific second primary cancer rates in patients not treated with radiation (e.g., rates for a population of cancer survivors following postsurgical cancer removal who did not receive radiotherapy). Furthermore, radiation risk models fitted to data on second primary cancers are often only based on a low number of cases, and therefore, do not have sufficient statistical power for fitting coefficients associated with effect modifiers, such as the age at which radiotherapy was administered. There are also uncertainties and reliability issues connected with directly applying estimates from studies on patients treated with older radiation techniques to recently treated patients receiving radiotherapy via contemporary modern radiation techniques. In addition, risk models based on radiation epidemiology require the knowledge of the dose at the site of the second primary cancer. Dose-response curves derived from epidemiological studies, however, are usually based on approximate dose estimates at the site of the second primary cancer, because of the uncertainties in the dose reconstruction and the macroscopic size of the diagnosed second primary cancer.

427. Risk predictions models for application after radiation treatment for breast cancer were developed by Simonetto et al. [S41] for second primary cancer of the lung, contralateral breast, and leukaemia. For the high-dose region, meta-analyses from relevant radiotherapy studies were performed. These were combined with risk estimates from the LSS of the cohort of Japanese atomic bombing survivors, to construct dose-response relationships for the full dose range as applied in breast cancer radiotherapy. The resulting estimates of long-term health risks were shown to be largely compatible with rate ratios observed in randomized breast cancer radiotherapy trials [T7].

428. A summary of the pooled results of the meta-analyses of ERRs per unit dose at the site of the second primary cancers estimated in this evaluation are presented in table 23, which gives values of ERRs per unit dose estimated in the main analyses and for the subgroup analysis limited to exposures in childhood or adolescence. The values estimated in the meta-analyses were included only if they were based on three or more studies. It should be noted that the ERRs per unit dose given in table 23 do not allow for a reliable characterization of specific risks of second primary cancer in other relevant subgroups, for example in patients who were also treated with chemotherapy.

Table 23. Pooled excess relative risks per unit dose of second primary cancer, for the main analyses and for the subgroup analysis restricted to exposures in childhood or adolescence

| *Meta-analysis cancer site* | *Number of study results* | *Pooled ERR per unit dose (95% CI)* |
|---|---|---|
| **MAIN STUDY** | | |
| Haematological | 11 | 0.01 (0.00, 0.03) |
| Sarcoma | 11 | 0.33 (0.09, 0.63) |
| Breast | 12 | 0.12 (0.04, 0.21) |
| Lung | 6 | 0.11 (0.04, 0.19) |
| Gastrointestinal | 19 | 0.04 (0.01, 0.07) |
| Thyroid, dose cut-off including categorical risks <15 Gy | 9 | 0.68 (0.09, 1.6) |
| Brain | 6 | 0.11 (0.03, 0.19) |
| **SUBGROUP (YOUNG PERSONS)** | | |
| Haematological | 4 | 0.00 (-0.04, 0.05) |
| Sarcoma | 9 | 0.56 (0.15, 1.13) |
| Breast | 3 | 0.23 (0.08, 0.4) |
| Gastrointestinal | 4 | 0.13 (0.04, 0.24) |
| Thyroid, dose cut-off including categorical risks <15 Gy | 6 | 0.42 (0.0, 0.98) |
| Brain | 6 | 0.11 (0.03, 0.19) |

429. One measure of organ sensitivity with respect to the development of second primary cancer that does not require explicit consideration of age-specific cancer rates from the general population, is the assigned share (AS) [K25, T7] also known as the probability of causation [L2] or the attributable risk [K25, T7]. The assigned share usually represents the probability that an observed cancer was caused by a preceding radiation exposure and can be calculated from the excess relative risk using the equation:

$$\mathrm{AS} = \frac{ERR}{1 + ERR}$$

430. This equation can be used to calculate the AS for a specific type of second primary cancer for the seven sites considered in the meta-analyses carried out in this evaluation from the mean dose to a specific organ at risk ($\bar{D}_{OAR}$) delivered during radiotherapy and the meta-analysis pooled ERR per unit dose to the organ at risk ($\mathrm{ERR_{pd}}$) yields:

$$AS(SPCa_i, \bar{D}_{OAR}) = \frac{ERR_{pd}(SPCa_i) \times \bar{D}_{OAR}}{1 + ERR_{pd}(SPCa_i) \times \bar{D}_{OAR}}$$

AS probabilities, for doses to the organ at risk of 100 mGy, 1, 5 and 10 Gy, are presented in table 24. Values are based on the ERRs from the primary meta-analyses and for the subgroup analyses restricted to exposure in childhood given in table 22.

431. Table 24 shows that the highest AS values for all ages at treatment are those for thyroid cancer ranging from 0.06 t0.87 for organ doses of 100 mGy and 10 Gy, respectively. The second highest AS values for all ages at treatment is for sarcoma, ranging from 0.03 to 0.77 for organ doses of 100 mGy and 10 Gy, respectively. The assigned shares for the other cancer sites investigated are lower. A similar pattern is observed for the AS of patients treated as young persons (table 24). The ASs for thyroid cancer

for all ages at treatment vary between 0.04 and 0.81for organ dose of 100 mGy and 10 Gy, respectively. The highest AS for treatment at young ages is for sarcoma, and it increases from a value of 0.05 to 0.85 between organ doses of 100 mGy and 10 Gy, respectively. The AS values, for patients treated as young persons, for the other cancer sites investigated are lower than those for thyroid and sarcoma. At a dose of 10 Gy the AS varies between 0.52 and 0.70 for the other cancer types, with the exception of leukaemia for which the AS remains very small (0.02).

Table 24. Assigned share for four potential doses to the site of the organ at risk

| *Second primary cancer site* | *Assigned share (95% CI)* [a] | | | |
|---|---|---|---|---|
| | *100 mGy* | *1 Gy* | *5 Gy* | *10 Gy* |
| **ALL AGES AT RADIATION TREATMENT** | | | | |
| Thyroid, dose cut-off for including categorical risks 15 <Gy | 0.06(0.00, 0.13) | 0.41 (0.15, 0.67) | 0.77 (0.58, 0.96) | 0.87 (0.75, 0.99) |
| Breast | 0.01(0.00, 0.02) | 0.11(0.04, 0.17) | 0.37(0.21, 0.54) | 0.54(0.37, 0.72) |
| Lung | 0.01(0.00, 0.02) | 0.10(0.04, 0.16) | 0.36(0.21, 0.51) | 0.53(0.37, 0.69) |
| Brain | 0.01 (0.00, 0.02) | 0.10 (0.03, 0.16) | 0.35 (0.17, 0.52) | 0.52 (0.33, 0.71) |
| Sarcoma | 0.03 (0.01, 0.06) | 0.25 (0.10, 0.40) | 0.62 (0.43, 0.81) | 0.77 (0.63, 0.91) |
| Gastrointestinal | 0.004 (0.001, 0.007) | 0.04 (0.01, 0.07) | 0.17 (0.07, 0.27) | 0.29 (0.14, 0.44) |
| Haematological | 0.001 (0.000, 0.003) | 0.01 (0.00, 0.03) | 0.06 (0.00, 0.12) | 0.12 (0.01, 0.22) |
| **TREATED WITH RADIATION AS YOUNG PERSONS** | | | | |
| Thyroid, dose cut-off for including categorical risks <15 Gy | 0.04 (0.00, 0.08) | 0.29 (0.06, 0.53) | 0.68 (0.42, 0.93) | 0.81 (0.63, 0.99) |
| Breast | 0.02 (0.01, 0.04) | 0.19 (0.08, 0.29) | 0.53 (0.36, 0.71) | 0.70 (0.55, 0.84) |
| Brain | 0.01 (0.00, 0.02) | 0.10 (0.03, 0.16) | 0.35 (0.17, 0.52) | 0.52 (0.33, 0.71) |
| Sarcoma | 0.05 (0.01, 0.10) | 0.36 (0.16, 0.56) | 0.74 (0.57, 0.90) | 0.85 (0.74, 0.96) |
| Gastrointestinal | 0.013 (0.003, 0.023) | 0.12 (0.04, 0.20) | 0.40 (0.22, 0.58) | 0.57 (0.39, 0.76) |
| Haematological | 0.00 (0.000, 0.005) | 0.00 (0.00, 0.05) | 0.01 (0.00, 0.23) | 0.02 (0.00, 0.45) |

[a] The 95% Cis in the table were calculated using propagation of errors.

432. The AS methodology assumes a linear dose-response relationship, since it is based on a linear pooled estimate of the ERR per unit dose from the meta-analysis. Justification for assuming linearity can be found in the review by Berrington de Gonzalez et al. [B19] based on 28 studies of high-dose radiotherapy patients in which it is stated that "there was little evidence that the dose-response curve was non-linear in the direction of a down-turn in risk, even at organ doses of ≥60 Gy. Thyroid cancer was the only exception with evidence of a downturn after 20 Gy". However, the review by Berrington de Gonzalez et al. did not present results for second primary cancers of the haematopoietic and lymphoid system, and only analysed radiotherapy studies with organ doses above 5 Gy, and therefore, no inferences regarding linearity at lower doses could be made. As shown in this annex, as well as in the review of Berrington de Gonzalez et al. [B19] and in other studies, the estimates of ERR per unit dose from radiotherapy studies tend to be significantly lower than those from other types of epidemiological studies of non-radiotherapy low doses. For the outcome first or second primary cancers of the haematopoietic and lymphoid system, and for the whole dose range from 0 to 10 Gy, such a discrepancy could be indicative of a non-linear dose-response relationship [S41].

433. A disadvantage of the AS methodology is the insensitivity of the AS values to the actual timing of occurrence of the second primary cancer. Two methods are suggested to deal with this problem in future research. The first method would involve collecting results on latency, time frames for second primary cancer occurrence, and age at treatment effects from the literature and apply those to the pooled ERR per unit dose from the meta-analysis. Another method to characterize effects of age at exposure and attained age on radiation-induced second primary cancer risk in cohorts of patients treated with radiotherapy would rely on modifying the pooled ERR per unit dose model using the risk modification from the Japanese atomic bombing survivors cohort [S19].

434. As far as the first method is concerned, some considerations could be given to the minimum latency period for the occurrence of second primary cancers and to the total time frame after radiotherapy, during which radiation-induced second primary cancers have been reported in the literature. This information could then be built into the calculation of the AS. For example, based on the information given in chapter V.B.2, most of the radiation-induced haematological cancers are known to occur in the period between 2 and 20 years after exposure. Some publications [L27, M19, T19] report a latency period for onset of second primary non-leukaemia cancers after radiotherapy of 10 years or more after the initial treatment.

435. Although, the AS methodology based on ERR per unit dose from meta-analyses lacks information on risk modification by sex and age at treatment, reliable pooled estimates of the risks per unit dose at some sites of second primary cancer can still be provided. Some studies provide strong epidemiological evidence for risk modification by age at exposure. Notably, the LSS of Japanese atomic bombing survivors estimated that the ERR for solid cancers decreased by about 17% (90% CI: 7, 25) per decade increase in age at exposure [P12]. This estimate was later confirmed by Grant et al. [G18] to be 19% (95% CI: 12, 27) with an extended follow-up. While not allowing firm conclusions on the age at exposure effects, the sensitivity of children treated by radiotherapy has been demonstrated in the CCSS [B22, M29], which reported a cumulative incidence of 7.9%, notably 10% with radiotherapy and 5% without radiotherapy, for invasive cancers at 30 years from primary cancer diagnosis [F16]. Similarly, studies on second primary cancer after treatment for Hodgkin lymphoma in children indicated that the incidence was 7% at 15 years after treatment [W18]. Evidence on effect modification is rare for second primary cancer, but can be found in some studies. For example, Hancock et al., reported that breast cancer risk is strongly modified by age at treatment. In particular, radiation risk appears to have been overall high for treatment under age 30 years with the highest risk for treatment under age 15 years and lower risk for treatment in patients over age 30 years [H13].

436. The second method combines data on dose-response relationships from cohorts with information on effect modification from the LSS of the Japanese atomic bombing survivors. For this purpose, comparisons were made of patterns of effect modification by age, between models developed using data from the cohorts of Japanese atomic bombing survivors and models for cancer risks previously reported for radiotherapy patients [S19]. Schneider and Walsh reported that the projection models for the Japanese atomic bombing survivors predict risk well for the outcomes of testicular cancer survivors, but survivors of Hodgkin lymphoma show steeper variations of risk with both age at exposure and attained age than the models used for the Japanese atomic bombing survivors. They reported that the standard functions for risk effect modification by age, based on the Japanese atomic bombing survivors data, predict risks of second primary cancer reasonably well in radiotherapy patients for ages at exposure prior to 40 years and attained ages before 55 years. They also provided extended models which were better fitted than the simple models used for the Japanese atomic bombing survivors to predict solid cancer risk increase as a function of age at exposure beyond 40 years and the risk decrease as a function of attained age beyond 65 years.

437. Based on the information provided above, a generalized suggested risk projection method for most cancer sites, but excluding leukaemia and thyroid, could be to calculate mean doses to OAR from the treatment plan for a specific patient. One can then apply the relevant ERR per unit dose from the

meta-analysis or from informative single papers for the cancer sites not included in the meta-analyses results, and the risk effect modifiers by age at exposure and attained age. This approach would describe the enhanced risk for children and would allow the prediction of AS for a specific patient age at treatment and at a specific attained age. For leukaemia and thyroid, special consideration of the non-linear dose-response relationship and internal exposures for thyroid and the different risk effect modifiers for leukaemia would need to be accounted for.

438. A methodological limitation of the AS projection is the "Greenland effect" where radiation exposure may work to move forward the occurrence of cancers that would have occurred without radiation exposure[G22]. The alternative suggestion made by Greenland to report expected years of cancer-free-life lost due to radiation exposure is suitable to assess risk in the general population but cannot be used to assess risk in cancer survivors, because if the lack of baseline rates for second primary cancers in patients not treated with radiation, on which to base years of life lost statistics. A quantification of the Greenland effect has been undertaken in a simulation study with biologically-based models [E3]. Earlier occurrence of a given cancer may depend on the mechanism of radiation acting as initiator or promotor of preneoplastic clones. Further work that has been done in the past on modelling the risk of second primary cancers after radiotherapy can be found in the works of Schneider and Taylor et al. [S16, T8].

## D. Comparison of second primary cancer risks to cancer risks from other radiation exposure

439. *Haematopoietic and lymphoid malignancies*: The pooled ERR per unit dose estimated in the main meta-analysis was 0.01 (95% CI: 0.00, 0.03) $Gy^{-1}$. For comparison, Hsu et al. [H49] examined 312 cases of leukaemia, other than CLL or ATL, from the LSS of Japanese atomic bombing survivors The sex-averaged ERR per unit dose to the ABM at age 70 after exposure at age 30 years, estimated using a linear dose-response model, was 0.79 (95% CI: 0.03, 1.93) $Gy^{-1}$, which is much higher than the ERR estimated in the meta-analysis performed in this annex; the two estimates have non-overlapping CIs that are coincident on one side. The ERR at 1 Gy estimated for patients in the LSS with ages at exposure between 0 and 19 years was 6.5 (95% CI: 4.0, 10.3). Comparisons of ERRs are limited by a more homogenous outcome definition in the LSS than in the meta-analysis and the unknown effect of age-related effect modifiers on the pooled estimates of the meta-analysis. It is also important to note that the conditions of the dose delivery to the atomic bombings survivors was very different from those encountered in clinical radiotherapy. Other limitations of the pooled estimates in a meta-analysis include the small numbers of cases for rare types of cancer, which is only partially linked to the attained age at the end of follow-up, uncertainty in the estimates of organ doses, heterogeneity in the definition of study outcomes and missing or incomplete information on competing risk factors, in particular on chemotherapy and on cancer predisposing syndromes.

440. *Sarcoma*: The pooled ERR per unit dose estimated in the main meta-analysis was 0.33 (95% CI: 0.09, 0.63) $Gy^{-1}$, while the pooled estimate of the ERR per unit dose, based on nine studies related to exposures in childhood or early adolescence, was 0.56 (95% CI: 0.15, 1.13) $Gy^{-1}$. However, the uncertainties are large, and the study design is too heterogeneous in terms of the age-related modification of the effect on the risk factor, to draw any conclusions regarding effects of age at exposure. For comparison, a study by Preston et al. [P12] reported 149 cases of sarcomas among the Japanese atomic bombing survivors and estimates of the ERR per unit dose that were statistically compatible. The estimate of the sex-averaged ERR per unit dose was 0.49 (90% CI: 0.07, 1.4) $Gy^{-1}$; the ERR per unit dose for men at 0.76 (90% CI: 0.08, 2.3) $Gy^{-1}$ was not significantly higher than that for women (0.20 (90% CI: 0.02,

0.80) $Gy^{-1}$). Preston et al. further reported that "the ERR increased with decreasing age at exposure and attained age, with the effect of attained age being marginally significant" [P12]. Samartzis et al. reported 104 cases of soft-tissue sarcomas among the Japanese atomic bombing survivors with a mean colon dose of 0.18 Gy [S3]; mean ages at the time of the atomic bombings and sarcoma diagnosis were 26.8 and 63.6 years, respectively. A linear dose-response model with an ERR per unit dose of 1.01 (95% CI: 0.13, 2.46; $p$=0.019) $Gy^{-1}$ was reported for soft-tissue sarcoma. Given the large uncertainties in the risks from both the meta-analysis and the Japanese atomic bombing survivors, the risks appear to be statistically compatible, but this comparison is unable to support strong conclusions on risk effect modifiers, such as age at exposure.

441. *Breast cancer*: The pooled ERR per unit dose estimated in the main meta-analysis was 0.12 (95% CI: 0.04, 0.21) $Gy^{-1}$. A pooled analysis of eight cohorts, including the Japanese atomic bombing survivors [P11], reported that no simple unified summary model adequately described the excess risks in all groups. However, a simple pooled ERR per unit dose model was presented and the central estimate was 0.86 (95% CI: 0.7, 1.04) $Gy^{-1}$ for women exposed at age 25 years, except for patients in three of the eight cohorts, for whom the age at exposure was six months. This risk appears to be about seven times higher than the value estimated in this evaluation, but a comprehensive assessment of the influence of the risk effect modifiers in this comparison is made difficult by the lack of detailed information in the papers that contributed to the meta-analysis. A study on the Stockholm cohort of patients treated for skin haemangioma reported an ERR per unit dose to the breast at age 50 years of 0.48 (95% CI: 0.28, 0.69) $Gy^{-1}$ [E2]. This risk is about four times higher than the main pooled value estimated in this evaluation, but a complete assessment of the influence of the risk effect modifiers is difficult.

442. *Lung cancer*: The pooled ERR per unit dose estimated in the main meta-analysis was 0.11 (95% CI: 0.04, 0.19) $Gy^{-1}$. A study on the Japanese atomic bombing survivors [C3] reported that for non-smokers, the sex-averaged ERR per unit dose to the lung at age 70 after radiation exposure at age 30 was 0.81 (95% CI: 0.51, 1.18) $Gy^{-1}$ with a female to male ratio of 2.83. Given the large uncertainties in the risks from both the meta-analysis and the Japanese atomic bombing survivors, the risks do not appear to be statistically compatible, and this comparison is unable to support strong conclusions on risk effect modifiers such as age at exposure, smoking status, and sex due to a general lack of information in the studies that contributed to the meta-analysis.

443. *Gastrointestinal cancers*: The pooled ERR per unit dose estimated in the main meta-analysis was 0.04 (95% CI: 0.01, 0.07) $Gy^{-1}$. Studies on non-radiotherapy cohorts, such as the LSS of the Japanese atomic bombing survivors, usually report data on gastrointestinal outcomes separately (e.g. stomach and colon cancer); therefore, a suitable comparison based on published risks for the combined gastrointestinal cancer is not currently possible without further analysis of the data. However, for comparison purposes, Sugiyama et al. [S54] reported a significant linear ERR per unit dose of 0.63 (95% CI: 0.34, 0.98) $Gy^{-1}$ for total colon cancer incidence and Sakata et al. [S2] reported a smoking adjusted ERR per unit dose of 0.33 (95% CI: 0.20, 0.47) $Gy^{-1}$ for stomach cancer incidence. Both ERR values are sex-averaged for an attained age of 70 years—the ERR for colon cancer is for an age at exposure of 30 years—and, statistically, are significantly larger than the pooled estimate of the ERR performed for this evaluation.

444. *Thyroid cancer*: The pooled ERR per unit dose estimated in the main meta-analysis, based on applying an upper cut-off for dose category risks of 15 Gy, was 0.68 (95% CI: 0.09, 1.60) $Gy^{-1}$. Furukawa et al. analysed 371 thyroid cancer cases from the LSS of Japanese atomic bombing survivors and using a linear dose-response model, estimated an ERR at a dose of 1 Gy of 1.28 (95% CI. 0.59, 2.70) $Gy^{-1}$ at age 60 after acute exposure at age 10, and is statistically compatible with the meta-analysis result [F17]. However no increased risk was found for those exposed at age 20 or older. Ron et al., carried out a pooled analysis of seven studies of thyroid cancer, which included 120,000 people, nearly 700 thyroid cancers and 3 million PY at risk [R10]. The study reported that linearity down to 0.10 Gy best described the dose-response relationship for exposures before age 15 years. At the very high doses commonly using in cancer

therapy (>10 Gy), the risk appeared to level off or decrease. For childhood exposures, the pooled ERR per unit dose to the thyroid was 7.7 (95% CI: 2.1, 28.7) $Gy^{-1}$, which is higher than, and not statistically compatible with, the meta-analysis result. However, it is unclear if the ERRs can be compared because of a lack of overall information on overall risk-effect modifier, such as age at exposure, in the studies contributing to the meta-analysis. In an update to this study Veiga et al. [V12] pooled data from 12 studies of thyroid cancer patients who were exposed to radiation before age 20 years. Estimates of ERR per unit dose to the thyroid increased with decreasing age at exposure. The estimated ERR per unit dose was 2.76 (95% CI: 0.94, 4.98) $Gy^{-1}$, and significant within 10 years of radiation exposure, based on 42 exposed cases, and remained elevated at 50 years and more after exposure. This linear risk for the dose-response relationship for thyroid cancer is over three times higher than, but statistically compatible with, the main meta-analysis result. A study on children treated for skin haemangioma [L19] reported an ERR per unit dose of 7.5 (95% CI: 0.4, 18.1) $Gy^{-1}$, for a mean age at exposure of 15 months, which is also higher than, but statistically compatible with, the meta-analysis result based on a much wider range of age at exposure. The meta-analysis estimate is a factor of two smaller than the EOR per unit dose of 1.36 (95% CI: 0.39, 4.15) $Gy^{-1}$ from exposure to $^{131}I$ about 30 years after the Chornobyl accident reported by Tronko et al. [T22].

445. *Brain cancer*: The pooled ERR per unit dose estimated in the main meta-analysis was 0.11 (95% CI: 0.03, 0.19) $Gy^{-1}$. The most recent paper on brain tumour cases diagnosed between 1958 and 2009 among the cohort of Japanese atomic bombings survivors [B43] reported 285 CNS tumours (67 gliomas, 107 meningiomas, 49 schwannomas, and 62 other or not otherwise specified tumours) among 105,444 individuals with a total dose to the brain estimated to contribute 3.1 million PY at risk. The paper reported that acute irradiation of the brain with a weighted dose of 1 Gy increased the risk of all CNS tumours as a group (over background rates) by 140%. Using a simple linear model without taking account of effect modification, the ERR per unit weighted dose to the brain was 1.40 (95% CI: 0.61, 2.57) $Gy^{-1}$, which is statistically, significantly higher than the ERR estimated in the meta-analysis, which also used a simple linear model without effect modification. A study on children treated for skin haemangioma [L19] reported an ERR per unit dose of 10.9 (95% CI: 3.6, 20.5) $Gy^{-1}$ for a mean age at exposure of 15 months, which is much higher, and statistically significantly different, than the value estimated by the meta-analysis on the basis of a wider age at exposure range. However, the risk of skin haemangioma was based on SIR and a low mean dose of 77 mGy, which may be important factors contributing to the high risk per unit dose reported.

446. A study by Li et al. [L12] identified 14,048 patients in the Japanese atomic bombings survivors LSS diagnosed with a first primary cancer, of whom 1,088 were diagnosed with a second primary cancer. A similar linear dose-response relationship for both first and second primary solid tumours was reported, with an ERR per unit dose of 0.65 (95% CI: 0.57, 0.74) $Gy^{-1}$ and 0.56 (95% CI: 0.33, 0.80) $Gy^{-1}$, respectively. The ERRs per unit dose for both first and second primary leukaemia were also similar, with values of 2.65 (95% CI: 1.78, 3.78) $Gy^{-1}$ and 3.65 (95% CI: 0.96, 10.70) $Gy^{-1}$, respectively.

## E. Limitations of existing evidence, major uncertainties, and gaps in knowledge

447. The main limitation in the studies on second primary cancer after radiotherapy included in the meta-analysis performed in this evaluation is that patients from many of the cohorts examined in, especially those of childhood cancer survivors, had not reached retirement age at end of the follow-up period. As such, their attained age was below the age at which the risk of cancer types more generally associated with old age, such as gastrointestinal cancer, is greatest.

448. There are initiatives dedicated to filling current gaps in knowledge, some concerning platforms and registries, such as "E$^2$-RADIatE" which is a platform that collects real-world data through prospective data registries in radiotherapy, and was developed through a collaboration between the European Organisation for Research and Treatment of Cancer and the European Society for Radiotherapy and Oncology [E6].A further initiative to improve knowledge gaps comes from the Dutch model-based approach which is used for the selection of patients for proton therapy. Although results are still pending, the application of this approach to breast cancer patients has been described in a recently published paper by Boersma et al. [B31].

449. Estimates of second primary cancer risks after receiving radiation treatment using new modalities such as IMRT, volumetric modulated arc-therapy, proton and heavy ion radiotherapy are of direct relevance for current risk projections. However, the groundwork for assessing the long-term risks from these novel radiation treatment techniques is only in its initial phase. Two studies are worth mentioning in this regard. The Pediatric Proton and Photon Therapy Comparison Cohort, has been recently established by the NCI in the United States to perform a retrospective cohort study with prospective follow-up including paediatric cancer patients treated with proton therapy and a comparison group of similar patients treated with photon therapy. This new cohort will build on the foundation created by the Pediatric Proton Consortium Registry [N15] which currently includes 20 proton therapy centres in the United States and has recruited more than 3,000 paediatric proton patients who have actively enrolled in the registry since 2012. The EU-HARMONIC project is a European-funded project with a consortium of 24 European partners that aims to better understand the long-term health effects of medical exposure to ionizing radiation in children, including those undergoing radiotherapy for cancer, in order to optimize treatment plans in young patients and reduce the risk of late toxicities[H16]. The expected impacts are the provision of much needed information on the effects of low to moderate doses of radiation on humans, and improvements in radiological protection in medicine. The project is based on building a close relationship with clinicians, such as paediatric oncologists and cardiologists, social scientists and patients, which will ensure the study is relevant not only in terms of clinical effectiveness, but also in terms of patient care and quality of life.

450. Although both projects are expected to help fill current knowledge gaps on second primary cancer risks after modern radiation treatment, their findings may not become available for many years due to the long latency period for the development of solid tumours. Furthermore, patients treated for a first primary cancer represent a unique group with very specific characteristics that may not have the same baseline risk of cancer as the rest of the general population. The meta-analyses presented in this annex rely solely on estimates of ERR, due to the unavailability of estimates of EAR in the literature reviewed. As a result, the comparison of these values with those associated with other non-radiotherapy exposure situations is highly dependent on the baseline cancer rate, posing aa limitation in such comparisons. This underscores the need for future research to incorporate EAR estimates per unit dose, ensuring a more comprehensive understanding of radiation-associated risks.

# F. Summary and discussion

451. An extensive literature search and meta-analysis were conducted to estimate risks of second primary cancer after radiotherapy with respect to the relevant organ doses at the affected cancer site. Data on cumulative risks and latency periods were also presented. Sufficient information was found in the open literature for the meta-analysis of seven second primary cancer sites: haematopoietic and lymphoid malignancies, sarcoma, breast cancer, lung cancer, gastrointestinal cancer, thyroid cancer, and brain cancer. The highest ERRs per unit dose to the OAR estimated in the primary meta-analysis were for thyroid cancer (0.68 (95% CI: 0.09, 1.6) Gy$^{-1}$) and sarcoma (0.33 (95% CI: 0.09, 0.63) Gy$^{-1}$). For other

cancer sites, the ERRs per unit dose were estimated at or below 0.12 Gy. Insufficient data were found in the open literature to justify a meta-analysis of risks for second primary skin cancer and other second primary cancer sites (e.g. kidney cancer and cancers of the male and female genitalia). No firm conclusions could be drawn on the risk of second primary non-Hodgkin lymphoma because due to the small number of cases in most studies. Additionally, comparisons between studies for haematopoietic and lymphoid malignancies were made complicated by major shifts in the classification of neoplasms of the haematopoietic and lymphoid tissues, as some lymphomas and leukaemia are different manifestation of the same disease. Also it should be noted that the meta-analysis of gastrointestinal tumours combined organs of different radiosensitivity and the meta-analysis of brain tumours included both benign and malignant cancers.

452. A meta-analyses of studies reporting risks for all solid cancer risks as a single group could not be performed due to heterogeneity of the reference doses. However, an overall meta-analysis was conducted by combining all 63 risk estimates from the six solid cancer types considered in the meta-analysis. This resulted in an estimate of the overall ERR per unit dose for primary solid cancer of 0.08 (95% CI: 0.06, 0.11) $Gy^{-1}$. Limitations of this estimate are that *(a)* information available in the open literature for some second primary cancer sites was not sufficient to draw firm conclusions regarding the risk per unit organ dose; *(b)* most studies in the literature assumed a linear dose-response relationship, limiting firm conclusions on the shape of the dose-response relationship, except for a few studies about second primary thyroid and haematopoietic cancers; and *(c)* it was not feasible to draw firm overall conclusions on effect modification of risks of second primary cancer by age at first radiation treatment, age at risk, smoking status, or sex. However, with the exception of gastrointestinal cancer, most risks included in the primary meta-analyses were either adjusted to account for the effect of chemotherapy or based on patients not treated with chemotherapy. Additionally, there was a general lack of information regarding radiopharmaceuticals in the papers reviewed and further research is needed on this topic.

453. The weight of the Lönn et al. [L23] study in the meta-analysis depends on the standard error of its linear ERR per unit organ dose (Gy), which was derived from the width of the CI, in turn dependent on the inverse of the mean organ dose. Mean organ doses varied significantly across cancer sites; for example, the mean organ dose for breast and lung cancer were estimated to be 0.2 Gy while the mean organ dose for cancer of the haematopoietic and lymphoid tissues was estimated at 19 Gy. The highest organ doses were estimated for gastrointestinal cancer (i.e. 40 Gy for cancer of the small intestine, and 50 Gy for cancer of the rectum). As a result, the CIs of the ERR per unit dose for gastrointestinal cancers determined by Lönn et al. [L23] are much narrower than those for breast and lung cancer, giving these ERR estimates greater weight in the meta-analysis. Consequently, there is sufficient evidence to consider excluding data in the study by Lönn et al. from the meta-analysis.

454. Overall, data on the shape of the second primary cancer dose-response relationship is very limited in the studies included in the meta-analysis. For breast, lung, gastrointestinal, and brain cancers, most available information is based on linear models, suggesting that the dose-response relationship is linear. However, for second primary sarcoma, thyroid cancer, and haematopoietic cancers, studies report both linear and non-linear risks. Since statistical methods for incorporating non-linear risks from models with different parameters in meta-analyses are not well established, the risks reported in this annex did not account for potential non-linear effects.

455. In some studies included in the meta-analysis, it was not possible to clearly distinguish between the follow-up period, the cancer diagnosis period, the radiation treatment period, age at first radiation treatment or sex. Therefore, firm conclusions based on risk effect modification by these risk modification factors could not be drawn. Most studies reported risks for patients treated with both radiation and chemotherapy. However, the risks per unit organ dose that were provided in these studies, were mostly adjusted for chemotherapy or for patients who did not receive it. There was a general lack of information

about radiopharmaceuticals in the studies reviewed, highlighting the need for further research on this topic. Finally, it is important to emphasize that the risks and conclusions presented in this chapter should not discourage people and medical officials from considering of the lifesaving benefit of radiotherapy for childhood or adult cancer patients.

456. The ERRs per unit dose calculated in the meta-analysis were compared to those from other types of epidemiological studies on radiation effects such as those based on data from the cohort Japanese atomic bombing survivors. The ERR per unit dose estimated for the seven second primary cancer sites in this annex are generally lower than those reported by other radiation epidemiological studies on non-radiotherapy exposures. For sarcoma, the comparisons were statistically compatible. For thyroid, statistical compatibility depended on the choice of risks used for comparison. However, for haematological, lung, gastrointestinal, brain and breast cancers the pooled estimates of risks from radiotherapy cohorts ware in general statistically significantly smaller than those from other types of radiation epidemiological studies. One possible explanation is that the higher doses delivered during radiotherapy, particularly to the target organ and tissues, may trigger radiobiological effects (as discussed in chapter III), reduce risks associated with radiotherapy compared to those reported in epidemiological studies on other types of radiation exposures. The risks estimated in the meta-analysis were used to determine AS values which do not rely on unrealistic assumptions based on external cancer rate comparisons, for organ doses from radiotherapy of 0.1, 1, 5 and 10 Gy.

# VI. FUTURE RESEARCH NEEDS

457. As cancer survivors are living to much older ages than in the past the occurrence of second primary cancers has become increasingly significant. It is therefore important to continue initiating and supporting projects on long-term and large-scale follow-up registries as well as studies on current and future first primary cancer survivors to better understand the risk of developing second primary cancers after radiotherapy. These studies should also be designed to evaluate the impact of risk effect modifiers, such as age at exposure, attained age, sex, race and ethnicity.

458. Research should be directed at integrating knowledge from radiobiology, oncology, and epidemiology, to evaluate the lifelong burden of all morbidities, including second primary cancer, associated with prior treatment for cancer. Studies on patients treated with radiotherapy generally give a low estimate of risk, whereas data from the LSS of Japanese atomic bombings survivors shows elevated risks associated with exposures at low level of doses [L11]. Research should aim at investigating the causes of these differences, and their relationship with biological mechanisms for different dose ranges and outcomes. Similarly, the analysis of studies on risks of second primary cancers in breast, lung, brain, and gastrointestinal organs, shows that there is strong evidence that risk estimates vary significantly in the different dose regions. Epidemiological studies should harmonize the outcomes considered by integrating the WHO classification of neoplasms of haematopoietic and lymphoid tissues into the study design.

459. Research on genetic susceptibility to second primary cancers from naturally occurring cancers is still in its early stages. Future technologies may be able to identify genetic changes specific to radiation-induced cancers, but this diagnostic advancement has yet to occur. High-throughput technologies like microarray genotyping and next-generation sequencing hold promise for advancing our understanding of the genetic susceptibility to second primary cancer and providing valuable biological insights into their development. Further studies on single nucleotide polymorphisms within candidate genes are needed. Key research priorities include identifying novel susceptibility genes, analysing rare germline mutations,

investigating gene-environment interactions, developing polygenic risk score, characterizing identified variants functionally, validating findings across diverse populations, and integrating genomic data with other omics data.

460. Future research may reveal molecular signatures associated with radiation exposure enabling the identification of radiation-induced second primary cancers which would be beneficial for clinical decision-making. Current these diagnoses rely on information on the location of the tumour within the radiation field and its delayed development after exposure. Research priorities on molecular patterns linked to radiation aetiology of tumours include identifying specific molecular signatures, integrating multi-omics data, validating findings across diverse patient populations, conducting longitudinal studies, functional characterization of molecular patterns, and utilizing machine learning approaches to identify complex interactions between genes or proteins. Addressing these research needs could enhance understanding of radiation-induced second primary cancers and improve clinical decision-making for patients undergoing radiation treatments.

461. Additional research directed at examining factors contributing to second primary cancers should include performing long-term follow-up studies, integrating multi-omics data, identifying biomarkers for risk prediction, investigating the role of the microbiome, understanding the impact of socioeconomic status, exploring gene-environment/lifestyle interactions, developing preventive interventions, and conducting population-specific studies. A comprehensive approach to analyse these factors will help develop personalized prevention strategies and improve clinical management of second primary cancers.

462. Reliable biomarkers need to be identified to predict individual risks for second primary cancers based on host, lifestyle and environmental factors. Future studies in this area should focus on improving understanding of these risks to guide radiotherapy decisions and enhance risk stratification for long-term patient follow-up. Descriptive and analytical epidemiological studies should fully investigate non-linear effects in the dose-response relationship and potential risk effect modifiers and be integrated with molecular and genetic approaches to assess the risk of emerging toxicities and to improve the current understanding on the evolution of late effects, especially associated with the ageing of cancer survivors. Additionally, future research should aim at providing a more complete and comprehensive reporting of information in original studies since currently vital data, such as CIs on the risk, ranges of dates for follow-up and therapy and diagnosis, are often missing or incomplete and a robust organ dosimetry is lacking.

463. The establishment of prospective cohorts of cancer patients should be supported to allow for the evaluation of the lifelong risk of second primary cancer and other late effects. These cohorts should include:

*(a)* Patients treated with modern modalities, including radiopharmaceutical therapies. Treatment information should be documented according to the guidelines given in either the Euregional Computer Assisted Theragnostics project (EuroCAT) [D13] or the Australian Computer Assisted Theragnostics (AusCAT) network [F6].

*(b)* First primary cancer sites where field size and radiation dose have been reduced. In particular, for modern treatments, it should be established whether risks of second primary cancer have decreased or are merely delayed, and whether there is any effect on histologic types of second primary cancer.

*(c)* Selected cohorts of cancer survivors treated with surgery or non-genotoxic therapies to investigate the natural history of these cancers, especially the baseline risks of second primary cancer, for comparison with patients treated with radiotherapy.

*(d)* Collection of biological samples (e.g. lymphocytes from blood cells and DNA from tissue or blood cells) to facilitate future studies on genetic factors related to survival and development of second primary cancers.

464. Comprehensive organ dosimetry systems should be developed for in-field, near-field, and out-of-field areas for all radiotherapy patients. Current epidemiological studies face challenges in reconstructing organ doses at the sites of second primary cancer from past treatments, often decades old. Following the radiation treatment, CT or MR images coupled with Digital Imaging and Communications in Medicine (DICOM) radiation treatment files, should be used to perform a complete Monte Carlo radiation transport simulation of the radiation treatment (including all ionizing radiation imaging modalities for diagnosis, treatment guidance, and post-therapy monitoring) at a time when all detailed information on the tumour, the organ anatomy and the primary and scattered radiation fields are readily available. Absorbed doses to all relevant tissues and organs should be reported in the patient's electronic medical record together with associated uncertainties.

465. Monte Carlo radiation transport simulations may be performed using whole-body, nearly patient-specific computational phantoms. Libraries of existing phantoms allow to select a close representation of individual patients based on matching criteria, such as height and weight, effective diameter, or water equivalent diameter. Partial-body CT or MR imaging can then be used to perform further 3D-morphing of both outer body contours and internal organ structures of the phantom. This approach ensures that all tissues, including those far from the treatment field, are available for dosimetry calculations.

466. Analytic studies should continue to address the relation between radiation dose and the risk of second primary cancer risk, the shape of the dose-response relationship, and the role of modifying factors, considering histologic type. Particular attention should be given to adolescent and young adult cancer survivors (aged between 18 and 39 years), considering the lack of data and the particular needs of these patients.

467. It is important to develop dose references and calibration procedures tailored to new treatment modalities, ensuring traceability to absorbed dose in tissues and to develop a benchmark library which can be used to compare and validate TPS algorithms.

468. Treatment planning systems must account for tissue doses due to secondary radiation fields to improve the accuracy of dose reconstruction when using whole-body virtual patient models. For example, some existing Monte Carlo-based TPS algorithms for proton radiotherapy include doses delivered by primary and secondary protons, but they fail to consider (p, n) reactions in tissue elements resulting in internally produced neutron fields and other particles.

469. A better understanding of the interactions of radiotherapy with other risk factors is essential and requires large patient cohorts with well-characterized treatment information together with extended follow-up and systematic collection of data on long-term morbidities. Study sizes should be adequately powered to characterize interactions between radiotherapy and other risk factors as additive, multiplicative, or other formulations. The possible interaction of radiotherapy with other variables (chemotherapy, immunotherapy, age at exposure, attained age, sex, race, lifestyle factors), obesity, and genetic modifiers of treatment in the development of site-specific cancer risk also need to be addressed.

470. Future comprehensive risk prediction models for second primary cancer should incorporate genetic modifiers of late sequelae, as well as treatment variables and well-established risk factors. The results of risk prediction models can be used to stratify patients into risk groups in order to customize follow-up strategies and develop evidence-based interventions.

471. Finally, international agreements should be established to facilitate the development of international databases on whole-body dosimetry, clinical and biological data, and on approaches for extended patient follow-up. policymakers and other stakeholders should participate in this initiative to facilitate data sharing, sustainability and usability, and ensure patient privacy. The use of anthropomorphic whole-body phantoms in combination with other computational techniques should ensure that an adequate balance is maintained between concerns related to patient privacy rights and research needs.

# VII. CONCLUSIONS

472. The global cancer burden has both negative and positive dimensions. On the negative side, cancer incidence continues to rise worldwide impacting the populations of both developing and developed countries. Cancer is the leading cause of premature death in 57 countries around the world, and ranks second to cardiovascular disease mortality in 70 others. Current trends suggest that cancer may surpass cardiovascular disease as the leading cause of premature death worldwide over the next several decades.

473. This trend is, however, offset by significant advances in modern cancer treatments. Radiation therapy, primarily delivered through external beam technologies, accounts for approximately 50% of all cancer treatments, with approximately 40% of patients cured of cancer having received radiotherapy as part of their treatment. As a result, global cancer survival rates have increased, leading to a growing population of survivors. Planning for cancer radiation treatment, includes optimizing the radiation dose delivered to the tumour target—typically the planning target volume—in a geometrically conformal 3D field, while additionally ensuring that the doses delivered to the adjacent organs and tissues at risk are below the thresholds for tissue reactions. This process maximizes the level of cell killing in the tumour area while simultaneously avoiding acute or chronic toxicity to healthy organs.

474. While continuous improvements in TPS have led to increasing levels of patient survival, the current optimization process does not take account of longer-term stochastic risks to the patient such as the development of a second primary cancer following radiotherapy. As cancer patient populations grows and post-therapy life expectancy increases, the risk of second primary cancer should now be considered in both the radiation treatment planning optimization process and the multidisciplinary decisions on the use of genotoxic treatments, such as some forms of chemotherapy. In this annex, radiation-related risks of second primary cancer following radiotherapy were investigated across four specialist areas: oncology, radiobiology, dosimetry, and epidemiology. The major conclusions from the review of relevant literature and evaluations for each specialist area are summarized in the following sections.

## A. Oncology

475. As cancer survival rates improve, a growing number of cancer patients are living long after completing their cancer treatments. While the main concern in the years immediately following treatment is relapse or metastasis of the cancer, the risk of developing a second primary cancer becomes a significant issue over the long term. Cancer survivors have a higher risk of developing a new primary cancer than is expected for the general population. Factors contributing to this higher risk include genetic susceptibility to cancer development, lifestyle and environmental factors, and treatment modalities used in the treatment of the first cancer, such as chemotherapy and radiotherapy.

476. Diagnosing a second primary cancer requires a review of the medical history of the patient, combined with the use of diagnostic imaging, and tissue biopsy. A biopsy is essential to confirm that the second primary cancer is histologically distinct from the first one. However, histological heterogeneity and complexity sometimes make histopathological decision challenging, especially if the cancer tissues share similar characteristics or if there are histological changes due to prior treatments. Recent technological advances in genomic and mutational profiling can assist pathologists and clinicians in reaching a more accurate diagnosis and classification of cancer types.

477. Cancer treatment involves multiple therapeutic modalities, often used in combination. As a result, it is a challenging task—and generally not required in current medical practice—to establish if a second primary cancer has been directly caused by radiotherapy or another form of cytotoxic therapy. However, in cases where radiotherapy was used for treatment of the first cancer, the development of a second primary cancer correlates the distribution of the dose in the tissue, the size of the irradiated volume, the total dose, and the dose rate. In order to improve the outcomes for the patients, technical developments focus on enhancing the precision of radiotherapy, maximizing its effectiveness against cancer cells while minimizing damage to surrounding tissues.

478. Raising awareness among healthcare providers about second primary cancers and establishing long-standing care programmes are crucial steps in reducing the risk of second primary cancers. Nevertheless, multidisciplinary patient care programmes following the completion of cancer treatment are still lacking. Most research on second primary cancers to date has been retrospective and focussed on specific cancer types in within isolated fields of expertise. There is a need to undertake coordinated multidisciplinary studies to provide insights on optimal survivorship care and enhance the quality of life of cancer survivors. Identifying patients with high susceptibility of developing second primary cancers and offering personalized management options is expected to become a standard practice in the future.

## B. Radiobiology

479. Radiotherapy works by causing lethal DNA damage to tumour cells. However, current techniques often result in a range of doses received by patients, with high doses targeting the tumour area and lower doses affecting surrounding tissues. In addition to the mechanisms of radiation-induced cancer formation at low doses, higher doses also contribute to cancer formation through mechanisms such as tissue reorganization, cell death and repopulation by surviving stem cells, cellular senescence or differentiation, inflammation-driven oncogenic stimuli, and changes in the tumour microenvironment.

480. While genetic susceptibility to certain cancers is well established knowledge about genetic susceptibility to radiation-induced second primary cancers remains limited. It is known that heterozygous carriers of certain hereditary disease-related mutations have a higher risk of radiation-associated second primary cancers. However, the potential link between common genetic variations in the general population and the risk of second primary cancer risk after radiotherapy remains unclear.

481. Several patient-related factors, such as genetic background, age, sex, race, hormonal and oestrogen levels are correlated with the risks for second primary cancer. Additionally, chronic exposure to carcinogens, such as alcohol, tobacco, environmental stress, and occupational hazards, further increases the risk of cancer survivors to develop second primary cancers. There is strong epidemiological evidence that one of the most important factors for determining the likelihood of developing a second primary cancer is the patient's age at the time of radiotherapy, with younger

patients having a higher probability of developing a second cancer, particularly when combined with other genotoxic treatments such as some forms of chemotherapy.

482. Given the multitude of factors that promote cancer formation, the activity of chemotherapy and its interaction with radiotherapy can have various effects on the normal tissue of patients, depending on the type and dose of the therapy, the location of the first primary cancer and the individual patient response. Chemotherapy may promote the development of second primary cancer by inducing mutations that lead to cell transformation or by inhibiting the immune system, weakening its ability to cancer surveillance. Conversely, chemotherapy can also reduce the risk of second primary cancer by eliminating radiation-damaged cells or in some case, such as breast cancer, by altering the hormonal status of the survivors though ovarian ablation. Ultimately, there is evidence that chemotherapy can either increase or reduce the risk of radiotherapy-induced second primary cancer, depending on the specific therapy regime.

483. The current evidence on the effectiveness of biomarkers for predicting second primary cancer risk is incomplete. An ideal biomarker would serve multiple purposes, including acting as a prognostic indicator, comparing radiotherapy techniques and their impact on healthy tissues, serving as biodosimetry tools to monitor absorbed doses and potential cancer development, and monitoring critical organs during radiotherapy. Studies using oncoproteomics and gene expression analysis have identified potential genes that can act as prognostic indicators, while genome-wide association studies have also identified genomic variants linked to increased risk of second primary cancer risk. Additionally, specific mutational signatures provide supporting evidence of a causal relationship between ionizing radiation and the development of second primary cancers such as sarcomas, particularly in high-dose radiation fields. Nonetheless, other contributing factors including genetic susceptibility play roles in the development of second primary cancers.

## C. Dosimetry

484. Epidemiological and clinical studies on second primary cancer after radiotherapy rely on the accurate estimate of the absorbed dose delivered during radiotherapy of the first primary cancer at the site of appearance of the second primary cancer. Since the second primary cancer can occur far from the planned treatment volume, absorbed doses at the site of the second primary cancer are not generally calculated. The consequence is that crucial information to derive risks for second primary cancer after radiotherapy is not available.

485. Treatment planning software is used during a course of radiotherapy to compute the distribution of doses delivered by the irradiating device in the PTV and the surrounding tissues including those delineated as OARs. This calculation is performed using the CT scan image of the patient, which, in most cases, is partial-body in extent and only includes the PTV and the OARs. Limitations on the TPS and the lack of whole-body CT images of patients hinder the possibility of computing the distribution of whole-body doses in routine clinical practice.

486. Current TPS based on Monte Carlo codes or linear Boltzmann transport equation solvers are capable of computing with high accuracy the distribution of the absorbed dose in the PTV and OARs. Nevertheless, most codes do not provide accurate results far from the treated volume. This limitation concerns photon treatments as well as radiotherapy using either protons or heavy ions.

487. Analytical and Monte Carlo programs aimed at the computation of the distribution of whole-body doses during either photon or proton treatments are available. However, these programs can currently be used in routine clinical practice, because of their limited capability to calculate dose distributions with

modern radiotherapy techniques, their lack of availability to the clinical community, and the exceedingly long simulation time required to perform the calculation for a single patient.

488. Comparisons between different beam qualities necessarily requires the application of biological weighting factors to derive RBE values. In the absence of a more adequate approach, the weighting factors recommended by the ICRP are generally adopted for this purpose. However, the applicability of these factors to radiotherapy is questionable since they were derived for radiological protection purposes.

489. Recent studies have indicated that the distribution of out-of-field doses produced from imaging procedures undertaken for diagnostic, planning, and patient-positioning purposes, can be comparable, in some cases, to the distribution of out-of-field doses produced during therapy with protons or heavy ions. It is therefore necessary to consider the distribution of whole-body doses from medical imaging systems in the calculation of cancer risks.

490. The validity of the predictions of risk projection models that utilize dose distributions for their calculations requires further analysis by comparison with epidemiological studies.

491. Efforts are needed to develop TPSs capable of routinely computing distributions of whole-body doses. Additionally, treatment planning should seek to optimize risk by incorporating the probability of second primary cancer occurrence into the process. To support future clinical and epidemiological research, standardized methods for storing and exchanging whole-body dosimetry data should be established.

492. Access to traceable dosimetry and validated radiobiological dosimetry quantities is essential for epidemiology, radiobiology, and radiation oncology. Ensuring metrological traceability through experimental validation and international comparisons guarantees the consistency of measured quantities. However, in radiation oncology and epidemiology, most current treatment planning systems and dose reconstruction programmes disrupt this traceability due to multiple assumptions and the reporting of results as integrated quantities with significant uncertainties and errors. Efforts to validate and verify these programmes, along with greater accuracy in radiobiological weighting factors for beam qualities used in both existing and new treatment modalities, are essential for improving radiobiology studies and treatment planning. In the mid- and long-term, these improvements are also necessary to enhance the accuracy of epidemiological research.

## D. Epidemiology

493. An extensive literature search and meta-analysis were conducted in this evaluation to gain information on the risks of secondary primary cancer following radiotherapy. Information on radiation related cumulative second primary cancer rates and associated latency periods was also presented. Sufficient information was found in the open literature for seven second primary cancer sites to justify their inclusion in the meta-analysis: haematopoietic and lymphoid malignancies; sarcoma; breast; lung; gastrointestinal organs; thyroid, and brain. A meta-analysis of studies that reported risks for all solid cancer combined was not considered, because of the problems associated with the heterogeneity of the reference dose for this cancer grouping. However, data used to perform individual meta-analyses of the risks for the six solid cancers mentioned above, were grouped together in an overall meta-analysis. The overall excess relative risk per unit organ dose for all second primary solid cancers was estimated to be 0.08 (95% CI: 0.06, 0.11) $Gy^{-1}$. This level of excess risk suggests that while patients who have undergone radiotherapy have a slightly increased risk of developing a second primary cancer, the overall magnitude of this risk remains small. This implies that radiotherapy contributes only a minor fraction to the total risk of second primary cancer in these patients.

494. The limitations of this approach are that:

*(a)* There was insufficient information available in the open literature on some second primary cancer sites which could be used to draw firm conclusions regarding the risk per unit organ dose;

*(b)* With the exception of a few studies for second primary thyroid cancer and second primary haematopoietic cancer, risks associated with radiotherapy reported in the literature were calculated on the assumption of a linear dose response relationship, and therefore no reliable inferences could be made about the shape of the dose-response relationship;

*(c)* The methodology to analyse non-linear dose response relationships by meta-analysis, is only currently in the development phase; and

*(d)* It was not considered currently feasible to draw firm conclusions on the effect modification of the risk of second primary cancer by age at first radiation treatment, age at risk, smoking status, sex and some other treatment related factors.

However, except for gastrointestinal cancer, the risks contributing to the primary meta-analyses were, in general, either adjusted for the co-treatment by chemotherapy or had been estimated for patients not treated with chemotherapy.

495. The ERRs per unit dose calculated in the meta-analysis were compared to cancer risks per unit dose from other types of radiation epidemiological studies such as those based on data from the Japanese atomic bombing survivors. The ERRs per unit dose estimated by the meta-analysis for the seven second primary cancer sites are generally lower than the risks reported by other types of radiation epidemiological studies such as those for the Japanese atomic bombing survivors. For sarcoma, such comparisons are statistically compatible; for thyroid cancer, statistical compatibility depends on the choice of comparison risks; while for haematological and lymphoid malignancies, and lung, gastrointestinal, brain and breast cancers, the pooled estimates from the cohorts of patients treated with radiotherapy are generally, statistically, and significantly smaller than the corresponding risks from other types of epidemiological studies of radiation effects. Conclusions drawn on whether or not these comparisons are based on approximately similar age at exposure ranges, attained ages, and proportion of males and females, are uncertain due to a general lack of overall information on risk effect modifiers.

496. The Committee is aware that this evaluation was limited to studies available in the open literature for which organ specific absorbed doses at the site of the second primary cancer were reported. As a result, there are limitations in the current analysis that can be attributed to the inherent difficulties in pooling heterogeneous studies. The sensitivity and subgroup analyses included in the meta-analysis attempted to account for some of the possible effect modifiers such as age at exposure (e.g., in large categories: young persons at exposure; adults at exposure; or all ages at exposure), prior exposure to genotoxic chemotherapy, internal radiopharmaceutical exposures, and imaging modalities using ionizing radiation. However, the meta-analysis was not able to address the effects on second primary cancer risks attributable to dose rate effects, attained age, sex, smoking status, and time-since-exposure, in a similar way, because the literature cannot be divided into broad subsets of studies by these parameters in a manner sufficient to justify meta-analysis calculations. These limitations are emphasized because, from radiation epidemiological knowledge in general, the evidence ranges from weak to strong, depending on cancer type, that these are factors that influence radiation-related cancer risk.

# ACKNOWLEDGEMENTS

The Committee acknowledges with gratitude the work of the members of the expert group involved in preparing this annex and especially to W. Bolch (United States) who chaired the expert group and L. Lebaron-Jacobs (France) who served as senior technical advisor. The Committee also thanks, in particular, J. K. Lee (Republic of Korea), M. Di Giorgio (Argentina), J.-M. Bordy (France), L. Walsh (Germany), for acting as lead writers for the chapters II to IV, respectively. Further, the Committee acknowledges the contribution of U. Schneider (Switzerland) in his role as quality assurance lead writer. The Committee also expresses its appreciation to A. Buhler (United States), research librarian, for her support in setting up the literature review. Finally, the Committee would like to thank Q. Sun (China), W.-U. Müller (Germany), A. L. Lachos Davila (Peru), D. Bartlett (United Kingdom), O. Belyakov and O. Holmberg (IAEA), A. Kesminiene (IARC/WHO) and W. Small (ICRP) for critically reviewing the manuscript. The views expressed in this annex remain those of the Committee and do not necessarily represent the views of the United Nations or its Member States.

## Expert group

*Chair*: W. Bolch (United States); *Members*: Z. Mokrani (Algeria), M. Di Giorgio, D. Dubner, N. Lendoiro and A. Rossini (Argentina), M. Sanagou (Australia), R. Wilkins (Canada), Y. Li and Q. Sun (China), E. Attalla (Egypt), J.-M. Bordy, V. Dedieu, C. Huet, L. Lebaron-Jacobs, J. Mazurier, F. Menetrier and J. Thariat (France), L. Brualla, W.-U. Müller, L. Walsh and D. Wollschläger (Germany), S. Yoshinaga (Japan), J. K. Lee, W. J. Lee and E. K. Paik (Republic of Korea), A. L. Lachos Davila (Peru), I. Toma-Dasu and A. Wojcik (Sweden), U. Schneider (Switzerland), O. El-Koha (United Arab Emirates), D. Bartlett (United Kingdom), and G. Woloschak (United States); *Observers*: O. Belyakov and O. Holmberg (IAEA), A. Kesminiene (IARC/WHO), and W. Small (ICRP).

# GLOSSARY

**Absolute risk:** Excess risk attributed to an agent and usually expressed as the numeric difference between exposed and unexposed populations (e.g., five cancer deaths over a lifetime per 100 people, each irradiated with 1 Sv).

**Absorbed dose (radiation):** The average energy imparted to matter in an element of volume by ionizing radiation divided by the mass of that element of volume. The SI unit for absorbed dose is joule per kilogramme (J/kg) and its special name is gray (Gy).

**Activity:** The rate at which spontaneous transformations occur in a given amount of radioactive material. Strictly, the expectation value of the number of nuclear transformations occurring in a given quantity of material per unit time. It is measured in becquerels (Bq); 1 Bq equals one transformation per second.

**Acute exposure:** Exposure received within a short time period.

The temporary modulation (usually reduction) by small "priming" doses of the response to subsequent high radiation doses.

**Adjustment:** The process of statistically accounting for effects of differences between groups or populations under comparison in order to control confounding. Adjustment is frequently performed when estimating effect measures from epidemiological data, for example by stratification or multivariate regression analysis.

**Alkylating agents:** Compounds that transfer an alkyl group to DNA.

**Apoptosis:** Cell death caused by an intracellular pathway which can be activated by external stimuli.

**Assigned share:** Assigned share is the probability that an observed health effect in an individual was caused by a specific radiation exposure. It is equal to the fraction of the total number of cases of a specific type of cancer diagnosed among individuals which is in excess to the baseline number of cases for persons who share the same attributes, such as absorbed organ dose, age, time since last exposure, sex, smoking history. The AS is often referred to as the "attributable fraction" or "probability of causation" assuming that the calculated excess relative risk represents the net consequences of mechanisms of disease manifestation for a given individual diagnosed with disease.

**Attained age:** Age at observation or during the follow-up, usually refers to age as time-dependent characteristic in a cohort study.

**Attribution:** The ascribing of an actual or manifest outcome to a cause. In the context of this annex, attribution refers to the ascribing of an outcome—in particular, a health effect—to radiation exposure:

*(a)* An outcome may be an individual outcome (such as the occurrence of a health effect in an individual);

*(b)* An outcome may also be a collective outcome (such as a change in the frequency of occurrence of health effects in a population or a group);

*(c)* Attribution may require expert assessment (or the consensus of a number of expert assessments), which may be made on the basis of a combination of evidence (facts) and argument (explanation).

**Baseline (hazard) rate (frequency):** Rate of an event of interest for a population in the absence of exposure. See also "Hazard rate (function)" and "Excess (absolute) rate".

**Baseline risk:** Baseline risk refers to the probability that an event of interest (e.g., diagnosis of cancer) will occur in an individual among an unexposed population over a given time period (e.g., lifetime following exposure). See "Risk".

**Becquerel (Bq):** The SI unit of activity, equal to one transformation per second. As the unit is so small, multiples are frequently used such as megabecquerels (MBq) which is $10^6$ or a million becquerels (1 GBq is $10^9$ Bq; 1 TBq is $10^{12}$ Bq; and 1 PBq is $10^{15}$ Bq).

**Bias:** Systematic deviation of results or inferences from the truth. A statistical estimation procedure is "biased" if the expected value of the estimate of the quantity of interest is not equal to the true value of the quantity. The "bias" of the procedure is the difference between the expected and true values (see also "selection bias", "information bias", "survivor bias").

**Biodosimetry:** The use of biological (or chemical, physiological) markers of exposure to reconstruct acute or protracted radiation doses.

**Biological indicator:** Measurable biological effect that is clearly, specifically, and quantifiably related to an exposure.

**Case-control study:** An epidemiological study type where cases with the health outcome of interest are selected from the population and compared in terms of exposure with controls free of outcomes under the condition that cases and controls derive from the same population (i.e., controls are comparable to cases except for the lack of outcomes). The effect of exposure on the outcome in a case-control study is expressed as an odds ratio (OR).

**Causation:** The relationship between cause (radiation exposure) and effect (radiogenic health effect or associated disease/mortality risk). The term can also mean the action of causing something.

**Chromosomal aberration:** Induced chromosomal aberrations can be divided into two main classes: chromosome-type aberrations involving both chromatids of a chromosome; and chromatid-type aberrations involving only one of the two chromatids. Ionizing radiation induces chromosome-type aberrations in the G0 or G1 stage of the cell cycle (i.e., prior to replication), while chromatid-type aberrations are induced during the S or G2 stage (i.e., during or after replication of the affected chromatid segment). In peripheral lymphocytes, most of which are in the G0 stage of the cell cycle, ionizing radiation induces mainly chromosome-type aberrations.

**Cohort study:** An epidemiological study type where people free of the health outcome of interest are selected and subdivided into groups based on exposure of interest. The occurrence of the health outcome (typically incidence or mortality) is recorded for the entire study population and the effect of exposure on the risk of the outcome is expressed as a rate ratio, risk ratio or hazard ratio. Follow-up for development of disease cases since start of study period can be prospective or retrospective (collection of information of the cases that have already occurred).

**Combined effect:** The joint effects of two or more agents on the level of molecules, cells, organs, and organisms in the production of a biological effect.

**Confidence interval (CI):** In "frequentist inference", a confidence interval is an interval defined in terms of the sampling distribution of a statistic of interest (i.e., the distribution of estimates of the statistic that would arise from repeated—generally hypothetical—realizations of data generated from the same underlying distribution as the observed data) such that, for example, the probability that a 95% confidence interval for a given parameter contains the true value of that parameter is 0.95 (compare this with "credible intervals" used in "Bayesian inference").

**Confounder/Confounding factor:** A variable that can cause or prevent the outcome of interest, is not an intermediate variable, and is associated with the factor under investigation. Such a variable must be controlled in order to obtain an undistorted estimate of the effect of the study factor(s) on risk.

**Confounding:** A situation in which the result of a causal relationship between the factor of interest and the outcome is distorted by the other factor (i.e., confounding factor) that causes the outcome and also is non-causally but correlated to the factor of interest.

**Deletion:** Loss of a portion of a gene or chromosome; a type of mutation; a synonym for deficiency.

**DNA sequence:** The relative order of base pairs, whether in a fragment of DNA, a gene, a chromosome, or an entire genome.

**Dose:** A measure of the energy deposited by radiation in a target. Dose can be used as a shorthand for absorbed dose and effective dose when the context is clear.

*Radiation:* See "Absorbed dose radiation".

*Chemicals:* The amount of a chemical administered to an organism. See also "absorbed dose".

**Dose rate:** Dose relative to time during which the dose is received. In experimental studies, dose rate is often expressed per minute, while in epidemiological studies of long-term exposure, dose per year is frequently used.

**Dose-response relationship:** The relationship between the magnitude of exposure to a chemical, biological, or physical agent (dose) and the magnitude or frequency and/or severity of associated adverse effects (response).

**Effect modification:** Effect modification (also called statistical "interaction") occurs when the magnitude of the influence of the dependent variable of interest (e.g., radiation dose or exposure) on an independent (outcome) variable (e.g., lung cancer) depends on the magnitude of some other factor (the effect modifier). For example, analyses of the atomic bombing's survivors suggest that the magnitude of the outcome of ionizing radiation exposure on the excess relative risk of lung cancer depends on smoking intensity, with a larger outcome from unit dose among moderate smokers and much smaller outcome among heavy smokers. An effect modifier may or may not be a confounding factor, depending on whether or not there is a correlation between the dependent variable of interest and the effect modifying factor.

**Effective dose:** Fundamental dosimetric quantity for the purposes of radiological protection defined as the tissue-weighted sum of the equivalent doses in all specified tissues and organs of the body; measured in sieverts (Sv).

**Epigen, epigenetic:** Changes in an organism brought about by alterations in the expression of genetic information without any change in the genome itself; the genotype is unaffected by such a change, but the phenotype is altered.

**Equivalent dose:** Fundamental dosimetric quantity for the purposes of radiological protection defined as the product of the absorbed dose in the tissue or organ and the appropriate radiation weighting factor for the type of radiation giving rise to the dose; measured in sieverts (Sv).

**Equivalent dose, $H_T$:** The averaged absorbed dose in tissue or organ *T*, modified by the radiation weighting factor, $w_R$. The unit of equivalent dose is joule per kilogram, which is given the name sievert (Sv). The recommended values of $w_R$ are given in the ICRP Publication 103.

**Excess absolute risk/rate (EAR):** The difference between the disease rate in an exposed population and the baseline rate in that population. This is often a function of dose, age (or some other measure of time) and other factors (effect modifiers). Excess absolute rate is often called "excess absolute risk" or "excess rate".

**Excess relative risk/rate (ERR):** The excess relative risk is the proportional increase above the baseline rate of the disease or condition under study. Numerically, it is the "relative risk/rate" minus one. The excess relative rate is strictly a statistic calculated from observed frequencies/rates, while the excess relative risk is a prospective estimate inferred from the data and reasoning. The ERR is often considered as a function of dose and other factors.

**Error:** The difference between an observed or estimated value and the true (but unknown) value. Since the true value is unknowable, error cannot be quantified. Error can be contrasted with uncertainty, which can be estimated by repeated measurements or Bayesian inference.

**Excess risk/rate:** A measure of the statistical relationship between a given risk factor and a specific outcome. Depending on the context it can refer to some characterization of the influence on rates such as the "relative risk", "excess relative risk", or "excess (absolute) rate", or, perhaps most appropriately, to estimates of the risk over some period of time (lifetime risk) associated with an exposure of interest. The excess rate is strictly a statistic calculated from observed frequencies/rates, while the excess risk is a prospective estimate inferred from the observations and reasoning.

**Exposure:** The act or condition of being subject to irradiation. External exposure is exposure to radiation from a source outside of the body. Internal exposure is exposure to radiation from a source within the body. In the radiation field, exposure may also denote the electrical charge of ions caused by X- or gamma rays per unit mass of air; however, the term is used in its more general sense as described here.

**Frequency (of occurrence or existence of disease):** The number of new or existing cases of the disease under study divided by the number of people in a population over a defined time period.

**Gene:** The fundamental physical and functional unit of heredity. A gene is an ordered sequence of nucleotides located in a particular position on a particular chromosome that encodes a specific functional product.

**Genome:** The complete DNA sequence of an organism.

**Genomic instability:** An all-embracing term to describe the increased rate of acquisition of alterations in the genome. As compared with the direct actions of radiation, i.e., those consequence directly induced by energy deposition, radiation-induced instability is observed in cells at delayed times after irradiation and manifests in the progeny of exposed cells multiple generations after the initial insult.

**Genotype:** *(a)* the genetic constitution of an individual; *(b)* the types of alleles found at a locus in an individual.

**Gray (Gy):** Unit of absorbed dose, equal to 1 joule per kilogram (J/kg).

**Hazard:** Set of inherent properties of a substance, mixture of substances, or a process involving substances that, under production, usage, or disposal conditions, make it capable of having adverse effects on organisms or the environment, depending on the degree of exposure; in other words, a source of danger.

**Hazard rate (function):** A hazard rate refers to the limiting probability of occurrence of an event of interest in a short time interval $(t, t + \Delta t)$ as the length of the interval approaches 0 given that the event can occur after time *t*. In adiation risk estimation, *t* generally refers to age or time since exposure. Since hazard rates generally vary with time and other factors (e.g., sex or dose), the concept of a hazard function is often used.

**Hazard ratio:** A measure of association used in prospective studies describing an instantaneous risk over the study time period, or some subset thereof. It is the ratio of two hazard rates (functions) corresponding to the conditions described by two levels of a variable. Hazard ratio differs from relative risk and odds ratio as it is not cumulative over the entire study time and can indicate risks that happen before the endpoint.

**Health effects (of radiation)** (also radiogenic health effects)**:**

> **Deterministic (health) effect:** A health effect of radiation for which generally a threshold level of dose exists, which varies with person and circumstance, above which the severity of the health effect is greater for a higher dose. The ICRP has introduced a term "tissue reactions" to describe a group of health effects comprising deterministic effects and some health effects (such as cataracts and fibrosis) that are not determined solely at the time of irradiation but can be modified after radiation exposure.
>
> **Stochastic (health) effect:** A radiation-related health effect, the probability of occurrence of which depends on radiation dose and the severity of which (if it occurs) is independent of dose. In radiation protection the so-called LNT-model is used (linear non-threshold) meaning that the assumption is made that one can linearly extrapolate from moderate/high doses to low and very low doses without a threshold. Note that this should not be confused with the stochastic actions or processes of radiation interaction at the molecular or cellular level.

**In vitro measurement:** A procedure used to determine the nature, activity, location, or retention of radionuclides in the body by analysis of material excreted or otherwise removed from the body.

**In vivo measurement:** A procedure used to determine the nature, activity, location, or retention of radionuclides in the body by direct measurement.

**Inference:** The process of drawing conclusions from scientific observations, evidence, and reasoning in the presence of uncertainty. While this annex is focussed on prospectively inferring risk, note that estimating an assigned share (or probability of causation) is also inference, but retrospective.

**Initiator:** In the multistage model of carcinogenesis, initiators are defined by their ability to induce persistent changes (probably due to genotoxic effects) in the cell (initiation). If there is subsequent promotion, these changes may result in tumour formation.

**Interaction:** Combined, mutual effects between agents on a molecular and/or cellular level within a short time.

**Latency (period):** The period between exposure and manifestation of a health effect. This is also the period after which statistically significant increases in frequency of occurrence of the health effect in a population have been seen; theoretically, there might be an undetectable increased frequency of occurrence of the health effect in an exposed population during the presumed latency period, but this possibility becomes vanishingly small in the period shortly after exposure because there is a finite time required for damaged cells to replicate in an uncontrolled manner and manifest as a cancerous growth.

**Lifetime risk:** An estimate of the risk associated with an exposure of interest over the (remaining) lifetime of an exposed population. Technically, the lifetime risk following an exposure to dose *D* is given by:

$$\int_{t_0}^{\infty} h_c(u,D)\ S(u,D)\,\mathrm{d}u$$

where $h_c(u, D)$ is the assumed (hazard) rate for the cause of interest at time *u* given exposure to dose *D* and $S(u, D)$ is the probability of still being alive and at risk of having the event at time *u* given dose *D*. The excess lifetime risk is defined as

$$\int_{t_0}^{\infty} [h_c(u,D) - h_c(u,0)]\,S(u,D)\,\mathrm{d}u$$

In the literature on radiogenic health effects and radiation protection, this quantity is often called the REID or REIC (risk of exposure-induced excess deaths or cases). It is sometimes approximated by the LAR (lifetime attributable risk), which is defined as

$$\int_{t_0}^{\infty} [h_c(u,D) - h_c(u,0)]\,S(u,0)\,\mathrm{d}u$$

or the ELR (excess lifetime risk) which is defined as

$$\int_{t_0}^{\infty} h_c(u,D)\,S(u,D)\,\mathrm{d}u - \int_{t_0}^{\infty} h_c(u,0)\,S(u,0)\,\mathrm{d}u$$

If $h_c(u, D) > h_c(u, 0)$, the ELR understates the impact of the exposure because it does not account for deaths (cases) that might occur earlier than they would have in the absence of exposure, while the LAR tends to overestimate the number of exposure-related events at higher doses. By limiting the range of integration in the above expressions, excess risks for periods of interest other than the full lifetime after exposure can be computed.

**Linear energy transfer (LET):** The average linear rate of energy loss of charged particle radiation in a medium, i.e., the radiation energy lost per unit length of path through a material. That is, the quotient of d*E* by d*l*, where d*E* is the mean energy lost by a charged particle owing to collisions with electrons in traversing a distance d*l* in matter.

$$L = \frac{\mathrm{d}E}{\mathrm{d}l}$$

The unit of *L* is J/m, often given in keV/µm.

**Meta-analysis:** In statistics, a meta-analysis combines the results of two or more studies that address a set of related research hypotheses, for example by estimating a certain unknown parameter or parametric function common to two or more data sets, while controlling or adjusting for differences in other parameters unrelated to those of immediate interest. The purpose is to obtain an estimate that is more informative than any obtainable from a single study, while ensuring that other data do not corrupt or bias the result. The term "meta-analysis" is also used for statistical pooling of results available in published form, in contrast with a pooled analysis of original data from two or more studies.

**Misclassification:** Measurement error for discrete variables (sex, ethnicity etc.) where individuals are wrongly classified into a group to which they do not belong. If the classification error depends on the value of other variables (e.g., it differs by exposure group), it is called differential misclassification. If it is independent of the value of other variables, the term non-differential misclassification is used. The latter generally introduces a bias towards the null value, but there are exceptions.

**Mitochondria:** Sub-cellular organelles that are the main site of energy production. Mitochondria contain a small circular DNA molecule that encodes some of their constituent proteins.

**Model:** An analytical or physical representation or quantification of a real system and the ways in which phenomena occur within that system, used to predict or assess the behaviour of the real system under specified (often hypothetical) conditions. This is in contrast to a relationship, which is simply the way two or more factors are connected or related (i.e., relationships include causal relationships, those based on models, and those that simply fit the observed data).

**Monte Carlo uncertainty propagation (also called Monte Carlo analysis):** Computation of a probability distribution of an output of a model on the basis of repeated calculations using random sampling of values of uncertain input variables specified as probability distributions. The numerical sampling strategy used may be Simple Random Sampling, or a form of stratified sampling such as Latin Hypercube Sampling.

**Mutagen:** A substance that can induce heritable changes (mutations) of the genotype in a cell as a consequence of alteration or loss of genes or chromosomes (or parts thereof).

**Mutation:** A hereditary change in genetic material. A mutation can be a change in a single base (point mutation) or a single gene or it can involve larger chromosomal rearrangements such as deletions and translocations.

**Non-genotoxic effect:** Effect of an agent at the cellular, organ, or organism level without direct effects on the genome such as DNA damage.

**Non-homologous end joining:** A DNA double-strand-break repair process that is the predominant pathway of repair in mammalian cells, predominately active in G1.

**Nucleotide:** A subunit of DNA or RNA consisting of a nitrogenous base: adenine (A), guanine (G), thymine (T) or cytosine (C) in DNA; adenine, guanine, uracil (U) or cytosine in RNA.

**Odds ratio (OR):** The effect measure used in case-control studies. Its interpretation is similar to that of a risk ratio or rate ratio, but its numerical value is more extreme (further from the null), particularly when the studied disease is common. An odds ratio is obtained as the ratio of the number of exposed cases to non-exposed cases relative to the ratio of exposed controls to non-exposed controls. It can also be computed from studies using other designs, with slightly different interpretation.

**Oncogene:** A gene, or more forms of which is associated with cancer; many oncogenes are involved, directly or indirectly, in controlling the rate of cell growth.

**Penetrance:** The frequency of expression of a trait or genotype; the proportion of individuals observed to show a particular phenotypic effect of a mutant gene compared with that expected on the basis of Mendelian inheritance.

**Phenotype:** The physical characteristics of a cell or organism as determined by the genetic constitution.

**Point estimate:** The most supported single value of the property estimated from the observed data such as rate ratio or odds ratio in an epidemiological study. The results of a study are, however, best interpreted as consistent with a range of values, such as that in the confidence interval or supported range, though the data generally lend more support to the central values within that range.

**Pooled analysis:** A combined analysis of original data from two or more data sets bearing on a common question of interest. The analysis may include parameters that distinguish between the different data sets. (Contrast with "meta-analysis").

**Progression:** Increase in autonomous growth and malignancy; used in particular to describe the transition from benign to malignant tumours and the progression of malignancy. There are probably numerous stages of progression during neoplastic development. The process of progression features in the general model of carcinogenesis as well as in the multistage model.

**Quantity of interest:** The true but unknown value being estimated, measured, or observed (also known formally as the "measurand"). It can be a single value, such as the dose or risk to a specific individual, or a set of true values, such as the set of doses to each individual in a defined cohort.

**Radionuclide:** A radioactive isotope of an element. Different isotopes of an element have the same number of protons but different numbers of neutrons and hence, different atomic masses. If there are too many or too few neutrons, the nuclei of the isotope tend to be unstable and transform into the nuclei of another element and in the process emit radiation.

**Radiosensitivity:** Susceptibility of cells, tissues and organisms to effects induced by ionizing radiation, especially, but not exclusively, in relation to radiation-associated carcinogenesis. The term "radiosensitivity" with regard to cancer induction refers to the rate of radiogenic tumour induction and does not refer to the degree of malignancy.

**Rate:** In general, the ratio of two quantities, where the denominator is usually a function of the period of time at risk. Rates of interest in epidemiology (and radiation risk estimation) are of the form $\frac{e}{\mathrm{PY}}$, where $e$ is the number of events over some time period of interest in a study population and PY (expressed as person-years) is the sum of time at risk during this time period for each person in the study population. Rates for an event of interest can vary with such factors as age, sex, and dose. Characterization of how rates depend on dose (or exposure) is central to radiation risk estimation. See also "Hazard function", "Risk", "Relative risk", "Excess rate", "Relative risk", and "Excess relative risk".

**Relationship:** The way two or more factors are connected or related (i.e., relationships include causal relationships, those based on models, and those that simply fit the observed data).

**Relative biological effectiveness (RBE):** For a specified radiation, RBE is the ratio of the *(a)* absorbed dose of a reference radiation required to produce a specific level of a response in a biological system to the *(b)* absorbed dose of the specified radiation required to produce an equal response, with all physical and biological variables, except radiation quality, being held as constant as possible. The reference radiation with a defined RBE of unity normally is gamma rays (often those from $^{60}$Co) or X-rays (most commonly 180–250 kVp X-rays). The RBE generally depends on dose, dose per fraction (if the dose is fractionated), dose rate, and biological end point. Note that RBE is an experimentally determined quantity and is not appropriate for the inference of risk and risk assessment, where a modifying factor that quantifies differences in risk for the same value of dose for a high-energy gamma ray compared with other types of radiation is more appropriate.

**Relative risk (RR):** Ratio between the cancer cases in the exposed population to the number of cases in the unexposed population. A relative risk of 1.5 indicates a 50% increase in cancer due to the agent under consideration. Excess relative risk (ERR) is numerically given as RR – 1.

**Relative risk/rate ratio:** The ratio of disease rates in different groups (e.g., an exposed and unexposed group) or for different exposure conditions (e.g., people exposed at high-dose-rates and people exposed at low dose-rates). It is often useful to view the relative risk as a function of variables, such as dose, sex, or age. Note that while this ratio is commonly called a relative risk, this is a misnomer; it is actually a ratio of rates, as are statistics derived from it (such as excess relative risk).

**Risk:** In the context of (radiation-related) health effects, risk refers to the probability that an event of interest (e.g., onset of cancer) will occur (i.e., it is prospective) during a given time period (e.g., the rest of life following an exposure). Risks can be estimated using evidence from epidemiological investigations of disease rates in previously exposed populations (i.e., based on past observations). The results from such retrospective analyses often are used, with appropriate modifying and adjustment factors, to make inferences about the risk for other exposure situations involving different populations for which direct epidemiological data on the dose-response relationship is not available.

**Sievert (Sv):** Unit of equivalent dose and of effective dose, equal to 1 joule per kilogram (J/kg).

**Signalling:** Transmission of activatory or inhibitory signals in the form of association and/or dissociation and/or modification of molecules that affect biological processes within a cell. Signalling can be intra-cellular or inter-cellular. Disruption of signalling can contribute to disease pathogenesis.

**Single nucleotide polymorphisms:** Variants of the genome at individual base pairs which can be useful in genetic mapping of diseases or susceptibles.

**Standardized incidence ratio (SIR):** An estimate of the number of cancer cases in a given population compared to what might be expected based on a comparison with the cancer experience in a larger population.

**Statistical significance:** Statistical significance testing is a technique originally developed for decision making, where the compatibility of the observations is contrasted with that expected in the absence of any difference (null hypothesis). An alternative or study hypothesis needs to be specified positing a difference, usually evaluating difference to any direction (called two-sided test), or to a specific direction (one-sided test). Conventionally, 0.05 has been used as the limit (critical value, alpha level or type I error) for significance, corresponding to a frequency of 1/20 for occurrence of a difference equally large or larger as observed in the absence of any true difference (under the null hypothesis). The choice of the cut-off level is, however, arbitrary, and statistical significance should not be seen as the sole criterion for interpretation of the findings.

**Stratification:** The process of stratification leads to the separation or grouping of a sample into several subsamples defined by the level of stratification variables such as socioeconomic status, age group or sex. Stratification is the basic principle of confounder control in the analysis of epidemiological data.

**Target (biological):** Any organism, organ, tissue, or cell that is subject to the action of a pollutant or other chemical, physical, or biological agent.

**Threshold dose:** The minimum dose that will produce a biological effect. Dose below which no effects occur ("true", mechanistically derived threshold) or are measurable (apparent threshold). For a given agent there can be multiple threshold doses, in essence one for each definable effect.

**Tissue dose:** Local dose in an organ or a functional or structural entity of an organ. See also "Absorbed dose".

**Toxicity:** Capacity of an agent to cause injury to a living organism. Toxicity can only be defined in quantitative terms with reference to the quantity of substance administered or absorbed, the way in which this quantity is administered (e.g., inhalation, ingestion, or injection) and distributed in time (e.g., single or repeated doses), the type and severity of injury, and the time needed to produce the injury.

**Transcriptome:** The complete collection of transcribed RNAs in a cell, investigated using the techniques of transcriptomics.

**Uncertainty:** Expression of having doubt, or being unsure about study results, hypotheses, model-based estimations, or results of measurements, and specifically, the true value of a quantity of interest. This may be due to lack of complete knowledge about true values for an individual or to a lack of complete knowledge of factors explaining the inter-individual variability of true values in a defined subgroup or population. Unlike error, uncertainties can be quantified. Estimates of uncertainty represent the amount or percentage by which an observed or calculated value might differ from its true value. For a quantity of interest that has a true fixed value, uncertainty is defined here as a degree of belief probability distribution comprising many realizations of possibly true values. For a quantity of interest that is a group or population of true values, uncertainty can be characterized as many alternative realizations of sets of true values.

**Variability:** Heterogeneity, diversity, or a range which characterizes variation in estimated, measured, or true values of a quantity of interest. The term variability is often used to describe differences in measured or true values among individuals in a population or cohort. Examples include inter-individual differences in body weight and/or dose or inter-cohort differences in exposure–response due to differences in sensitivity to a hazardous agent. Further study cannot reduce variability but may provide additional information to explain reasons why some of this variability occurs. This additional information can reduce the fraction of inter-individual variability initially treated as stochastic.

# REFERENCES


A1 Acharya, S., K. Sarafoglou, M. LaQuaglia et al. Thyroid neoplasms after therapeutic radiation for malignancies during childhood or adolescence. Cancer 97(10): 2397-2403 (2003).

A2 ACR. The burden of cancer. American Cancer Society. [Internet] Available from (https://canceratlas.cancer.org/the-burden/the-burden-of-cancer) on July 2022.

A3 Adhikari, J., P. Sharma and V.R. Bhatt. Risk of secondary solid malignancies after allogeneic hematopoietic stem cell transplantation and preventive strategies. Future Oncol 11(23): 3175-3185 (2015).

A4 Adjadj, E., C. Rubino, A. Shamsaldim et al. The risk of multiple primary breast and thyroid carcinomas. Cancer 98(6): 1309-1317 (2003).

A5 AGIR. Human radiosensitivity. Report of the independent Advisory Group on Ionising Radiation. RCE-21. Health Protection Agency, Chilton, 2013.

A6 Agostinelli, S., J. Allison, K. Amako et al. Geant4—a simulation toolkit. NIM-A 506(3): 250-303 (2003).

A7 Agrawal, C., L. Guthrie, M.S. Sturm et al. Comparison of thyroid nodule prevalence by ultrasound in childhood cancer survivors with and without thyroid radiation exposure. J Pediatr Hematol Oncol 38(1): 43-48 (2016).

A8 Ahdida, C., D. Bozzato, D. Calzolari et al. New capabilities of the FLUKA multi-purpose code. Front Phys 9 (2022).

A9 Aherne, N.J. and B.M. Murphy. Radiation-induced gliomas. Crit Rev Oncog 23(1-2): 113-118 (2018).

A10 Ahnesjo, A. and M.M. Aspradakis. Dose calculations for external photon beams in radiotherapy. Phys Med Biol 44(11): R99-155 (1999).

A11 Akuwudike, P., A. Tartas, M. Lopez-Riego et al. Cell type-specific patterns in the accumulation of DNA damage following multifractional radiation exposure. Int J Mol Sci 23(21) (2022).

A12 Akuwudike, P., M. Lopez-Riego, C. Dehours et al. Impact of fractionated cisplatin and radiation treatment on cell growth and accumulation of DNA damage in two normal cell types differing in origin. Sci Rep 13(1): 14891 (2023).

A13 Al-Juhaishi, T., A. Khurana and D. Shafer. Therapy-related myeloid neoplasms in lymphoma survivors: Reducing risks. Best Pract Res Clin Haematol 32(1): 47-53 (2019).

A14 Alazhri, J., C. Saclarides and E. Avisar. A rare complication resulting in a rare disease: radiation-induced male breast cancer. BMJ Case Rep 2016 (2016).

A15 Aldoss, I. and V. Pullarkat. Therapy-related acute myeloid leukemia with favorable cytogenetics: still favorable? Leuk Res 36(12): 1547-1551 (2012).

A16 Alexandrov, L.B., S. Nik-Zainal, D.C. Wedge et al. Signatures of mutational processes in human cancer. Nature 500(7463): 415-421 (2013).

A17 Alghamdi, S. and A. Tajaldeen. Evaluation of dose calculation algorithms using different density materials for in-field and out-of-field conditions. Exp Oncol 41(1): 46-52 (2019).

A18 Allan, J.M., C.P. Wild, S. Rollinson et al. Polymorphism in glutathione S-transferase P1 is associated with susceptibility to chemotherapy-induced leukemia. Proc Natl Acad Sci U S A 98(20): 11592-11597 (2001).

A19 Allard, A., N. Haddy, M.C. Le Deley et al. Role of radiation dose in the risk of secondary leukemia after a solid tumor in childhood treated between 1980 and 1999. Int J Radiat Oncol Biol Phys 78(5): 1474-1482 (2010).

A20 Allemani, C., T. Matsuda, V. Di Carlo et al. Global surveillance of trends in cancer survival 2000-14 (CONCORD-3): analysis of individual records for 37 513 025 patients diagnosed with one of 18 cancers from 322 population-based registries in 71 countries. Lancet 391(10125): 1023-1075 (2018).

A21 Allodji, R.S., B. Schwartz, C. Veres et al. Risk of subsequent leukemia after a solid tumor in childhood: Impact of bone marrow radiation therapy and chemotherapy. Int J Radiat Oncol Biol Phys 93(3): 658-667 (2015).

A22 Allodji, R.S., N. Haddy, G. Vu-Bezin et al. Risk of subsequent colorectal cancers after a solid tumor in childhood: Effects of radiation therapy and chemotherapy. Pediatr Blood Cancer 66(2): e27495 (2019).

A23 Allodji, R.S., M.A. Tucker, M.M. Hawkins et al. Role of radiotherapy and chemotherapy in the risk of leukemia after childhood cancer: an international pooled analysis. Int J Cancer: 2079-2089 (2020).

A24 APT. MCNPX™ User's Manual. (L.S. Waters, ed.) Version 2.3.0: LA-UR-02-2607. Los Alamos National Laboratory, Accelerator Production of Tritium, 2002.

A25 Arai, A., S. Ozawa, S. Kinoshita et al. Radiation-induced angiosarcoma of the parotid gland after postoperative radiotherapy for hypopharyngeal carcinoma. Auris Nasus Larynx 46(6): 940-945 (2019).

A26 Arber, D.A., A. Orazi, R. Hasserjian et al. The 2016 revision to the World Health Organization classification of myeloid neoplasms and acute leukemia. Blood 127(20): 2391-2405 (2016).

A27 Armstrong, G.T., Q. Liu, Y. Yasui et al. Long-term outcomes among adult survivors of childhood central nervous system malignancies in the Childhood Cancer Survivor Study. J Natl Cancer Inst 101(13): 946-958 (2009).

A28 Avgerinos, K.I., N. Spyrou, C.S. Mantzoros et al. Obesity and cancer risk: Emerging biological mechanisms and perspectives. Metabolism 92: 121-135 (2019).

B1 Baek, H.J., T.H. Kim, D. Shin et al. Radiobiological characterization of proton beam at the National Cancer Center in Korea. J Radiat Res 49(5): 509-515 (2008).

B2 Baker, K.S., W.M. Leisenring, P.J. Goodman et al. Total body irradiation dose and risk of subsequent neoplasms following allogeneic hematopoietic cell transplantation. Blood 133(26): 2790-2799 (2019).

B3 Barletta, J.A., A.M. Bellizzi and J.L. Hornick. Immunohistochemical staining of thyroidectomy specimens for PTEN can aid in the identification of patients with Cowden syndrome. Am J Surg Pathol 35(10): 1505-1511 (2011).

B4 Barnett, G.C., C.M. West, A.M. Dunning et al. Normal tissue reactions to radiotherapy: towards tailoring treatment dose by genotype. Nat Rev Cancer 9(2): 134-142 (2009).

B5 Baskar, R., K.A. Lee, R. Yeo et al. Cancer and radiation therapy: current advances and future directions. Int J Med Sci 9(3): 193-199 (2012).

B6 Basu, S.K., C. Schwartz, S.G. Fisher et al. Unilateral and bilateral breast cancer in women surviving pediatric Hodgkin's disease. Int J Radiat Oncol Biol Phys 72(1): 34-40 (2008).

B7 Bateni, S.B., N.G. Coburn, C. Law et al. Second primary cancers and survival among neuroendocrine tumor patients. Endocr Relat Cancer 30(8) (2023).

B8 Batho, H.F. Lung corrections in Cobalt-60 beam therapy. J Can Assoc Radiol 15: 79-83 (1964).

B9 Bazire, L., Y. De Rycke, B. Asselain et al. Risks of second malignancies after breast cancer treatment: Long-term results. Cancer Radiother 21(1): 10-15 (2017).

B10 Becher, H. and V. Winkler. Estimating the standardized incidence ratio (SIR) with incomplete follow-up data. BMC Med Res Methodol 17(1): 55 (2017).

B11 Bechler, J.R., W.W. Robertson, Jr., A.T. Meadows et al. Osteosarcoma as a second malignant neoplasm in children. J Bone Joint Surg Am 74(7): 1079-1083 (1992).

B12 Behjati, S., G. Gundem, D.C. Wedge et al. Mutational signatures of ionizing radiation in second malignancies. Nat Commun 7: 12605 (2016).

B13 Belzile-Dugas, E. and M.J. Eisenberg. Radiation-induced cardiovascular disease: Review of an underrecognized pathology. J Am Heart Assoc 10(18): e021686 (2021).

B14 Bentzen, S.M., D.I. Rosenthal, E.A. Weymuller et al. Increasing toxicity in nonoperative head and neck cancer treatment: investigations and interventions. Int J Radiat Oncol Biol Phys 69(2 Suppl): S79-82 (2007).

B15 Benveniste, M.F., D. Gomez, B.W. Carter et al. Recognizing radiation therapy-related complications in the chest. Radiographics 39(2): 344-366 (2019).

B16 Bergom, C., C.M. West, D.S. Higginson et al. The implications of genetic testing on radiation therapy decisions: A guide for radiation oncologists. Int J Radiat Oncol Biol Phys 105(4): 698-712 (2019).

B17 Bernstein, J.L., WECARE Study Collaborative Group and P. Concannon. ATM, radiation, and the risk of second primary breast cancer. Int J Radiat Biol 93(10): 1121-1127 (2017).

B18 Berrington, A. and D.R. Cox. Generalized least squares for the synthesis of correlated information. Biostatistics 4(3): 423-431 (2003).

B19 Berrington de Gonzalez, A., E. Gilbert, R. Curtis et al. Second solid cancers after radiation therapy: a systematic review of the epidemiologic studies of the radiation dose-response relationship. Int J Radiat Oncol Biol Phys 86(2): 224-233 (2013).

B20 Best, T., D. Li, A.D. Skol et al. Variants at 6q21 implicate PRDM1 in the etiology of therapy-induced second malignancies after Hodgkin's lymphoma. Nat Med 17(8): 941-943 (2011).

B21 Bettega, D., P. Calzolari, P. Hessel et al. Neoplastic transformation induced by carbon ions. Int J Radiat Oncol Biol Phys 73(3): 861-868 (2009).

B22 Bhatia, S., L.L. Robison, O. Oberlin et al. Breast cancer and other second neoplasms after childhood Hodgkin's disease. N Engl J Med 334(12): 745-751 (1996).

B23 Bhatia, S., H.N. Sather, O.B. Pabustan et al. Low incidence of second neoplasms among children diagnosed with acute lymphoblastic leukemia after 1983. Blood 99(12): 4257-4264 (2002).

B24 Bhatia, S. and C. Sklar. Second cancers in survivors of childhood cancer. Nat Rev Cancer 2(2): 124-132 (2002).

B25 Bhatia, S., M.D. Krailo, Z. Chen et al. Therapy-related myelodysplasia and acute myeloid leukemia after Ewing sarcoma and primitive neuroectodermal tumor of bone: A report from the Children's Oncology Group. Blood 109(1): 46-51 (2007).

B26 Bhatla, D., R.B. Gerbing, T.A. Alonzo et al. DNA repair polymorphisms and outcome of chemotherapy for acute myelogenous leukemia: a report from the Children's Oncology Group. Leukemia 22(2): 265-272 (2008).

B27 Bhatti, P., L.H. Veiga, C.M. Ronckers et al. Risk of second primary thyroid cancer after radiotherapy for a childhood cancer in a large cohort study: an update from the childhood cancer survivor study. Radiat Res 174(6): 741-752 (2010).

B28 Blagosklonny, M.V. Carcinogenesis, cancer therapy and chemoprevention. Cell Death Differ 12(6): 592-602 (2005).

B29 Blettner, M. and J.D. Boice, Jr. Radiation dose and leukaemia risk: general relative risk techniques for dose-response models in a matched case-control study. Stat Med 10(10): 1511-1526 (1991).

B30 Bluhm, E.C., C. Ronckers, R.J. Hayashi et al. Cause-specific mortality and second cancer incidence after non-Hodgkin lymphoma: a report from the Childhood Cancer Survivor Study. Blood 111(8): 4014-4021 (2008).

B31 Boersma, L.J., M.G.A. Sattler, J.H. Maduro et al. Model-based selection for proton therapy in breast cancer: Development of the national indication protocol for proton therapy and first clinical experiences. Clin Oncol (R Coll Radiol) 34(4): 247-257 (2022).

B32 Boice, J.D., Jr., M. Blettner, R.A. Kleinerman et al. Radiation dose and leukemia risk in patients treated for cancer of the cervix. J Natl Cancer Inst 79(6): 1295-1311 (1987).

B33 Boice, J.D., Jr., G. Engholm, R.A. Kleinerman et al. Radiation dose and second cancer risk in patients treated for cancer of the cervix. Radiat Res 116(1): 3-55 (1988).

B34 Boice, J.D., Jr., M. Blettner, R.A. Kleinerman et al. Radiation dose and breast cancer risk in patients treated for cancer of the cervix. Int J Cancer 44(1): 7-16 (1989).

B35 Boice, J.D., Jr., E.B. Harvey, M. Blettner et al. Cancer in the contralateral breast after radiotherapy for breast cancer. N Engl J Med 326(12): 781-785 (1992).

B36 Boivin, J.F., G.B. Hutchison, A.G. Zauber et al. Incidence of second cancers in patients treated for Hodgkin's disease. J Natl Cancer Inst 87(10): 732-741 (1995).

B37 Bonita, R., R. Beaglehole, T. Kjellström et al. Basic Epidemiology. 2nd edition. World Health Organization, Geneva, 2006.

B38 Boterberg, T., C. Dunlea, S. Harrabi et al. Contemporary paediatric radiation oncology. Arch Dis Child 108(5): 332-337 (2022).

B39 Boukheris, H., M. Stovall, E.S. Gilbert et al. Risk of salivary gland cancer after childhood cancer: a report from the Childhood Cancer Survivor Study. Int J Radiat Oncol Biol Phys 85(3): 776-783 (2013).

B40 Brada, M., D. Ford, S. Ashley et al. Risk of second brain tumour after conservative surgery and radiotherapy for pituitary adenoma. BMJ 304(6838): 1343-1346 (1992).

B41 Bray, F., M. Laversanne, E. Weiderpass et al. The ever-increasing importance of cancer as a leading cause of premature death worldwide. Cancer 127(16): 3029-3030 (2021).

B42 Brenner, A.V., D.L. Preston, R. Sakata et al. Incidence of breast cancer in the Life Span Study of Atomic Bomb Survivors: 1958-2009. Radiat Res 190(4): 433-444 (2018).

B43 Brenner, A.V., H. Sugiyama, D.L. Preston et al. Radiation risk of central nervous system tumors in the Life Span Study of atomic bomb survivors, 1958-2009. Eur J Epidemiol 35(6): 591-600 (2020).

B44 Brenner, A.V., D.L. Preston, R. Sakata et al. Comparison of all solid cancer mortality and incidence dose-response in the Life Span Study of Atomic Bomb Survivors, 1958-2009. Radiat Res 197(5): 491-508 (2022).

B45 Brenner, D.J., R.E. Curtis, E.J. Hall et al. Second malignancies in prostate carcinoma patients after radiotherapy compared with surgery. Cancer 88(2): 398-406 (2000).

B46 Breslow, N.E., J.R. Takashima, J.A. Whitton et al. Second malignant neoplasms following treatment for Wilm's tumor: a report from the National Wilms' Tumor Study Group. J Clin Oncol 13(8): 1851-1859 (1995).

B47 Bright, C.J., M.M. Hawkins, D.L. Winter et al. Risk of soft-tissue sarcoma among 69 460 five-year survivors of childhood cancer in Europe. J Natl Cancer Inst 110(6): 649-660 (2018).

B48 Broeks, A., L.M. Braaf, A. Huseinovic et al. Identification of women with an increased risk of developing radiation-induced breast cancer: a case only study. Breast Cancer Res 9(2): R26 (2007).

B49 Brooks, J.D., S.N. Teraoka, A.S. Reiner et al. Variants in activators and downstream targets of ATM, radiation exposure, and contralateral breast cancer risk in the WECARE study. Hum Mutat 33(1): 158-164 (2012).

B50 Brown, C.A., D.R. Youlden, J.F. Aitken et al. Therapy-related acute myeloid leukemia following treatment for cancer in childhood: A population-based registry study. Pediatr Blood Cancer 65(12): e27410 (2018).

B51 Brualla, L., M. Rodriguez and A.M. Lallena. Monte Carlo systems used for treatment planning and dose verification. Strahlenther Onkol 193(4): 243-259 (2017).

B52 Brusamolino, E., A.P. Anselmo, C. Klersy et al. The risk of acute leukemia in patients treated for Hodgkin’s disease is significantly higher after combined modality programs than after chemotherapy alone and is correlated with the extent of radiotherapy and type and duration of chemotherapy: a case-control study. Haematologica 83(9): 812-823 (1998).

B53 Bryant, A.K., M.P. Banegas, M.E. Martinez et al. Trends in radiation therapy among cancer survivors in the United States, 2000-2030. Cancer Epidemiol Biomarkers Prev 26(6): 963-970 (2017).

B54 Burt, L.M., J. Ying, M.M. Poppe et al. Risk of secondary malignancies after radiation therapy for breast cancer: Comprehensive results. Breast 35: 122-129 (2017).

B55 Bushberg, J.T., J.A. Seiberg, E.M. Leidholdt et al. The Essential Physics of Medical Imaging. Wolters Kluwer Health, 2020.

C1 Cabanas, M.L., C. Yan, R.J. Lalonde et al. Which dose specification should be used for NRG radiation therapy trials: Dose-to-medium or dose-to-water? Pract Radiat Oncol 10(2): e103-e110 (2020).

C2 Cahan, W.G., H.Q. Woodard, N.L. Higinbotham et al. Sarcoma arising in irradiated bone: report of 11 cases. Cancer 82(1): 3-29 (1948).

C3 Cahoon, E.K., D.L. Preston, D.A. Pierce et al. Lung, laryngeal and other respiratory cancer incidence among Japanese Atomic Bomb Survivors: An updated analysis from 1958 through 2009. Radiat Res 187(5): 538-548 (2017).

C4 Cardous-Ubbink, M.C., R.C. Heinen, P.J. Bakker et al. Risk of second malignancies in long-term survivors of childhood cancer. Eur J Cancer 43(2): 351-362 (2007).

C5 Carrasco, P., N. Jornet, M.A. Duch et al. Comparison of dose calculation algorithms in phantoms with lung equivalent heterogeneities under conditions of lateral electronic disequilibrium. Med Phys 31(10): 2899-2911 (2004).

C6 Casagranda, L., M. Oriol, F. Freycon et al. Second malignant neoplasm following childhood cancer: A nested case-control study of a recent cohort (1987-2004) from the Childhood Cancer Registry of the Rhone-Alpes region in France. Pediatr Hematol Oncol 33(6): 371-382 (2016).

C7 Chaplain, G., C. Milan, C. Sgro et al. Increased risk of acute leukemia after adjuvant chemotherapy for breast cancer: a population-based study. J Clin Oncol 18(15): 2836-2842 (2000).

C8 Chaturvedi, A.K., E.A. Engels, E.S. Gilbert et al. Second cancers among 104,760 survivors of cervical cancer: evaluation of long-term risk. J Natl Cancer Inst 99(21): 1634-1643 (2007).

C9 Choi, G., B. Huang, E. Pinarbasi et al. Genetically mediated Nf1 loss in mice promotes diverse radiation-induced tumors modeling second malignant neoplasms. Cancer Res 72(24): 6425-6434 (2012).

C10 Chow, E.J., D.L. Friedman, M. Stovall et al. Risk of thyroid dysfunction and subsequent thyroid cancer among survivors of acute lymphoblastic leukemia: a report from the Childhood Cancer Survivor Study. Pediatr Blood Cancer 53(3): 432-437 (2009).

C11 Chung, C.C. and S.J. Chanock. Current status of genome-wide association studies in cancer. Hum Genet 130(1): 59-78 (2011).

C12 Chung, C.S., T.I. Yock, K. Nelson et al. Incidence of second malignancies among patients treated with proton versus photon radiation. Int J Radiat Oncol Biol Phys 87(1): 46-52 (2013).

C13 Clarkson, J.R. A note on depth doses in fields of irregular shape. Br J Radiol 14(164): 265-268 (1941).

C14 Coleman, C.N. Adverse effects of cancer therapy. Risk of secondary neoplasms. Am J Pediatr Hematol Oncol 4(1): 103-111 (1982).

C15 Colnot, J., V. Barraux, C. Loiseau et al. A new Monte Carlo model of a Cyberknife((R)) system for the precise determination of out-of-field doses. Phys Med Biol 64(19): 195008 (2019).

C16 Colnot, J., S. Zefkili, R. Gschwind et al. Out-of-field doses from radiotherapy using photon beams: A comparative study for a pediatric renal treatment. J Appl Clin Med Phys 22(3): 94-106 (2021).

C17 Cooke, R., M.E. Jones, D. Cunningham et al. Breast cancer risk following Hodgkin lymphoma radiotherapy in relation to menstrual and reproductive factors. Br J Cancer 108(11): 2399-2406 (2013).

C18 COSMIC. Cancer gene census. Catalogue of Somatic Mutations in Cancer. [Internet] Available from (https://cancer.sanger.ac.uk/census) on 26 October 2023.

C19 Croce, O., S. Hachem, E. Franchisseur et al. Contact radiotherapy using a 50kV X-ray system: Evaluation of relative dose distribution with the Monte Carlo code PENELOPE and comparison with measurements. Radiat Phys Chem 81(6): 609-617 (2012).

C20 Cunningham, J.R. Scatter-air ratios. Phys Med Biol 17(1): 42-51 (1972).

C21 Curtis, R.E., J.D. Boice, Jr., M. Stovall et al. Leukemia risk following radiotherapy for breast cancer. J Clin Oncol 7(1): 21-29 (1989).

C22 Curtis, R.E., J.D. Boice, Jr., M. Stovall et al. Risk of leukemia after chemotherapy and radiation treatment for breast cancer. N Engl J Med 326(26): 1745-1751 (1992).

C23 Curtis, R.E., J.D. Boice, Jr., M. Stovall et al. Relationship of leukemia risk to radiation dose following cancer of the uterine corpus. J Natl Cancer Inst 86(17): 1315-1324 (1994).

D1 D'Angio, G.J., A. Meadows, V. Mike et al. Decreased risk of radiation-associated second malignant neoplasms in actinomycin-D-treated patients. Cancer 37(2 Suppl): 1177-1185 (1976).

D2 Daniels, R.D. and M.K. Schubauer-Berigan. A meta-analysis of leukaemia risk from protracted exposure to low-dose gamma radiation. Occup Environ Med 68(6): 457-464 (2011).

D3 De Bruin, M.L., J. Sparidans, M.B. van't Veer et al. Breast cancer risk in female survivors of Hodgkin's lymphoma: lower risk after smaller radiation volumes. J Clin Oncol 27(26): 4239-4246 (2009).

D4 De Saint-Hubert, M., F. Suesselbeck, F. Vasi et al. Experimental validation of an analytical program and a Monte Carlo simulation for the computation of the far out-of-field dose in external beam photon therapy applied to pediatric patients. Front Oncol 12: 882506 (2022).

D5 De Saint-Hubert, M., N. Verbeek, C. Baumer et al. Validation of a Monte Carlo framework for out-of-field dose calculations in proton therapy. Front Oncol 12: 882489 (2022).

D6 De Saint-Hubert, M., G. Boissonnat, U. Schneider et al. Complete patient exposure during paediatric brain cancer treatment for photon and proton therapy techniques including imaging procedures. Front Oncol 13: 1222800 (2023).

D7 de Vathaire, F., M. Schlumberger, M.J. Delisle et al. Leukaemias and cancers following iodine-131 administration for thyroid cancer. Br J Cancer 75(5): 734-739 (1997).

D8 de Vathaire, F., C. Hardiman, A. Shamsaldin et al. Thyroid carcinomas after irradiation for a first cancer during childhood. Arch Intern Med 159(22): 2713-2719 (1999).

D9 de Vathaire, F., M. Hawkins, S. Campbell et al. Second malignant neoplasms after a first cancer in childhood: temporal pattern of risk according to type of treatment. Br J Cancer 79(11-12): 1884-1893 (1999).

D10 de Vathaire, F., N. Haddy, R.S. Allodji et al. Thyroid radiation dose and other risk factors of thyroid carcinoma following childhood cancer. J Clin Endocrinol Metab 100(11): 4282-4290 (2015).

D11 de Vathaire, F., B. Scwhartz, C. El-Fayech et al. Risk of a second kidney carcinoma following childhood cancer: Role of chemotherapy and radiation dose to kidneys. J Urol 194(5): 1390-1395 (2015).

D12 Debniak, T. and J. Lubinski. Principles of genetic predisposition to malignancies. Hered Cancer Clin Pract 6(2): 69-72 (2008).

D13 Deist, T.M., A. Jochems, J. van Soest et al. Infrastructure and distributed learning methodology for privacy-preserving multi-centric rapid learning health care: euroCAT. Clin Transl Radiat Oncol 4: 24-31 (2017).

D14 Delaney, G., S. Jacob, C. Featherstone et al. The role of radiotherapy in cancer treatment: estimating optimal utilization from a review of evidence-based clinical guidelines. Cancer 104(6): 1129-1137 (2005).

D15 Denekamp, J. Changes in the rate of repopulation during multifraction irradiation of mouse skin. Br J Radiol 46(545): 381-387 (1973).

D16 Diallo, I., A. Lamon, A. Shamsaldin et al. Estimation of the radiation dose delivered to any point outside the target volume per patient treated with external beam radiotherapy. Radiother Oncol 38(3): 269-271 (1996).

D17 Diallo, I., N. Haddy, E. Adjadj et al. Frequency distribution of second solid cancer locations in relation to the irradiated volume among 115 patients treated for childhood cancer. Int J Radiat Oncol Biol Phys 74(3): 876-883 (2009).

D18 Diamandidou, E., A.U. Buzdar, T.L. Smith et al. Treatment-related leukemia in breast cancer patients treated with fluorouracil-doxorubicin-cyclophosphamide combination adjuvant chemotherapy: the University of Texas M.D. Anderson Cancer Center experience. J Clin Oncol 14(10): 2722-2730 (1996).

D19 DiMarzio, P., R. Peila, O. Dowling et al. Smoking and alcohol drinking effect on radiotherapy associated risk of second primary cancer and mortality among breast cancer patients. Cancer Epidemiol 57: 97-103 (2018).

D20 Dineen, S.P., C.L. Roland, R. Feig et al. Radiation-associated undifferentiated pleomorphic sarcoma is associated with worse clinical outcomes than sporadic lesions. Ann Surg Oncol 22(12): 3913-3920 (2015).

D21 Divgi, C., J.A. Carrasquillo, R. Meredith et al. Overcoming barriers to radiopharmaceutical therapy (RPT): An overview from the NRG-NCI working group on dosimetry of radiopharmaceutical therapy. Int J Radiat Oncol Biol Phys 109(4): 905-912 (2021).

D22 Dogan, A., P. Kern, B. Schultheis et al. Radiogenic angiosarcoma of the breast: case report and systematic review of the literature. BMC Cancer 18(1): 463 (2018).

D23 Doi, K., M.N. Mieno, Y. Shimada et al. Risk of second malignant neoplasms among childhood cancer survivors treated with radiotherapy: meta-analysis of nine epidemiological studies. Paediatr Perinat Epidemiol 23(4): 370-379 (2009).

D24 Doi, K., M.N. Mieno, Y. Shimada et al. Methodological extensions of meta-analysis with excess relative risk estimates: application to risk of second malignant neoplasms among childhood cancer survivors treated with radiotherapy. J Radiat Res 55(5): 885-901 (2014).

D25 Dores, G.M., R.E. Curtis, F.E. van Leeuwen et al. Pancreatic cancer risk after treatment of Hodgkin lymphoma. Ann Oncol 25(10): 2073-2079 (2014).

D26 Dorr, W. and T. Herrmann. Cancer induction by radiotherapy: dose dependence and spatial relationship to irradiated volume. J Radiol Prot 22(3A): A117-121 (2002).

D27 Dosani, M., K.A. Schrader, A. Nichol et al. Severe late toxicity after adjuvant breast radiotherapy in a patient with a germline ataxia telangiectasia mutated gene: Future treatment decisions. Cureus 9(7): e1458 (2017).

D28 Drooger, J., D. Akdeniz, J.P. Pignol et al. Adjuvant radiotherapy for primary breast cancer in BRCA1 and BRCA2 mutation carriers and risk of contralateral breast cancer with special attention to patients irradiated at younger age. Breast Cancer Res Treat 154(1): 171-180 (2015).

D29 Dunst, J., S. Ahrens, M. Paulussen et al. Second malignancies after treatment for Ewing's sarcoma: a report of the CESS-studies. Int J Radiat Oncol Biol Phys 42(2): 379-384 (1998).

D30 Dupuy, A., A. Shamsaldin, E. Quiniou et al. Risk of melanoma following adulthood cancer: a case-control study. Eur J Cancer 41(18): 2904-2910 (2005).

E1 Ehrhardt, M.J., C.R. Howell, K. Hale et al. Subsequent breast cancer in female childhood cancer survivors in the St Jude Lifetime Cohort Study (SJLIFE). J Clin Oncol 37(19): 1647-1656 (2019).

E2 Eidemuller, M., E. Holmberg, P. Jacob et al. Breast cancer risk and possible mechanisms of radiation-induced genomic instability in the Swedish hemangioma cohort after reanalyzed dosimetry. Mutat Res 775: 1-9 (2015).

E3 Eidemuller, M., J. Becker, J.C. Kaiser et al. Concepts of association between cancer and ionising radiation: accounting for specific biological mechanisms. Radiat Environ Biophys 62(1): 1-15 (2023).

E4 El Meddeb, H.A., J.M. Chehata, S. Kanoun et al. Bilateral breast cancer after prophylactic bilateral pulmonary irradiation. Gulf J Oncolog 1(23): 63-66 (2017).

E5 Elkind, M.M., H. Sutton-Gilbert, W.B. Moses et al. Sub-lethal and lethal radiation damage. Nature 214(5093): 1088-1092 (1967).

E6 EORTC. E$^2$-RADIatE The platform for radiotherapy. European Organisation for Research and Treatment of Cancer. [Internet] Available from (https://project.eortc.org/e2-radiate/) on November 2023.

E7 Estey, E., R.P. Hasserjian and H. Dohner. Distinguishing AML from MDS: A fixed blast percentage may no longer be optimal. Blood 139(3): 323-332 (2022).

E8 Eterovic, A.K., O.M. Maher, J. Chandra et al. Metachronous medulloblastoma in a child with successfully treated neuroblastoma: Case report and novel findings of DNA sequencing. J Natl Compr Canc Netw 16(6): 683-691 (2018).

E9 Evans, D.G., E.R. Woodward, S. Bajalica-Lagercrantz et al. Germline TP53 testing in breast cancers: Why, when and how? Cancers (Basel) 12(12): 3762 (2020).

E10 Evans, D.G. and E.R. Woodward. New surveillance guidelines for Li-Fraumeni and hereditary TP53 related cancer syndrome: implications for germline TP53 testing in breast cancer. Fam Cancer 20(1): 1-7 (2021).

F1 Farran, Y., O. Padilla, K. Chambers et al. Atypical presentation of radiation-associated breast angiosarcoma: A case report and review of literature. Am J Case Rep 18: 1347-1350 (2017).

F2 Feller, A., K.L. Matthes, A. Bordoni et al. The relative risk of second primary cancers in Switzerland: a population-based retrospective cohort study. BMC Cancer 20(1): 51 (2020).

F3 Fernandes, C.M., N. Hamerschlak, R.K. Kishimoto et al. Radiotherapy-related acute myeloid leukaemia with KMT2A amplification. Br J Haematol 172(5): 653 (2016).

F4 Fernandez-Antoran, D., G. Piedrafita, K. Murai et al. Outcompeting p53-mutant cells in the normal esophagus by redox manipulation. Cell Stem Cell 25(3): 329-341 (2019).

F5 Ferrarini, A., A. Auteri-Kaczmarek, A. Pica et al. Early occurrence of lung adenocarcinoma and breast cancer after radiotherapy of a chest wall sarcoma in a patient with a de novo germline mutation in TP53. Fam Cancer 10(2): 187-192 (2011).

F6 Field, M., S. Vinod, N. Aherne et al. Implementation of the Australian Computer-Assisted Theragnostics (AusCAT) network for radiation oncology data extraction, reporting and distributed learning. J Med Imaging Radiat Oncol 65(5): 627-636 (2021).

F7 Finke, I., P. Scholz-Kreisel, U. Hennewig et al. Radiotherapy and subsequent thyroid cancer in German childhood cancer survivors: A nested case-control study. Radiat Oncol 10: 219 (2015).

F8 Fippel, M. Fast Monte Carlo dose calculation for photon beams based on the VMC electron algorithm. Med Phys 26(8): 1466-1475 (1999).

F9 Fisher, B., H. Rockette, E.R. Fisher et al. Leukemia in breast cancer patients following adjuvant chemotherapy or postoperative radiation: The NSABP experience. J Clin Oncol 3(12): 1640-1658 (1985).

F10 Fogliata, A., G. Nicolini, A. Clivio et al. Dosimetric evaluation of Acuros XB Advanced Dose Calculation algorithm in heterogeneous media. Radiat Oncol 6: 82 (2011).

F11 Followill, D., P. Geis and A. Boyer. Estimates of whole-body dose equivalent produced by beam intensity modulated conformal therapy. Int J Radiat Oncol Biol Phys 38(3): 667-672 (1997).

F12 Followill, D.S., M.S. Stovall, S.F. Kry et al. Neutron source strength measurements for Varian, Siemens, Elekta, and General Electric linear accelerators. J Appl Clin Med Phys 4(3): 189-194 (2003).

F13 Fouad, Y.A. and C. Aanei. Revisiting the hallmarks of cancer. Am J Cancer Res 7(5): 1016-1036 (2017).

F14 Francois, P., C. Beurtheret and A. Dutreix. Calculation of the dose delivered to organs outside the radiation beams. Med Phys 15(6): 879-883 (1988).

F15 Frebourg, T., S. Bajalica Lagercrantz, C. Oliveira et al. Guidelines for the Li-Fraumeni and heritable TP53-related cancer syndromes. Eur J Hum Genet 28(10): 1379-1386 (2020).

F16 Friedman, D.L., J. Whitton, W. Leisenring et al. Subsequent neoplasms in 5-year survivors of childhood cancer: The Childhood Cancer Survivor Study. J Natl Cancer Inst 102(14): 1083-1095 (2010).

F17 Furukawa, K., D. Preston, S. Funamoto et al. Long-term trend of thyroid cancer risk among Japanese atomic-bomb survivors: 60 years after exposure. Int J Cancer 132(5): 1222-1226 (2013).

G1 Garcia Yip, F., J. Šolc, R. Behrens et al. Dose distributions of the INTRABEAM electronic brachytherapy system measured with traceable detectors. JINST 18(07): P07049 (2023).

G2 Gazit, E., R. Gil and R. Altshuler. Lymphocytes modulated by beta interferon express new non-HLA-A,B,C class I antigens. Hum Immunol 18(1): 53-63 (1987).

G3 Ghareeb, E.R., R. Bhargava, J.A. Vargo et al. Primary and radiation-induced breast angiosarcoma: Clinicopathologic predictors of outcomes and the impact of adjuvant radiation therapy. Am J Clin Oncol 39(5): 463-467 (2016).

G4 Giannoula, E., A. Gkantaifi and I. Iakovou. Radiation treatment of head and neck carcinomas as a risk factor for thyroid carcinomas. Hell J Nucl Med 19(1): 65-74 (2016). (Greek).

G5 Giefing, M., M. Rydzanicz, K. Szukala et al. Second primary tumors (SPT) of head and neck: distinguishing of "true" SPT from micrometastasis by LOH analysis of selected chromosome regions. Neoplasma 52(5): 374-380 (2005).

G6 Gilbert, E.S., M. Stovall, M. Gospodarowicz et al. Lung cancer after treatment for Hodgkin's disease: Focus on radiation effects. Radiat Res 159(2): 161-173 (2003).

G7 Gilbert, E.S., R.E. Curtis, M. Hauptmann et al. Stomach cancer following Hodgkin lymphoma, testicular cancer and cervical cancer: A pooled analysis of three international studies with a focus on radiation effects. Radiat Res 187(2): 186-195 (2017).

G8 Gillespie, C.S., A.I. Islim, B.A. Taweel et al. The growth rate and clinical outcomes of radiation induced meningioma undergoing treatment or active monitoring. J Neurooncol 153(2): 239-249 (2021).

G9 Giunta, S. Decoding human cancer with whole genome sequencing: A review of PCAWG Project studies published in February 2020. Cancer Metastasis Rev 40(3): 909-924 (2021).

G10 Glazebrook, K.N., M.J. Magut and C. Reynolds. Angiosarcoma of the breast. AJR Am J Roentgenol 190(2): 533-538 (2008).

G11 Gold, D.G., J.P. Neglia and K.E. Dusenbery. Second neoplasms after megavoltage radiation for pediatric tumors. Cancer 97(10): 2588-2596 (2003).

G12 Gold, E.B. The timing of the age at which natural menopause occurs. Obstet Gynecol Clin North Am 38(3): 425-440 (2011).

G13 Golemis, E.A., P. Scheet, T.N. Beck et al. Molecular mechanisms of the preventable causes of cancer in the United States. Genes Dev 32(13-14): 868-902 (2018).

G14 Gomarteli, K., J. Fleckenstein, S. Kirschner et al. Radiation-induced malignancies after intensity-modulated versus conventional mediastinal radiotherapy in a small animal model. Sci Rep 9(1): 15489 (2019).

G15 Gonin-Laurent, N., A. Gibaud, M. Huygue et al. Specific TP53 mutation pattern in radiation-induced sarcomas. Carcinogenesis 27(6): 1266-1272 (2006).

G16 Gonzalez-Vasconcellos, I., T. Domke, V. Kuosaite et al. Differential effects of genes of the Rb1 signalling pathway on osteosarcoma incidence and latency in alpha-particle irradiated mice. Radiat Environ Biophys 50(1): 135-141 (2011).

G17 Goske, M.J., K.E. Applegate, J. Boylan et al. The Image Gently campaign: Working together to change practice. AJR Am J Roentgenol 190(2): 273-274 (2008).

G18 Grant, E.J., A. Brenner, H. Sugiyama et al. Solid cancer incidence among the Life Span Study of Atomic Bomb Survivors: 1958-2009. Radiat Res 187(5): 513-537 (2017).

G19 Grantzau, T., M.S. Thomsen, M. Vaeth et al. Risk of second primary lung cancer in women after radiotherapy for breast cancer. Radiother Oncol 111(3): 366-373 (2014).

G20 Graves, S.A. and R.F. Hobbs. Dosimetry for optimized, personalized radiopharmaceutical therapy. Semin Radiat Oncol 31(1): 37-44 (2021).

G21 Green, D.M., A. Hyland, M.P. Barcos et al. Second malignant neoplasms after treatment for Hodgkin's disease in childhood or adolescence. J Clin Oncol 18(7): 1492-1499 (2000).

G22 Greenland, S. Relation of probability of causation to relative risk and doubling dose: A methodologic error that has become a social problem. Am J Public Health 89(8): 1166-1169 (1999).

G23 Guerin, S., A. Dupuy, H. Anderson et al. Radiation dose as a risk factor for malignant melanoma following childhood cancer. Eur J Cancer 39(16): 2379-2386 (2003).

G24 Guerin, S., C. Guibout, A. Shamsaldin et al. Concomitant chemo-radiotherapy and local dose of radiation as risk factors for second malignant neoplasms after solid cancer in childhood: A case-control study. Int J Cancer 120(1): 96-102 (2007).

G25 Guibout, C., E. Adjadj, C. Rubino et al. Malignant breast tumors after radiotherapy for a first cancer during childhood. J Clin Oncol 23(1): 197-204 (2005).

H1 Haddy, N., M.C. Le Deley, A. Samand et al. Role of radiotherapy and chemotherapy in the risk of secondary leukaemia after a solid tumour in childhood. Eur J Cancer 42(16): 2757-2764 (2006).

H2 Hadj-Hamou, N.S., N. Ugolin, C. Ory et al. A transcriptome signature distinguished sporadic from postradiotherapy radiation-induced sarcomas. Carcinogenesis 32(6): 929-934 (2011).

H3 Halg, R.A., J. Besserer and U. Schneider. Systematic measurements of whole-body dose distributions for various treatment machines and delivery techniques in radiation therapy. Med Phys 39(12): 7662-7676 (2012).

H4 Halg, R.A., J. Besserer and U. Schneider. Systematic measurements of whole-body imaging dose distributions in image-guided radiation therapy. Med Phys 39(12): 7650-7661 (2012).

H5 Halg, R.A., J. Besserer, M. Boschung et al. Measurements of the neutron dose equivalent for various radiation qualities, treatment machines and delivery techniques in radiation therapy. Phys Med Biol 59(10): 2457-2468 (2014).

H6 Hall, E.J., S.G. Martin, H. Amols et al. Photoneutrons from medical linear accelerators-radiobiological measurements and risk estimates. Int J Radiat Oncol Biol Phys 33(1): 225-230 (1995).

H7 Hall, E.J. Radiation, the two-edged sword: Cancer risks at high and low doses. Cancer J 6(6): 343-350 (2000).

H8 Hallquist, A., L. Hardell and P.O. Lofroth. External radiotherapy prior to thyroid cancer: A case-control study. Int J Radiat Oncol Biol Phys 27(5): 1085-1089 (1993).

H9 Han, T., J.K. Mikell, M. Salehpour et al. Dosimetric comparison of Acuros XB deterministic radiation transport method with Monte Carlo and model-based convolution methods in heterogeneous media. Med Phys 38(5): 2651-2664 (2011).

H10 Hanahan, D. and R.A. Weinberg. The hallmarks of cancer. Cell 100(1): 57-70 (2000).

H11 Hanahan, D. and R.A. Weinberg. Hallmarks of cancer: The next generation. Cell 144(5): 646-674 (2011).

H12 Hanahan, D. Hallmarks of cancer: New dimensions. Cancer Discov 12(1): 31-46 (2022).

H13 Hancock, S.L., M.A. Tucker and R.T. Hoppe. Breast cancer after treatment of Hodgkin's disease. J Natl Cancer Inst 85(1): 25-31 (1993).

H14 Harada, H., Y. Harada, H. Tanaka et al. Implications of somatic mutations in the AML1 gene in radiation-associated and therapy-related myelodysplastic syndrome/acute myeloid leukemia. Blood 101(2): 673-680 (2003).

H15 Harkenrider, M.M., F. Alite, S.R. Silva et al. Image-based brachytherapy for the treatment of cervical cancer. Int J Radiat Oncol Biol Phys 92(4): 921-934 (2015).

H16 HARMONIC. Health effects of cardiac fluoroscopy and modern radiotherapy in paediatrics. [Internet] Available from (https://harmonicproject.eu/) on 8 July 2022.

H17 Hartel, C., A. Nikoghosyan, M. Durante et al. Chromosomal aberrations in peripheral blood lymphocytes of prostate cancer patients treated with IMRT and carbon ions. Radiother Oncol 95(1): 73-78 (2010).

H18 Hasenbalg, F., H. Neuenschwander, R. Mini et al. Collapsed cone convolution and analytical anisotropic algorithm dose calculations compared to VMC++ Monte Carlo simulations in clinical cases. Phys Med Biol 52(13): 3679-3691 (2007).

H19 Hauptmann, M., S.D. Fossa, M. Stovall et al. Increased stomach cancer risk following radiotherapy for testicular cancer. Br J Cancer 112(1): 44-51 (2015).

H20 Hauptmann, M., T. Borge Johannesen, E.S. Gilbert et al. Increased pancreatic cancer risk following radiotherapy for testicular cancer. Br J Cancer 115(7): 901-908 (2016).

H21 Hauptmann, M., R.D. Daniels, E. Cardis et al. Epidemiological studies of low-dose ionizing radiation and cancer: Summary bias assessment and meta-analysis. J Natl Cancer Inst Monogr 2020(56): 188-200 (2020).

H22 Hauri, P., R.A. Halg, J. Besserer et al. A general model for stray dose calculation of static and intensity-modulated photon radiation. Med Phys 43(4): 1955 (2016).

H23 Hauri, P., S. Radonic, F. Vasi et al. Development of whole-body representation and dose calculation in a commercial treatment planning system. Z Med Phys 32(2): 159-172 (2022).

H24 Hawkins, C.J. and M.A. Miles. Mutagenic consequences of sublethal cell death signaling. Int J Mol Sci 22(11): 6144 (2021).

H25 Hawkins, M.M., L.M. Wilson, M.A. Stovall et al. Epipodophyllotoxins, alkylating agents, and radiation and risk of secondary leukaemia after childhood cancer. BMJ 304(6832): 951-958 (1992).

H26 Hawkins, M.M., L.M. Wilson, H.S. Burton et al. Radiotherapy, alkylating agents, and risk of bone cancer after childhood cancer. J Natl Cancer Inst 88(5): 270-278 (1996).

H27 Hegemann, N.S., A. Schlesinger-Raab, U. Ganswindt et al. Risk of second cancer following radiotherapy for prostate cancer: A population-based analysis. Radiat Oncol 12(1): 2 (2017).

H28 Henderson, T.O., J. Whitton, M. Stovall et al. Secondary sarcomas in childhood cancer survivors: A report from the Childhood Cancer Survivor Study. J Natl Cancer Inst 99(4): 300-308 (2007).

H29 Henderson, T.O., A. Amsterdam, S. Bhatia et al. Systematic review: Surveillance for breast cancer in women treated with chest radiation for childhood, adolescent, or young adult cancer. Ann Intern Med 152(7): 444-455; W144-454 (2010).

H30 Henderson, T.O., K.C. Oeffinger, J. Whitton et al. Secondary gastrointestinal cancer in childhood cancer survivors: A cohort study. Ann Intern Med 156(11): 757-766, W-260 (2012).

H31 Henderson, T.O., P. Rajaraman, M. Stovall et al. Risk factors associated with secondary sarcomas in childhood cancer survivors: A report from the Childhood Cancer Survivor Study. Int J Radiat Oncol Biol Phys 84(1): 224-230 (2012).

H32 Higgins, J.P., S.G. Thompson, J.J. Deeks et al. Measuring inconsistency in meta-analyses. BMJ 327(7414): 557-560 (2003).

H33 Hill, D.A., E. Gilbert, G.M. Dores et al. Breast cancer risk following radiotherapy for Hodgkin lymphoma: Modification by other risk factors. Blood 106(10): 3358-3365 (2005).

H34 Hinnen, K.A., M. Schaapveld, M. van Vulpen et al. Prostate brachytherapy and second primary cancer risk: A competitive risk analysis. J Clin Oncol 29(34): 4510-4515 (2011).

H35 Hiraoka, S., K. Sakanaka, T. Iwai et al. Therapy-related acute myeloid leukemia 2 months after chemoradiotherapy for esophageal cancer: A case report. Case Rep Oncol 13(1): 299-303 (2020).

H36 Hird, A.E., D.E. Magee, R. Matta et al. Assessment of secondary sarcomas among patients with cancer of the abdomen or pelvis who received combinations of surgery, radiation, and chemotherapy versus surgery alone. JAMA Netw Open 3(10): e2013929 (2020).

H37 Hodapp, N. The ICRU Report 83: prescribing, recording and reporting photon-beam intensity-modulated radiation therapy (IMRT). Strahlenther Onkol 188(1): 97-99 (2012).

H38 Hodgson, D.C., E.S. Koh, T.H. Tran et al. Individualized estimates of second cancer risks after contemporary radiation therapy for Hodgkin lymphoma. Cancer 110(11): 2576-2586 (2007).

H39 Holowaty, E.J., G.A. Darlington, C.K. Gajalakshmi et al. Leukemia after irradiation for endometrial cancer in Ontario. Cancer 76(4): 644-649 (1995).

H40 Holsti, L.R. Development of clinical radiotherapy since 1896. Acta Oncol 34(8): 995-1003 (1995).

H41 Holt, J.G., J.S. Laughlin and J.P. Moroney. The extension of the concept of tissue-air ratios (TAR) to high-energy x-ray beams. Radiology 96(2): 437-446 (1970).

H42 Hooning, M.J., B.M. Aleman, M. Hauptmann et al. Roles of radiotherapy and chemotherapy in the development of contralateral breast cancer. J Clin Oncol 26(34): 5561-5568 (2008)

H43 Hoskin, P. The price of anticancer intervention. Secondary malignancies after radiotherapy. Lancet Oncol 3(9): 577-578 (2002).

H44 Hosmann, A., F. Hinker, C. Dorfer et al. Management of choroid plexus tumors–An institutional experience. Acta Neurochir (Wien) 161(4): 745-754 (2019).

H45 Howard, R., E. Gilbert, C.F. Lynch et al. Risk of leukemia among survivors of testicular cancer: A population-based study of 42,722 patients. Ann Epidemiol 18(5): 416-421 (2008).

H46 Howell, R.M., S.F. Kry, E. Burgett et al. Secondary neutron spectra from modern Varian, Siemens, and Elekta linacs with multileaf collimators. Med Phys 36(9): 4027-4038 (2009).

H47 Howell, R.M., S.B. Scarboro, S.F. Kry et al. Accuracy of out-of-field dose calculations by a commercial treatment planning system. Phys Med Biol 55(23): 6999-7008 (2010).

H48 Howell, R.M., S.B. Scarboro, P.J. Taddei et al. Methodology for determining doses to in-field, out-of-field and partially in-field organs for late effects studies in photon radiotherapy. Phys Med Biol 55(23): 7009-7023 (2010).

H49 Hsu, W.L., D.L. Preston, M. Soda et al. The incidence of leukemia, lymphoma and multiple myeloma among atomic bomb survivors: 1950-2001. Radiat Res 179(3): 361-382 (2013).

H50 Hu, Y.A., H. Song, Z. Chen et al. Evaluation of an electron Monte Carlo dose calculation algorithm for electron beam. J Appl Clin Med Phys 9(3): 1-15 (2008).

H51 Huang, J. and W.J. Mackillop. Increased risk of soft tissue sarcoma after radiotherapy in women with breast carcinoma. Cancer 92(1): 172-180 (2001).

H52 Huang, J.Y., D.S. Followill, X.A. Wang et al. Accuracy and sources of error of out-of field dose calculations by a commercial treatment planning system for intensity-modulated radiation therapy treatments. J Appl Clin Med Phys 14(2): 4139 (2013).

H53 Huang, K.L., R.J. Mashl, Y. Wu et al. Pathogenic germline variants in 10,389 adult cancers. Cell 173(2): 355-370 e314 (2018).

I1 IAEA. Radiotherapy in Cancer Care: Facing the Global Challenge. International Atomic Energy Agency, Vienna, 2017.

I2 IARC. IARC monographs on the evaluation of carcinogenic risks to humans. Radiation: A review of human carcinogens. Volume 100D. International Agency for Research on Cancer, Lyon, 2012.

I3 IARC. Global cancer observatory. International Agency for Research on Cancer. [Internet] Available from (https://gco.iarc.fr/en) on 8 May 2022.

I4 ICRP. 1990 Recommendations of the International Commission on Radiological Protection. ICRP Publication 60. Annals of the ICRP 21(1-3). International Commission on Radiological Protection, Oxford, 1991.

I5 ICRP. The 2007 recommendations of the International Commission on Radiological Protection. ICRP Publication 103. Annals of the ICRP 37(2-4). International Commission on Radiological Protection, Elsevier Ltd., 2007.

I6 ICRU. Prescribing, recording, and reporting photon beam therapy. ICRU Report 62. International Commission on Radiation Units and Measurements, Bethesda, 1999.

I7 ICRU. Dosimetry-guided radiopharmaceutical therapy. ICRU Report 96. International Commission on Radiation Units and Measurements, Bethesda, 2021.

I8 Inskip, P.D., M. Stovall and J.T. Flannery. Lung cancer risk and radiation dose among women treated for breast cancer. J Natl Cancer Inst 86(13): 983-988 (1994).

I9 Inskip, P.D. and R.E. Curtis. New malignancies following childhood cancer in the United States, 1973-2002. Int J Cancer 121(10): 2233-2240 (2007).

I10 Inskip, P.D., L.L. Robison, M. Stovall et al. Radiation dose and breast cancer risk in the childhood cancer survivor study. J Clin Oncol 27(24): 3901-3907 (2009).

I11 Inskip, P.D., A.J. Sigurdson, L. Veiga et al. Radiation-related new primary solid cancers in the Childhood Cancer Survivor Study: Comparative radiation dose response and modification of treatment effects. Int J Radiat Oncol Biol Phys 94(4): 800-807 (2016).

I12 Iqbal, F.M., B. Ahmed and R. Vidya. Double-edged sword of radiotherapy: a cause of secondary angiosarcoma after breast conservation therapy. BMJ Case Rep 2016 (2016).

I13 Iranzo, J., I. Martincorena and E.V. Koonin. Cancer-mutation network and the number and specificity of driver mutations. Proc Natl Acad Sci U S A 115(26): E6010-E6019 (2018).

I14 Ishida, Y., T. Takabatake, S. Kakinuma et al. Genomic and gene expression signatures of radiation in medulloblastomas after low-dose irradiation in Ptch1 heterozygous mice. Carcinogenesis 31(9): 1694-1701 (2010).

I15 Izumi, M., J. Oyanagi, K. Sawa et al. Mutational landscape of multiple primary lung cancers and its correlation with non-intrinsic risk factors. Sci Rep 11(1): 5680 (2021).

J1 Jagetic, L.J. and W.D. Newhauser. A simple and fast physics-based analytical method to calculate therapeutic and stray doses from external beam, megavoltage x-ray therapy. Phys Med Biol 60(12): 4753-4775 (2015).

J2 Jenkinson, H.C., D.L. Winter, H.B. Marsden et al. A study of soft tissue sarcomas after childhood cancer in Britain. Br J Cancer 97(5): 695-699 (2007).

J3 Joung, J.Y., J. Lim, C.M. Oh et al. Risk of second primary cancer among prostate cancer patients in Korea: A population-based cohort study. PLoS One 10(10): e0140693 (2015).

J4 Journy, N., I. Mansouri, R.S. Allodji et al. Volume effects of radiotherapy on the risk of second primary cancers: A systematic review of clinical and epidemiological studies. Radiother Oncol 131: 150-159 (2019).

K1 Kaatsch, P., I. Reinisch, C. Spix et al. Case-control study on the therapy of childhood cancer and the occurrence of second malignant neoplasms in Germany. Cancer Causes Control 20(6): 965-980 (2009).

K2 Kaldor, J.M., N.E. Day, J. Bell et al. Lung cancer following Hodgkin's disease: A case-control study. Int J Cancer 52(5): 677-681 (1992).

K3 Kalra, S., R.J. Grimer, D. Spooner et al. Radiation-induced sarcomas of bone: Factors that affect outcome. J Bone Joint Surg Br 89(6): 808-813 (2007).

K4 Karlsson, P., E. Holmberg, A. Samuelsson et al. Soft tissue sarcoma after treatment for breast cancer – A Swedish population-based study. Eur J Cancer 34(13): 2068-2075 (1998).

K5 Karzmark, C.J., A. Deubert and R. Loevinger. Correspondence: Tissue-phantom ratios – An aid to treatment planning. Br J Radiol 38(446): 158-159 (1965).

K6 Kase, K.R., G.K. Svensson, A.B. Wolbarst et al. Measurements of dose from secondary radiation outside a treatment field. Int J Radiat Oncol Biol Phys 9(8): 1177-1183 (1983).

K7 Kaufman, E.L., J.S. Jacobson, D.L. Hershman et al. Effect of breast cancer radiotherapy and cigarette smoking on risk of second primary lung cancer. J Clin Oncol 26(3): 392-398 (2008).

K8 Kawrakow, I., M. Fippel and K. Friedrich. 3D electron dose calculation using a Voxel based Monte Carlo algorithm (VMC). Med Phys 23(4): 445-457 (1996).

K9 Kawrakow, I. Accurate condensed history Monte Carlo simulation of electron transport. I. EGSnrc, the new EGS4 version. Med Phys 27(3): 485-498 (2000).

K10 Kawrakow, I. and M. Fippel. VMC++, a MC algorithm optimized for electron and photon beam dose calculations for RTP. Proceedings of the 22nd Annual International Conference of the IEEE Engineering in Medicine and Biology Society (Cat. No.00CH37143). Chicago, IL, USA, 2000.

K11 Kayser, S., K. Dohner, J. Krauter et al. The impact of therapy-related acute myeloid leukemia (AML) on outcome in 2853 adult patients with newly diagnosed AML. Blood 117(7): 2137-2145 (2011).

K12 Kenney, L.B., Y. Yasui, P.D. Inskip et al. Breast cancer after childhood cancer: a report from the childhood cancer survivor study. Ann Intern Med 141(8): 590-597 (2004).

K13 Kern, W., T. Haferlach, S. Schnittger et al. Prognosis in therapy-related acute myeloid leukemia and impact of karyotype. J Clin Oncol 22(12): 2510-2511 (2004).

K14 Kerns, S.L., L. Fachal, L. Dorling et al. Radiogenomics consortium genome-wide association study meta-analysis of late toxicity after prostate cancer radiotherapy. J Natl Cancer Inst 112(2): 179-190 (2020).

K15 Khan, F.M., S.H. Levitt, V.C. Moore et al. Computer and approximation methods of calculating depth dose in irregularly shaped fields. Radiology 106(2): 433-436 (1973).

K16 Khanna, L., S.R. Prasad, S. Yedururi et al. Second malignancies after radiation therapy: Update on pathogenesis and cross-sectional imaging findings. Radiographics 41(3): 876-894 (2021).

K17 Khoury, J.D., E. Solary, O. Abla et al. The 5th edition of the World Health Organization classification of haematolymphoid tumours: Myeloid and histiocytic/dendritic neoplasms. Leukemia 36(7): 1703-1719 (2022).

K18 Kim, E., D.J. Han, B.H. Kim et al. Whole-genome sequencing reveals mutational signatures related to radiation-induced sarcomas and DNA-damage-repair pathways. Mod Pathol 36(1): 100004 (2023).

K19 Kirova, Y.M., D. Stoppa-Lyonnet, A. Savignoni et al. Risk of breast cancer recurrence and contralateral breast cancer in relation to BRCA1 and BRCA2 mutation status following breast-conserving surgery and radiotherapy. Eur J Cancer 41(15): 2304-2311 (2005).

K20 Kleinerman, R.A., M.A. Tucker, D.H. Abramson et al. Risk of soft tissue sarcomas by individual subtype in survivors of hereditary retinoblastoma. J Natl Cancer Inst 99(1): 24-31 (2007).

K21 Kleinerman, R.A., S.J. Schonfeld and M.A. Tucker. Sarcomas in hereditary retinoblastoma. Clin Sarcoma Res 2(1): 15 (2012).

K22 Kleinerman, R.A., S.A. Smith, E. Holowaty et al. Radiation dose and subsequent risk for stomach cancer in long-term survivors of cervical cancer. Int J Radiat Oncol Biol Phys 86(5): 922-929 (2013).

K23 Knežević, Ž., L. Stolarczyk, I. Ambrožová et al. Out-of-field doses produced by a proton scanning beam inside pediatric anthropomorphic phantoms and their comparison with different photon modalities. Front Oncol 12: 904563 (2022).

K24 Knight, J.A., A.D. Skol, A. Shinde et al. Genome-wide association study to identify novel loci associated with therapy-related myeloid leukemia susceptibility. Blood 113(22): 5575-5582 (2009).

K25 Kocher, D.C., A.I. Apostoaei, R.W. Henshaw et al. Interactive RadioEpidemiological Program (IREP): A web-based tool for estimating probability of causation/assigned share of radiogenic cancers. Health Phys 95(1): 119-147 (2008).

K26 Kroon, P.S., S. Hol and M. Essers. Dosimetric accuracy and clinical quality of Acuros XB and AAA dose calculation algorithm for stereotactic and conventional lung volumetric modulated arc therapy plans. Radiat Oncol 8: 149 (2013).

K27 Krul, I.M., A.W.J. Opstal-van Winden, B.M.P. Aleman et al. Breast cancer risk after radiation therapy for Hodgkin Lymphoma: Influence of gonadal hormone exposure. Int J Radiat Oncol Biol Phys 99(4): 843-853 (2017).

K28 Kry, S.F., M. Salehpour, D.S. Followill et al. The calculated risk of fatal secondary malignancies from intensity-modulated radiation therapy. Int J Radiat Oncol Biol Phys 62(4): 1195-1203 (2005).

K29 Kry, S.F., M. Salehpour, D.S. Followill et al. Out-of-field photon and neutron dose equivalents from step-and-shoot intensity-modulated radiation therapy. Int J Radiat Oncol Biol Phys 62(4): 1204-1216 (2005).

K30 Kry, S.F., B. Bednarz, R.M. Howell et al. AAPM TG 158: Measurement and calculation of doses outside the treated volume from external-beam radiation therapy. Med Phys 44(10): e391-e429 (2017).

K31 Kry, S.F., J. Lye, C.H. Clark et al. Report dose-to-medium in clinical trials where available: A consensus from the Global Harmonisation Group to maximize consistency. Radiother Oncol 159: 106-111 (2021).

K32 Kushner, C.J., W.T. Hwang, S. Wang et al. Long-term risk of second malignancies in women after breast conservation therapy for ductal carcinoma in situ or early-stage breast cancer. Breast Cancer Res Treat 170(1): 45-53 (2018).

K33 Kuttesch, J.F., Jr., L.H. Wexler, R.B. Marcus et al. Second malignancies after Ewing's sarcoma: Radiation dose-dependency of secondary sarcomas. J Clin Oncol 14(10): 2818-2825 (1996).

L1 Lae, M., A. Lebel, F. Hamel-Viard et al. Can c-myc amplification reliably discriminate postradiation from primary angiosarcoma of the breast? Cancer Radiother 19(3): 168-174 (2015).

L2 Land, C.E., E. Gilbert, J.M. Smith et al. Report of the NCI-CDC working group to revise the 1985 NIH radioepidemiological tables. NIH Publication 03-5387. National Cancer Institute, Bethesda, MD, 2003.

L3 Larson, R.A., Y. Wang, M. Banerjee et al. Prevalence of the inactivating 609C-->T polymorphism in the NAD(P)H:quinone oxidoreductase (NQO1) gene in patients with primary and therapy-related myeloid leukemia. Blood 94(2): 803-807 (1999).

L4 Lavey, R.S., N.L. Eby and L.R. Prosnitz. Impact of radiation therapy and/or chemotherapy on the risk for a second malignancy after breast cancer. Cancer 66(5): 874-881 (1990).

L5 Le Vu, B., F. de Vathaire, A. Shamsaldin et al. Radiation dose, chemotherapy and risk of osteosarcoma after solid tumours during childhood. Int J Cancer 77(3): 370-377 (1998).

L6 Leary, J.B., A. Anderson-Mellies and A.L. Green. Population-based analysis of radiation-induced gliomas after cranial radiotherapy for childhood cancers. Neurooncol Adv 4(1): vdac159 (2022).

L7 Lee, C., J.W. Jung, C. Pelletier et al. Reconstruction of organ dose for external radiotherapy patients in retrospective epidemiologic studies. Phys Med Biol 60(6): 2309-2324 (2015).

L8 Lehnhardt, M., J. Bohm, T. Hirsch et al. Radiation-induced angiosarcoma of the breast. Handchir Mikrochir Plast Chir 49(2): 103-110 (2017). (German).

L9 Lenz, G., M. Dreyling, E. Schiegnitz et al. Moderate increase of secondary hematologic malignancies after myeloablative radiochemotherapy and autologous stem-cell transplantation in patients with indolent lymphoma: results of a prospective randomized trial of the German Low Grade Lymphoma Study Group. J Clin Oncol 22(24): 4926-4933 (2004).

L10 Leone, G., E. Fabiani and M.T. Voso. De novo and therapy-related myelodysplastic syndromes: Analogies and differences. Mediterr J Hematol Infect Dis 14(1): e2022030 (2022).

L11 Leuraud, K., D.B. Richardson, E. Cardis et al. Ionising radiation and risk of death from leukaemia and lymphoma in radiation-monitored workers (INWORKS): An international cohort study. Lancet Haematol 2(7): e276-281 (2015).

L12 Li, C.I., N. Nishi, J.A. McDougall et al. Relationship between radiation exposure and risk of second primary cancers among atomic bomb survivors. Cancer Res 70(18): 7187-7198 (2010).

L13 Li, W., Q. Liu, X. Deng et al. Association between obesity and puberty timing: a systematic review and meta-analysis. Int J Environ Res Public Health 14(10): 1266 (2017).

L14 Li, Z., K. Wang, Y. Shi et al. Incidence of second primary malignancy after breast cancer and related risk factors-Is breast-conserving surgery safe? A nested case-control study. Int J Cancer 146(2): 352-362 (2020).

L15 Liang, F., S. Zhang, H. Xue et al. Risk of second primary cancers in cancer patients treated with cisplatin: a systematic review and meta-analysis of randomized studies. BMC Cancer 17(1): 871 (2017).

L16 Liao, Z., M.C. Rodrigues, J.N. Poynter et al. Risk of second malignant neoplasms in women and girls with germ cell tumors. Ann Oncol 28(2): 329-332 (2017).

L17 Lin, H.M. and M.A. Teitell. Second malignancy after treatment of pediatric Hodgkin disease. J Pediatr Hematol Oncol 27(1): 28-36 (2005).

L18 Lincoln, S.E., R.L. Nussbaum, A.W. Kurian et al. Yield and utility of germline testing following tumor sequencing in patients with cancer. JAMA Netw Open 3(10): e2019452 (2020).

L19 Lindberg, S., P. Karlsson, B. Arvidsson et al. Cancer incidence after radiotherapy for skin haemangioma during infancy. Acta Oncol 34(6): 735-740 (1995).

L20 Little, M.P., F. de Vathaire, A. Shamsaldin et al. Risks of brain tumour following treatment for cancer in childhood: modification by genetic factors, radiotherapy and chemotherapy. Int J Cancer 78(3): 269-275 (1998).

L21 Little, M.P. Comparison of the risks of cancer incidence and mortality following radiation therapy for benign and malignant disease with the cancer risks observed in the Japanese A-bomb survivors. Int J Radiat Biol 77(4): 431-464 (2001).

L22 Liu, A.P.Y., G. Wu, B.A. Orr et al. Outcome and molecular analysis of young children with choroid plexus carcinoma treated with non-myeloablative therapy: Results from the SJYC07 trial. Neurooncol Adv 3(1): vdaa168 (2021).

L23 Lönn, S., E.S. Gilbert, E. Ron et al. Comparison of second cancer risks from brachytherapy and external beam therapy after uterine corpus cancer. Cancer Epidemiol Biomarkers Prev 19(2): 464-474 (2010).

L24 Lonski, P., T. Kron, M. Taylor et al. Assessment of leakage dose in vivo in patients undergoing radiotherapy for breast cancer. Phys Imaging Radiat Oncol 5: 97-101 (2018).

L25 López, G.Y., J. Van Ziffle, C. Onodera et al. The genetic landscape of gliomas arising after therapeutic radiation. Acta Neuropathol 137(1): 139-150 (2019).

L26 Lorenz, E., P. Scholz-Kreisel, D. Baaken et al. Radiotherapy for childhood cancer and subsequent thyroid cancer risk: a systematic review. Eur J Epidemiol 33(12): 1139-1162 (2018).

L27 Lorigan, P., J. Radford, A. Howell et al. Lung cancer after treatment for Hodgkin's lymphoma: A systematic review. Lancet Oncol 6(10): 773-779 (2005).

L28 Low, D.A., W.B. Harms, S. Mutic et al. A technique for the quantitative evaluation of dose distributions. Med Phys 25(5): 656-661 (1998).

L29 Lubin, J.H. and J.D. Boice, Jr. Lung cancer risk from residential radon: meta-analysis of eight epidemiologic studies. J Natl Cancer Inst 89(1): 49-57 (1997).

M1 Ma, Y.P., F.E. van Leeuwen, R. Cooke et al. FGFR2 genotype and risk of radiation-associated breast cancer in Hodgkin lymphoma. Blood 119(4): 1029-1031 (2012).

M2 Mackie, T.R., A.F. Bielajew, D.W. Rogers et al. Generation of photon energy deposition kernels using the EGS Monte Carlo code. Phys Med Biol 33(1): 1-20 (1988).

M3 Majeed, H. and V. Gupta. Adverse Effects of Radiation Therapy. In: StatPearls. StatPearls Publishing, Treasure Island (FL), 2023.

M4 Majer, M., I. Ambrozova, M. Davidkova et al. Out-of-field doses in pediatric craniospinal irradiations with 3D-CRT, VMAT, and scanning proton radiotherapy: A phantom study. Med Phys 49(4): 2672-2683 (2022).

M5 Mancuso, M., E. Pasquali, S. Leonardi et al. Oncogenic bystander radiation effects in patched heterozygous mouse cerebellum. Proc Natl Acad Sci U S A 105(34): 12445-12450 (2008).

M6 Manem, V.S. and F. Taghizadeh-Hesary. Advances in personalized radiotherapy. BMC Cancer 24(1): 556 (2024).

M7 Marees, T., A.C. Moll, S.M. Imhof et al. Risk of second malignancies in survivors of retinoblastoma: More than 40 years of follow-up. J Natl Cancer Inst 100(24): 1771-1779 (2008).

M8 Martincorena, I., K.M. Raine, M. Gerstung et al. Universal patterns of selection in cancer and somatic tissues. Cell 171(5): 1029-1041.e1021 (2017).

M9 Matsumoto, S., S. Yonai and W.E. Bolch. Monte Carlo study of out-of-field exposure in carbon-ion radiotherapy: Organ doses in pediatric brain tumor treatment. Med Phys 46(12): 5824-5832 (2019).

M10 McCarthy, B.J., K.A. Schellinger, J.M. Propp et al. A case for the worldwide collection of primary benign brain tumors. Neuroepidemiology 33(3): 268-275 (2009).

M11 McDuff, S.G.R., J.R. Bellon, K.M. Shannon et al. ATM variants in breast cancer: Implications for breast radiation therapy treatment recommendations. Int J Radiat Oncol Biol Phys 110(5): 1373-1382 (2021).

M12 McGuire, V., D.Y. Lichtensztajn, L. Tao et al. Variation in patterns of second primary malignancies across U.S. race and ethnicity groups: A Surveillance, Epidemiology, and End Results (SEER) analysis. Cancer Causes Control 35(5): 799-815 (2024).

M13 McParland, B.J. and H.I. Fair. A method of calculating peripheral dose distributions of photon beams below 10 MV. Med Phys 19(2): 283-293 (1992).

M14 Meadows, G.G. and H. Zhang. Effects of alcohol on tumor growth, metastasis, immune response, and host survival. Alcohol Res 37(2): 311-322 (2015).

M15 Mee, T., N.F. Kirkby, N.N. Defourny et al. The use of radiotherapy, surgery and chemotherapy in the curative treatment of cancer: Results from the FORTY (Favourable Outcomes from RadioTherapY) project. Br J Radiol 96(1152): 20230334 (2023).

M16 Meeks, S.L., A.C. Paulino, E.C. Pennington et al. In vivo determination of extra-target doses received from serial tomotherapy. Radiother Oncol 63(2): 217-222 (2002).

M17 Menes, T.S., M.B. Terry, D. Goldgar et al. Second primary breast cancer in BRCA1 and BRCA2 mutation carriers: 10-year cumulative incidence in the Breast Cancer Family Registry. Breast Cancer Res Treat 151(3): 653-660 (2015).

M18 Menu-Branthomme, A., C. Rubino, A. Shamsaldin et al. Radiation dose, chemotherapy and risk of soft tissue sarcoma after solid tumours during childhood. Int J Cancer 110(1): 87-93 (2004).

M19 Metayer, C., C.F. Lynch, E.A. Clarke et al. Second cancers among long-term survivors of Hodgkin's disease diagnosed in childhood and adolescence. J Clin Oncol 18(12): 2435-2443 (2000).

M20 Meyniel, J.P., P.H. Cottu, C. Decraene et al. A genomic and transcriptomic approach for a differential diagnosis between primary and secondary ovarian carcinomas in patients with a previous history of breast cancer. BMC Cancer 10: 222 (2010).

M21 Michaeli, O. and U. Tabori. Pediatric high grade gliomas in the context of cancer predisposition syndromes. J Korean Neurosurg Soc 61(3): 319-332 (2018).

M22 Milas, Z.L., B. Neelands, S.J. Trufan et al. Survivorship care plans improve the identification of post-radiation hypothyroidism after head and neck cancer treatment. Anticancer Res 42(9): 4429-4437 (2022).

M23 Miralbell, R., A. Lomax, L. Cella et al. Potential reduction of the incidence of radiation-induced second cancers by using proton beams in the treatment of pediatric tumors. Int J Radiat Oncol Biol Phys 54(3): 824-829 (2002).

M24 Mirjolet, C., I. Diallo, A. Bertaut et al. Treatment related factors associated with the risk of breast radio-induced-sarcoma. Radiother Oncol 171: 14-21 (2022).

M25 Mohamad, O., T. Tabuchi, Y. Nitta et al. Risk of subsequent primary cancers after carbon ion radiotherapy, photon radiotherapy, or surgery for localised prostate cancer: a propensity score-weighted, retrospective, cohort study. Lancet Oncol 20(5): 674-685 (2019).

M26 Morton, L.M. and S.J. Chanock. A step toward slaying the hydra of second cancers. Nat Med 17(8): 924-925 (2011).

M27 Morton, L.M., E.S. Gilbert, P. Hall et al. Risk of treatment-related esophageal cancer among breast cancer survivors. Ann Oncol 23(12): 3081-3091 (2012).

M28 Morton, L.M., G.M. Dores, R.E. Curtis et al. Stomach cancer risk after treatment for Hodgkin lymphoma. J Clin Oncol 31(27): 3369-3377 (2013).

M29 Morton, L.M., K. Onel, R.E. Curtis et al. The rising incidence of second cancers: patterns of occurrence and identification of risk factors for children and adults. Am Soc Clin Oncol Educ Book: e57-67 (2014).

M30 Morton, L.M., A.J. Swerdlow, M. Schaapveld et al. Current knowledge and future research directions in treatment-related second primary malignancies. EJC Suppl 12(1): 5-17 (2014).

M31 Morton, L.M., J.N. Sampson, G.T. Armstrong et al. Genome-wide association study to identify susceptibility loci that modify radiation-related risk for breast cancer after childhood cancer. J Natl Cancer Inst 109(11) (2017).

M32 Moser, E.C., E.M. Noordijk, F.E. van Leeuwen et al. Risk of second cancer after treatment of aggressive non-Hodgkin's lymphoma: An EORTC cohort study. Haematologica 91(11): 1481-1488 (2006).

M33 Moskowitz, C.S., J.F. Chou, S.L. Wolden et al. Breast cancer after chest radiation therapy for childhood cancer. J Clin Oncol 32(21): 2217-2223 (2014).

M34 Mu, X., T. Bjork-Eriksson, S. Nill et al. Does electron and proton therapy reduce the risk of radiation induced cancer after spinal irradiation for childhood medulloblastoma? A comparative treatment planning study. Acta Oncol 44(6): 554-562 (2005).

M35 Murohashi, I. Leukemia in patients following radiotherapy for malignant neoplasms in the pelvic region. Leuk Res 9(9): 1201-1208 (1985).

M36 Murphy, M.J., J. Balter, S. Balter et al. The management of imaging dose during image-guided radiotherapy: report of the AAPM Task Group 75. Med Phys 34(10): 4041-4063 (2007).

N1 Nagle, P.W. and R.P. Coppes. Current and future perspectives of the use of organoids in radiobiology. Cells 9(12): 2649 (2020).

N2 Nandy, N. and C.A. Dasanu. Incidence of second primary malignancies in patients with esophageal cancer: A comprehensive review. Curr Med Res Opin 29(9): 1055-1065 (2013).

N3 Nassi, L., V. De Sanctis, G. Loseto et al. Second cancers in classical Hodgkin lymphoma and diffuse large B-cell lymphoma: A systematic review by the Fondazione Italiana Linfomi. Cancers (Basel) 14(3): 519 (2022).

N4 Navid, F., C. Billups, T. Liu et al. Second cancers in patients with the Ewing sarcoma family of tumours. Eur J Cancer 44(7): 983-991 (2008).

N5 NCRP. The relative biological effectiveness of radiations of different quality: Recommendations of the National Council on Radiation Protection and Measurements. NCRP Report No. 104. National Council on Radiation Protection and Measurements, Bethesda, 1990.

N6 NCRP. Second primary cancers and cardiovascular disease after radiation therapy. NCRP Report No. 170. National Council on Radiation Protection and Measurements, Bethesda, 2011.

N7 NCRP. Evaluating and communicating radiation risks for studies involving human subjects: Guidance for researchers and institutional review boards. NCRP Report No. 185. National Council on Radiation Protection and Measurements, Bethesda, 2020.

N8 Neglia, J.P., A.T. Meadows, L.L. Robison et al. Second neoplasms after acute lymphoblastic leukemia in childhood. N Engl J Med 325(19): 1330-1336 (1991).

N9 Neglia, J.P., D.L. Friedman, Y. Yasui et al. Second malignant neoplasms in five-year survivors of childhood cancer: Childhood cancer survivor study. J Natl Cancer Inst 93(8): 618-629 (2001).

N10 Neglia, J.P., L.L. Robison, M. Stovall et al. New primary neoplasms of the central nervous system in survivors of childhood cancer: A report from the Childhood Cancer Survivor Study. J Natl Cancer Inst 98(21): 1528-1537 (2006).

N11 Neugut, A.I., T. Murray, J. Santos et al. Increased risk of lung cancer after breast cancer radiation therapy in cigarette smokers. Cancer 73(6): 1615-1620 (1994).

N12 Newhauser, W.D. and M. Durante. Assessing the risk of second malignancies after modern radiotherapy. Nat Rev Cancer 11(6): 438-448 (2011).

N13 Ng, A.K., M.V. Bernardo, E. Weller et al. Second malignancy after Hodgkin disease treated with radiation therapy with or without chemotherapy: Long-term risks and risk factors. Blood 100(6): 1989-1996 (2002).

N14 Ng, J. and I. Shuryak. Minimizing second cancer risk following radiotherapy: Current perspectives. Cancer Manag Res 7: 1-11 (2015).

N15 NIH. US pediatric proton and photon therapy cohort. National Institute of Health. [Internet] Available from (https://sam.gov/opp/85d684890e1d49bf89cfdd67b7f4520e/view) on 9 July 2023.

N16 Nilsson, C., F. Linde, E. Hulegardh et al. Characterization of therapy-related acute myeloid leukemia: Increasing incidence and prognostic implications. Haematologica 108(4): 1015-1025 (2023).

N17 Nishimura, Y. Rationale for chemoradiotherapy. Int J Clin Oncol 9(6): 414-420 (2004).

N18 Nottage, K., J. McFarlane, M.J. Krasin et al. Secondary colorectal carcinoma after childhood cancer. J Clin Oncol 30(20): 2552-2558 (2012).

N19 NRC. Health risks from exposure to low levels of ionizing radiation. Committee to assess health risks from exposure to low levels of ionizing radiation, BEIR VII - Phase 2. National Research Council, T.N.A. Press, Washington D.C., 2006.

O1 O'Brien, M.M., S.S. Donaldson, R.R. Balise et al. Second malignant neoplasms in survivors of pediatric Hodgkin's lymphoma treated with low-dose radiation and chemotherapy. J Clin Oncol 28(7): 1232-1239 (2010).

O2 Olsen, J.H., T. Moller, H. Anderson et al. Lifelong cancer incidence in 47, 697 patients treated for childhood cancer in the Nordic countries. J Natl Cancer Inst 101(11): 806-813 (2009).

O3 Opstal-van Winden, A.W.J., H.G. de Haan, M. Hauptmann et al. Genetic susceptibility to radiation-induced breast cancer after Hodgkin lymphoma. Blood 133(10): 1130-1139 (2019).

O4 Ostrovnaya, I., A.B. Olshen, V.E. Seshan et al. A metastasis or a second independent cancer? Evaluating the clonal origin of tumors using array copy number data. Stat Med 29(15): 1608-1621 (2010).

O5 Otsuka, K. and T. Iwasaki. Insights into radiation carcinogenesis based on dose-rate effects in tissue stem cells. Int J Radiat Biol 99(10): 1503-1521 (2023).

P1 Pappo, A.S., G.T. Armstrong, W. Liu et al. Melanoma as a subsequent neoplasm in adult survivors of childhood cancer: a report from the childhood cancer survivor study. Pediatr Blood Cancer 60(3): 461-466 (2013).

P2 Pérez-Andújar, A., R. Zhang and W. Newhauser. Monte Carlo and analytical model predictions of leakage neutron exposures from passively scattered proton therapy. Med Phys 40(12): 121714 (2013).

P3 Perl, J., J. Shin, J. Schumann et al. TOPAS: An innovative proton Monte Carlo platform for research and clinical applications. Med Phys 39(11): 6818-6837 (2012).

P4 Picard, C., C. Faure-Conter, P. Leblond et al. Exploring heterogeneity of adrenal cortical tumors in children: The French pediatric rare tumor group (Fracture) experience. Pediatr Blood Cancer 67(2): e28086 (2020).

P5 Pierce, L.J., M. Strawderman, S.A. Narod et al. Effect of radiotherapy after breast-conserving treatment in women with breast cancer and germline BRCA1/2 mutations. J Clin Oncol 18(19): 3360-3369 (2000).

P6 Pierce, L.J., A.M. Levin, T.R. Rebbeck et al. Ten-year multi-institutional results of breast-conserving surgery and radiotherapy in BRCA1/2-associated stage I/II breast cancer. J Clin Oncol 24(16): 2437-2443 (2006).

P7 Pinheiro, J.C. and D. Bates. Mixed-Effects Models in S and S-PLUS. Springer, New York, 2000.

P8 Pollard, J.M. and R.A. Gatti. Clinical radiation sensitivity with DNA repair disorders: an overview. Int J Radiat Oncol Biol Phys 74(5): 1323-1331 (2009).

P9 Pondrom, M., G. Bougeard, M. Karanian et al. Rhabdomyosarcoma associated with germline TP53 alteration in children and adolescents: The French experience. Pediatr Blood Cancer 67(9): e28486 (2020).

P10 Porta, M. A Dictionary of Epidemiology. Oxford University Press, 2016.

P11 Preston, D.L., A. Mattsson, E. Holmberg et al. Radiation effects on breast cancer risk: A pooled analysis of eight cohorts. Radiat Res 158(2): 220-235 (2002).

P12 Preston, D.L., E. Ron, S. Tokuoka et al. Solid cancer incidence in atomic bomb survivors: 1958-1998. Radiat Res 168(1): 1-64 (2007).

P13 Prochazka, M., P. Hall, G. Gagliardi et al. Ionizing radiation and tobacco use increases the risk of a subsequent lung carcinoma in women with breast cancer: Case-only design. J Clin Oncol 23(30): 7467-7474 (2005).

P14 Pui, C.H., R.C. Ribeiro, M.L. Hancock et al. Acute myeloid leukemia in children treated with epipodophyllotoxins for acute lymphoblastic leukemia. N Engl J Med 325(24): 1682-1687 (1991).

R1 R Core Team. R project: A language and environment for statistical computing and graphics. R Foundation. [Internet] Available from (https://www.R-project.org)

R2 Reiner, A.S., M.E. Robson, L. Mellemkjaer et al. Radiation Treatment, ATM, BRCA1/2, and CHEK2*1100delC Pathogenic Variants and Risk of Contralateral Breast Cancer. J Natl Cancer Inst 112(12): 1275-1279 (2020).

R3 Relling, M.V., J.E. Rubnitz, G.K. Rivera et al. High incidence of secondary brain tumours after radiotherapy and antimetabolites. Lancet 354(9172): 34-39 (1999).

R4 Reulen, R.C., A.J. Taylor, D.L. Winter et al. Long term population-based risks of breast cancer after childhood cancer. Int J Cancer 123(9): 2156-2163 (2008).

R5 Reulen, R.C., C. Frobisher, D.L. Winter et al. Long-term risks of subsequent primary neoplasms among survivors of childhood cancer. JAMA 305(22): 2311-2319 (2011).

R6 Richards, S., N. Aziz, S. Bale et al. Standards and guidelines for the interpretation of sequence variants: A joint consensus recommendation of the American College of Medical Genetics and Genomics and the Association for Molecular Pathology. Genet Med 17(5): 405-424 (2015).

R7 Richardson, D.B., K. Abalo, M.O. Bernier et al. Meta-analysis of published excess relative risk estimates. Radiat Environ Biophys 59(4): 631-641 (2020).

R8 Richiardi, L., G. Scelo, P. Boffetta et al. Second malignancies among survivors of germ-cell testicular cancer: A pooled analysis between 13 cancer registries. Int J Cancer 120(3): 623-631 (2007).

R9 Rodriguez, M., J. Sempau and L. Brualla. PRIMO: A graphical environment for the Monte Carlo simulation of Varian and Elekta LINACS. Strahlenther Onkol 189(10): 881-886 (2013).

R10 Ron, E., J.H. Lubin, R.E. Shore et al. Thyroid cancer after exposure to external radiation: a pooled analysis of seven studies. Radiat Res 141(3): 259-277 (1995).

R11 Ronckers, C.M., A.J. Sigurdson, M. Stovall et al. Thyroid cancer in childhood cancer survivors: a detailed evaluation of radiation dose response and its modifiers. Radiat Res 166(4): 618-628 (2006).

R12 Rosemann, M., M. Lintrop, J. Favor et al. Bone tumorigenesis induced by alpha-particle radiation: mapping of genetic loci influencing predisposition in mice. Radiat Res 157(4): 426-434 (2002).

R13 Rosso, P., B. Terracini, T.R. Fears et al. Second malignant tumors after elective end of therapy for a first cancer in childhood: a multicenter study in Italy. Int J Cancer 59(4): 451-456 (1994).

R14 Ruben, J.D., C.M. Lancaster, P. Jones et al. A comparison of out-of-field dose and its constituent components for intensity-modulated radiation therapy versus conformal radiation therapy: implications for carcinogenesis. Int J Radiat Oncol Biol Phys 81(5): 1458-1464 (2011).

R15 Rubino, C., E. Adjadj, S. Guerin et al. Long-term risk of second malignant neoplasms after neuroblastoma in childhood: role of treatment. Int J Cancer 107(5): 791-796 (2003).

R16 Rubino, C., F. de Vathaire, M.E. Dottorini et al. Second primary malignancies in thyroid cancer patients. Br J Cancer 89(9): 1638-1644 (2003).

R17 Rubino, C., F. de Vathaire, A. Shamsaldin et al. Radiation dose, chemotherapy, hormonal treatment and risk of second cancer after breast cancer treatment. Br J Cancer 89(5): 840-846 (2003).

R18 Rubino, C., E. Adjadj, F. Doyon et al. Radiation exposure and familial aggregation of cancers as risk factors for colorectal cancer after radioiodine treatment for thyroid carcinoma. Int J Radiat Oncol Biol Phys 62(4): 1084-1089 (2005).

R19 Rubino, C., A. Shamsaldin, M.G. Le et al. Radiation dose and risk of soft tissue and bone sarcoma after breast cancer treatment. Breast Cancer Res Treat 89(3): 277-288 (2005).

S1 Sachs, R.K. and D.J. Brenner. Solid tumor risks after high doses of ionizing radiation. Proc Natl Acad Sci U S A 102(37): 13040-13045 (2005).

S2 Sakata, R., D.L. Preston, A.V. Brenner et al. Radiation-related risk of cancers of the upper digestive tract among Japanese atomic bomb survivors. Radiat Res 192(3): 331-344 (2019).

S3 Samartzis, D., N. Nishi, J. Cologne et al. Ionizing radiation exposure and the development of soft-tissue sarcomas in atomic-bomb survivors. J Bone Joint Surg Am 95(3): 222-229 (2013).

S4 Sampadi, B., S. Vermeulen, B. Mišovic et al. Divergent molecular and cellular responses to low and high-dose ionizing radiation. Cells 11(23): 3794 (2022).

S5 Sanchez-Nieto, B., M. Romero-Exposito, J.A. Terron et al. Uncomplicated and cancer-free control probability (UCFCP): a new integral approach to treatment plan optimization in photon radiation therapy. Phys Med 42: 277-284 (2017).

S6 Sanchez-Nieto, B., K.N. Medina-Ascanio, J.L. Rodriguez-Mongua et al. Study of out-of-field dose in photon radiotherapy: a commercial treatment planning system versus measurements and Monte Carlo simulations. Med Phys 47(9): 4616-4625 (2020).

S7 Sapkota, Y., L.M. Turcotte, M.J. Ehrhardt et al. Genome-wide association study in irradiated childhood cancer survivors identifies HTR2A for subsequent basal cell carcinoma. J Invest Dermatol 139(9): 2042-2045 e2048 (2019).

S8 Sato, T., Y. Iwamoto, S. Hashimoto et al. Features of Particle and Heavy Ion Transport Code System (PHITS) version 3.02. J Nucl Sci Technol 55(6): 684-690 (2018).

S9 Schaapveld, M., O. Visser, M.J. Louwman et al. Risk of new primary nonbreast cancers after breast cancer treatment: a Dutch population-based study. J Clin Oncol 26(8): 1239-1246 (2008).

S10 Schaapveld, M., B.M. Aleman, A.M. van Eggermond et al. Second cancer risk up to 40 years after treatment for Hodgkin's lymphoma. N Engl J Med 373(26): 2499-2511 (2015).

S11 Schellong, G., M. Riepenhausen, K. Ehlert et al. Breast cancer in young women after treatment for Hodgkin's disease during childhood or adolescence--an observational study with up to 33-year follow-up. Dtsch Arztebl Int 111(1-2): 3-9 (2014).

S12 Schneider, C.W., W.D. Newhauser, L.J. Wilson et al. A physics-based analytical model of absorbed dose from primary, leakage, and scattered photons from megavoltage radiotherapy with MLCs. Phys Med Biol 64(18): 185017 (2019).

S13 Schneider, U., A. Lomax and N. Lombriser. Comparative risk assessment of secondary cancer incidence after treatment of Hodgkin's disease with photon and proton radiation. Radiat Res 154(4): 382-388 (2000).

S14 Schneider, U. and L. Walsh. Cancer risk estimates from the combined Japanese A-bomb and Hodgkin cohorts for doses relevant to radiotherapy. Radiat Environ Biophys 47(2): 253-263 (2008).

S15 Schneider, U. Mechanistic model of radiation-induced cancer after fractionated radiotherapy using the linear-quadratic formula. Med Phys 36(4): 1138-1143 (2009).

S16 Schneider, U. Modeling the risk of secondary malignancies after radiotherapy. Genes (Basel) 2(4): 1033-1049 (2011).

S17 Schneider, U., M. Sumila and J. Robotka. Site-specific dose-response relationships for cancer induction from the combined Japanese A-bomb and Hodgkin cohorts for doses relevant to radiotherapy. Theor Biol Med Model 8: 27 (2011).

S18 Schneider, U., R.A. Halg, M. Hartmann et al. Accuracy of out-of-field dose calculation of tomotherapy and cyberknife treatment planning systems: a dosimetric study. Z Med Phys 24(3): 211-215 (2014).

S19 Schneider, U. and L. Walsh. Age at exposure and attained age variations of cancer risk in the Japanese A-bomb and radiotherapy cohorts. Med Phys 42(8): 4755-4761 (2015).

S20 Schneider, U., R.A. Halg and T. Lomax. Neutrons in active proton therapy: parameterization of dose and dose equivalent. Z Med Phys 27(2): 113-123 (2017).

S21 Schneider, U., L. Walsh and W. Newhauser. Tumour size can have an impact on the outcomes of epidemiological studies on second cancers after radiotherapy. Radiat Environ Biophys 57(4): 311-319 (2018).

S22 Schoch, C., W. Kern, S. Schnittger et al. Karyotype is an independent prognostic parameter in therapy-related acute myeloid leukemia (t-AML): an analysis of 93 patients with t-AML in comparison to 1091 patients with de novo AML. Leukemia 18(1): 120-125 (2004).

S23 Scholz, R. On the sensitivity of children to radiation. Med Glob Surv 1(1): 38-44 (1994).

S24 Schonfeld, S.J., R.M. Howell, S.A. Smith et al. Comparison of radiation dose reconstruction methods to investigate late adverse effects of radiotherapy for childhood cancer: a report from the Childhood Cancer Survivor Study. Radiat Res 193(2): 95-106 (2020).

S25 Schwartz, B., M.A. Benadjaoud, E. Clero et al. Risk of second bone sarcoma following childhood cancer: role of radiation therapy treatment. Radiat Environ Biophys 53(2): 381-390 (2014).

S26 Seedhouse, C., R. Bainton, M. Lewis et al. The genotype distribution of the XRCC1 gene indicates a role for base excision repair in the development of therapy-related acute myeloblastic leukemia. Blood 100(10): 3761-3766 (2002).

S27 Seedhouse, C., R. Faulkner, N. Ashraf et al. Polymorphisms in genes involved in homologous recombination repair interact to increase the risk of developing acute myeloid leukemia. Clin Cancer Res 10(8): 2675-2680 (2004).

S28 Seitz, H.K. and F. Stickel. Molecular mechanisms of alcohol-mediated carcinogenesis. Nat Rev Cancer 7(8): 599-612 (2007).

S29 Sempau, J., E. Acosta, J. Baro et al. An algorithm for Monte Carlo simulation of coupled electron-photon transport. NIM-B 132(3): 377-390 (1995).

S30 Sempau, J., S.J. Wilderman and A.F. Bielajew. DPM, a fast, accurate Monte Carlo code optimized for photon and electron radiotherapy treatment planning dose calculations. Phys Med Biol 45(8): 2263-2291 (2000).

S31 Seo, G.H., Y.Y. Cho, J.H. Chung et al. Increased risk of leukemia after radioactive iodine therapy in patients with thyroid cancer: a nationwide, population-based study in Korea. Thyroid 25(8): 927-934 (2015).

S32 Sgouros, G., L. Bodei, M.R. McDevitt et al. Radiopharmaceutical therapy in cancer: clinical advances and challenges. Nat Rev Drug Discov 19(9): 589-608 (2020).

S33 Sgouros, G., E. Frey, Y. Du et al. Imaging and dosimetry for alpha-particle emitter radiopharmaceutical therapy: improving radiopharmaceutical therapy by looking into the black box. Eur J Nucl Med Mol Imaging 49(1): 18-29 (2021).

S34 Shamsaldin, A., E. Grimaud, C. Hardiman et al. Dose distribution throughout the body from radiotherapy for Hodgkin's disease in childhood. Radiother Oncol 49(1): 85-90 (1998).

S35 Shaw, M., J. Lye, A. Alves et al. Measuring the dose in bone for spine stereotactic body radiotherapy. Phys Med 84: 265-273 (2021).

S36 Sherr, C.J. Ink4-Arf locus in cancer and aging. Wiley Interdiscip Rev Dev Biol 1(5): 731-741 (2012).

S37 Shuryak, I., P. Hahnfeldt, L. Hlatky et al. A new view of radiation-induced cancer: integrating short- and long-term processes. Part I: approach. Radiat Environ Biophys 48(3): 263-274 (2009).

S38 Siegel, R.L., K.D. Miller and A. Jemal. Cancer statistics, 2020. CA Cancer J Clin 70(1): 7-30 (2020).

S39 Siegel, R.L., K.D. Miller, H.E. Fuchs et al. Cancer statistics, 2022. CA Cancer J Clin 72(1): 7-33 (2022).

S40 Sigurdson, A.J., C.M. Ronckers, A.C. Mertens et al. Primary thyroid cancer after a first tumour in childhood (the Childhood Cancer Survivor Study): a nested case-control study. Lancet 365(9476): 2014-2023 (2005).

S41 Simonetto, C., D. Wollschlager, P. Kundrat et al. Estimating long-term health risks after breast cancer radiotherapy: merging evidence from low and high doses. Radiat Environ Biophys 60(3): 459-474 (2021).

S42 Singh, G.K., V. Yadav, P. Singh et al. Radiation-induced malignancies making radiotherapy a "two-edged sword": a review of literature. World J Oncol 8(1): 1-6 (2017).

S43 Slanina, J., F. Heinemann, K. Henne et al. Second malignancies after the therapy of Hodgkin's disease: the Freiburg collective 1940 to 1991. Strahlenther Onkol 175(4): 154-161 (1999).

S44 Smith, J.B., L.M. Cass, M.C. Simpson et al. Radiation-associated sarcoma of the head and neck: incidence, latency, and survival. Laryngoscope 132(5): 1034-1041 (2022).

S45 Sondka, Z., S. Bamford, C.G. Cole et al. The COSMIC Cancer Gene Census: describing genetic dysfunction across all human cancers. Nat Rev Cancer 18(11): 696-705 (2018).

S46 Sont, J.K., W.A. van Stiphout, E.M. Noordijk et al. Increased risk of second cancers in managing Hodgkins disease: the 20-year Leiden experience. Ann Hematol 65(5): 213-218 (1992).

S47 Steel, G., C. West and A. Nahum. Chapter 8: Dose fractionation in radiotherapy. In: Handbook of Radiotherapy Physics, second edition. CRC Press, Boca Raton, 2021.

S48 Storm, H.H. and J.D. Boice, Jr. Leukaemia after cervical cancer irradiation in Denmark. Int J Epidemiol 14(3): 363-368 (1985).

S49 Storm, H.H., M. Andersson, J.D. Boice, Jr. et al. Adjuvant radiotherapy and risk of contralateral breast cancer. J Natl Cancer Inst 84(16): 1245-1250 (1992).

S50 Stovall, M., S.A. Smith and M. Rosenstein. Tissue doses from radiotherapy of cancer of the uterine cervix. Med Phys 16(5): 726-733 (1989).

S51 Stovall, M., R. Weathers, C. Kasper et al. Dose reconstruction for therapeutic and diagnostic radiation exposures: use in epidemiological studies. Radiat Res 166(1 Pt 2): 141-157 (2006).

S52 Stovall, M., S.A. Smith, B.M. Langholz et al. Dose to the contralateral breast from radiotherapy and risk of second primary breast cancer in the WECARE study. Int J Radiat Oncol Biol Phys 72(4): 1021-1030 (2008).

S53 Strong, L.C., J. Herson, B.M. Osborne et al. Risk of radiation-related subsequent malignant tumors in survivors of Ewing's sarcoma. J Natl Cancer Inst 62(6): 1401-1406 (1979).

S54 Sugiyama, H., M. Misumi, A. Brenner et al. Radiation risk of incident colorectal cancer by anatomical site among atomic bomb survivors: 1958-2009. Int J Cancer 146(3): 635-645 (2020).

S55 Sung, H., N. Hyun, C.R. Leach et al. Association of first primary cancer with risk of subsequent primary cancer among survivors of adult-onset cancers in the United States. JAMA 324(24): 2521-2535 (2020).

S56 Sutton, A.J. and J.P. Higgins. Recent developments in meta-analysis. Stat Med 27(5): 625-650 (2008).

S57 Svahn-Tapper, G., S. Garwicz, H. Anderson et al. Radiation dose and relapse are predictors for development of second malignant solid tumors after cancer in childhood and adolescence: a population-based case-control study in the five Nordic countries. Acta Oncol 45(4): 438-448 (2006).

T1 Tabuchi, T., K. Ozaki, A. Ioka et al. Joint and independent effect of alcohol and tobacco use on the risk of subsequent cancer incidence among cancer survivors: a cohort study using cancer registries. Int J Cancer 137(9): 2114-2123 (2015).

T2 Takahashi, K. and T. Tango. Assignment of grouped exposure levels for trend estimation in a regression analysis of summarized data. Stat Med 29(25): 2605-2616 (2010).

T3 Tavernier, E., Q.H. Le, S. de Botton et al. Secondary or concomitant neoplasms among adults diagnosed with acute lymphoblastic leukemia and treated according to the LALA-87 and LALA-94 trials. Cancer 110(12): 2747-2755 (2007).

T4 Taylor, A.J., D.L. Winter, C.A. Stiller et al. Risk of breast cancer in female survivors of childhood Hodgkin's disease in Britain: a population-based study. Int J Cancer 120(2): 384-391 (2007).

T5 Taylor, A.J., D.L. Winter, K. Pritchard-Jones et al. Second primary neoplasms in survivors of Wilms' tumour--a population-based cohort study from the British Childhood Cancer Survivor Study. Int J Cancer 122(9): 2085-2093 (2008).

T6 Taylor, A.J., M.P. Little, D.L. Winter et al. Population-based risks of CNS tumors in survivors of childhood cancer: the British Childhood Cancer Survivor Study. J Clin Oncol 28(36): 5287-5293 (2010).

T7 Taylor, C., C. Correa, F.K. Duane et al. Estimating the risks of breast cancer radiotherapy: evidence from modern radiation doses to the lungs and heart and from previous randomized trials. J Clin Oncol 35(1): 1641-1649 (2017).

T8 Taylor, S., P. Lim, R. Ahmad et al. Risk of radiation-induced second malignant neoplasms from photon and proton radiotherapy in paediatric abdominal neuroblastoma. Phys Imaging Radiat Oncol 19: 45-52 (2021).

T9 Teepen, J.C., S.L. de Vroom, F.E. van Leeuwen et al. A systematic review: childhood cancer survivors and gastrointestinal cancer. Cancer Treat Rev 55: 210 (2017).

T10 Terenziani, M., M. Massimino, D. Magazzù et al. Management of breast cancer after Hodgkin's lymphoma and paediatric cancer. Eur J Cancer 51(13): 1667-1674 (2015).

T11 Thariat, J., F. Chevalier, D. Orbach et al. Avoidance or adaptation of radiotherapy in patients with cancer with Li-Fraumeni and heritable TP53-related cancer syndromes. Lancet Oncol 22(12): e562-e574 (2021).

T12 Travis, L.B., R.E. Curtis, J.D. Boice, Jr. et al. Second cancers following non-Hodgkin's lymphoma. Cancer 67(7): 2002-2009 (1991).

T13 Travis, L.B., R.E. Curtis, B. Glimelius et al. Second cancers among long-term survivors of non-Hodgkin's lymphoma. J Natl Cancer Inst 85(23): 1932-1937 (1993).

T14 Travis, L.B., R.E. Curtis, M. Stovall et al. Risk of leukemia following treatment for non-Hodgkin's lymphoma. J Natl Cancer Inst 86(19): 1450-1457 (1994).

T15 Travis, L.B., E.J. Holowaty, K. Bergfeldt et al. Risk of leukemia after platinum-based chemotherapy for ovarian cancer. N Engl J Med 340(5): 351-357 (1999).

T16 Travis, L.B., M. Andersson, M. Gospodarowicz et al. Treatment-associated leukemia following testicular cancer. J Natl Cancer Inst 92(14): 1165-1171 (2000).

T17 Travis, L.B., M. Gospodarowicz, R.E. Curtis et al. Lung cancer following chemotherapy and radiotherapy for Hodgkin's disease. J Natl Cancer Inst 94(3): 182-192 (2002).

T18 Travis, L.B., D.A. Hill, G.M. Dores et al. Breast cancer following radiotherapy and chemotherapy among young women with Hodgkin disease. JAMA 290(4): 465-475 (2003).

T19 Travis, L.B., D. Hill, G.M. Dores et al. Cumulative absolute breast cancer risk for young women treated for Hodgkin lymphoma. J Natl Cancer Inst 97(19): 1428-1437 (2005).

T20 Travis, L.B., C.S. Rabkin, L.M. Brown et al. Cancer survivorship--genetic susceptibility and second primary cancers: research strategies and recommendations. J Natl Cancer Inst 98(1): 15-25 (2006).

T21 Travis, L.B., A.K. Ng, J.M. Allan et al. Second malignant neoplasms and cardiovascular disease following radiotherapy. Health Phys 106(2): 229-246 (2014).

T22 Tronko, M., A.V. Brenner, T. Bogdanova et al. Thyroid neoplasia risk is increased nearly 30 years after the Chernobyl accident. Int J Cancer 141(8): 1585-1588 (2017).

T23 Tubiana, M. Can we reduce the incidence of second primary malignancies occurring after radiotherapy? A critical review. Radiother Oncol 91(1): 4-15; discussion 11-13 (2009).

T24 Tucker, M.A., G.J. D'Angio, J.D. Boice, Jr. et al. Bone sarcomas linked to radiotherapy and chemotherapy in children. N Engl J Med 317(10): 588-593 (1987).

T25 Tucker, M.A., A.T. Meadows, J.D. Boice, Jr. et al. Leukemia after therapy with alkylating agents for childhood cancer. J Natl Cancer Inst 78(3): 459-464 (1987).

T26 Tucker, M.A., P.H. Jones, J.D. Boice, Jr. et al. Therapeutic radiation at a young age is linked to secondary thyroid cancer. The Late Effects Study Group. Cancer Res 51(11): 2885-2888 (1991).

T27 Tukenova, M., I. Diallo, H. Anderson et al. Second malignant neoplasms in digestive organs after childhood cancer: a cohort-nested case-control study. Int J Radiat Oncol Biol Phys 82(3): e383-390 (2012).

T28 Turcotte, L.M., J.P. Neglia, R.C. Reulen et al. Risk, risk factors, and surveillance of subsequent malignant neoplasms in survivors of childhood cancer: a review. J Clin Oncol 36(21): 2145-2152 (2018).

T29 Turcotte, L.M., Q. Liu, Y. Yasui et al. Chemotherapy and risk of subsequent malignant neoplasms in the Childhood Cancer Survivor Study cohort. J Clin Oncol 37(34): 3310-3319 (2019).

T30 Turner, B.C., E. Harrold, E. Matloff et al. BRCA1/BRCA2 germline mutations in locally recurrent breast cancer patients after lumpectomy and radiation therapy: implications for breast-conserving management in patients with BRCA1/BRCA2 mutations. J Clin Oncol 17(10): 3017-3024 (1999).

U1 UNSCEAR. Ionizing Radiation: Levels and Effects. Volume I: Levels. UNSCEAR 1972 Report. United Nations Scientific Committee on the Effects of Atomic Radiation, 1972 Report to the General Assembly, with annexes. United Nations sales publication E.72.IX.17. United Nations, New York, 1972.

U2 UNSCEAR. Ionizing Radiation: Sources and Biological Effects. UNSCEAR 1982 Report. United Nations Scientific Committee on the Effects of Atomic Radiation, 1982 Report to the General Assembly, with annexes. United Nations sales publication E.82.IX.8. United Nations, New York, 1982.

U3 UNSCEAR. Sources and Effects of Ionizing Radiation. Volume II: Effects. UNSCEAR 2000 Report. United Nations Scientific Committee on the Effects of Atomic Radiation, 2000 Report to the General Assembly, with scientific annexes. United Nations sales publication E.00.IX.4. United Nations, New York, 2000.

U4 UNSCEAR. Effects of Ionizing Radiation. Volume I: Report to the General Assembly, Scientific Annexes A and B. UNSCEAR 2006 Report. United Nations Scientific Committee on the Effects of Atomic Radiation. United Nations sales publication E.08.IX.6. United Nations, New York, 2008.

U5 UNSCEAR. Sources, Effects and Risks of Ionizing Radiation. Volume II: Scientific Annex B. UNSCEAR 2013 Report. United Nations Scientific Committee on the Effects of Atomic Radiation. United Nations sales publication E.14.IX.2. United Nations, New York, 2013.

U6 UNSCEAR. Sources, Effects and Risks of Ionizing Radiation. Report to the General Assembly and Scientific Annexes A and B. UNSCEAR 2017 Report. United Nations Scientific Committee on the Effects of Atomic Radiation. United Nations sales publication E.18.IX.1. United Nations, New York, 2018.

U7 UNSCEAR. Sources, Effects and Risks of Ionizing Radiation. Report to the General Assembly and Scientific Annexes A and B. UNSCEAR 2019 Report. United Nations Scientific Committee on the Effects of Atomic Radiation. United Nations sales publication E.20.IX.5. United Nations, New York, 2020.

U8 UNSCEAR. Sources, Effects and Risks of Ionizing Radiation. Volume III: Scientific Annex C. UNSCEAR 2020/2021 Report. United Nations Scientific Committee on the Effects of Atomic Radiation. United Nations sales publication E.21.IX.3. United Nations, New York, 2021.

U9 Upton, A.C. The dose-response relation in radiation-induced cancer. Cancer Res 21: 717-729 (1961).

V1 van der Giessen, P.H. Peridose, a software program to calculate the dose outside the primary beam in radiation therapy. Radiother Oncol 58(2): 209-213 (2001).

V2 van der Meer, D.J., W.T.A. van der Graaf, D. van de Wal et al. Long-term second primary cancer risk in adolescent and young adult (15-39 years) cancer survivors: a population-based study in the Netherlands between 1989 and 2018. ESMO Open 9(1): 102203 (2024).

V3 van Leeuwen, F.E., A.M. Stiggelbout, A.W. van den Belt-Dusebout et al. Second cancer risk following testicular cancer: a follow-up study of 1,909 patients. J Clin Oncol 11(3): 415-424 (1993).

V4 van Leeuwen, F.E., W.J. Klokman, M. Stovall et al. Roles of radiotherapy and smoking in lung cancer following Hodgkin's disease. J Natl Cancer Inst 87(20): 1530-1537 (1995).

V5 van Leeuwen, F.E., W.J. Klokman, M.B. Veer et al. Long-term risk of second malignancy in survivors of Hodgkin's disease treated during adolescence or young adulthood. J Clin Oncol 18(3): 487-497 (2000).

V6 van Leeuwen, F.E., W.J. Klokman, M. Stovall et al. Roles of radiation dose, chemotherapy, and hormonal factors in breast cancer following Hodgkin's disease. J Natl Cancer Inst 95(13): 971-980 (2003).

V7 van Leeuwen, F.E. and A.K. Ng. Long-term risk of second malignancy and cardiovascular disease after Hodgkin lymphoma treatment. Hematology Am Soc Hematol Educ Program 2016(1): 323-330 (2016).

V8 Van Lierde, C., B. Gyselinck, J. Meulemans et al. The value of virtual chromoendoscopy in the workup of patients with head and neck squamous cell carcinoma. Curr Oncol Rep 22(12): 121 (2020).

V9 van Neerven, S.M. and L. Vermeulen. Cell competition in development, homeostasis and cancer. Nat Rev Mol Cell Biol 24(3): 221-236 (2023).

V10 Vassiliev, O.N., T.A. Wareing, J. McGhee et al. Validation of a new grid-based Boltzmann equation solver for dose calculation in radiotherapy with photon beams. Phys Med Biol 55(3): 581-598 (2010).

V11 Veiga, L.H., J.H. Lubin, H. Anderson et al. A pooled analysis of thyroid cancer incidence following radiotherapy for childhood cancer. Radiat Res 178(4): 365-376 (2012).

V12 Veiga, L.H., E. Holmberg, H. Anderson et al. Thyroid cancer after childhood exposure to external radiation: an updated pooled analysis of 12 studies. Radiat Res 185(5): 473-484 (2016).

V13 Veiga, L.H., R.E. Curtis, L.M. Morton et al. Association of breast cancer risk after childhood cancer with radiation dose to the breast and anthracycline use: a report from the Childhood Cancer Survivor Study. JAMA Pediatr 173(12): 1171-1179 (2019).

V14 Verbeek, N., J. Wulff, C. Baumer et al. Single pencil beam benchmark of a module for Monte Carlo simulation of proton transport in the PENELOPE code. Med Phys 48(1): 456-476 (2021).

V15 Verellen, D. and F. Vanhavere. Risk assessment of radiation-induced malignancies based on whole-body equivalent dose estimates for IMRT treatment in the head and neck region. Radiother Oncol 53(3): 199-203 (1999).

V16 Veres, C., R.S. Allodji, D. Llanas et al. Retrospective reconstructions of active bone marrow dose-volume histograms. Int J Radiat Oncol Biol Phys 90(5): 1216-1224 (2014).

V17 Viechtbauer, W. Conducting meta-analyses in R with the metafor package. J Stat Softw 36: 1-48 (2010).

V18 Vincenten, J.P.L., H.F. van Essen, B.I. Lissenberg-Witte et al. Clonality analysis of pulmonary tumors by genome-wide copy number profiling. PLoS One 14(10): e0223827 (2019).

V19 Virtanen, A., E. Pukkala and A. Auvinen. Incidence of bone and soft tissue sarcoma after radiotherapy: a cohort study of 295,712 Finnish cancer patients. Int J Cancer 118(4): 1017-1021 (2006).

W1 Wakeford, R. and M. Hauptmann. The risk of cancer following high, and very high, doses of ionising radiation. J Radiol Prot 42(2) (2022).

W2 Walsh, L., A. Ulanowski, J.C. Kaiser et al. Risk bases can complement dose bases for implementing and optimising a radiological protection strategy in urgent and transition emergency phases. Radiat Environ Biophys 58(4): 539-552 (2019).

W3 Wambersie, A., D. Jones and H. Suit. Presentation of the ICRU-IAEA joint report "Prescribing, recording, and reporting proton-beam therapy". Extended Abstracts from the Eighth International Meeting on Progress in Radio-Oncology ICRO/ÖGRO 8, Salzburg, 16–19 May 2007. Strahlenther Onkol 183(Suppl 2): 43–89 (2007).

W4 Wang, H., Y. Yin, R. Wang et al. Clinicopathological features, risk and survival in lung cancer survivors with therapy-related acute myeloid leukaemia. BMC Cancer 20(1): 1081 (2020).

W5 Wang, L. and G.X. Ding. The accuracy of the out-of-field dose calculations using a model based algorithm in a commercial treatment planning system. Phys Med Biol 59(13): N113-128 (2014).

W6 Wang, L. and G.X. Ding. Estimating the uncertainty of calculated out-of-field organ dose from a commercial treatment planning system. J Appl Clin Med Phys 19(4): 319-324 (2018).

W7 Wang, R., A.M. Zeidan, J.B. Yu et al. Myelodysplastic syndromes and acute myeloid leukemia after radiotherapy for prostate cancer: a population-based study. Prostate 77(5): 437-445 (2017).

W8 Wang, X., M. Zeng, X. Ju et al. Correlation between second and first primary cancer: systematic review and meta-analysis of 9 million cancer patients. Br J Surg 111(1) (2024).

W9 Wareing, T., J. McGhee, J. Morel et al. Discontinuous finite element SN methods on three-dimensional unstructured grids. Nucl Sci Eng 138: 256-268 (2001).

W10 Watson, A.P., B. Peterson, C. Lee et al. A pilot study evaluating chemotherapy tolerability for breast cancer patients who have received prior treatment and chest radiation for Hodgkin lymphoma. Cancer Treat Res Commun 15: 1-6 (2018).

W11 Watt, T.C., P.D. Inskip, K. Stratton et al. Radiation-related risk of basal cell carcinoma: a report from the Childhood Cancer Survivor Study. J Natl Cancer Inst 104(16): 1240-1250 (2012).

W12 Wegierek-Ciuk, A., A. Lankoff, H. Lisowska et al. Cisplatin reduces the frequencies of radiotherapy-induced micronuclei in peripheral blood lymphocytes of patients with gynaecological cancer: possible implications for the risk of second malignant neoplasms. Cells 10(10): 2709 (2021).

W13 Weinberg, O.K., A. Siddon, Y.F. Madanat et al. TP53 mutation defines a unique subgroup within complex karyotype de novo and therapy-related MDS/AML. Blood Adv 6(9): 2847-2853 (2022).

W14 WHO. WHO Report on cancer: setting priorities, investing wisely, and providing care for all. World Health Organization, Geneva, 2020.

W15 Wilson, L.J., W.D. Newhauser, C.W. Schneider et al. Method to quickly and accurately calculate absorbed dose from therapeutic and stray photon exposures throughout the entire body in individual patients. Med Phys 47(5): 2254-2266 (2020).

W16 Winter, G., R. Kirschner-Schwabe, S. Groeneveld-Krentz et al. Clinical and genetic characteristics of children with acute lymphoblastic leukemia and Li-Fraumeni syndrome. Leukemia 35(5): 1475-1479 (2021).